GATE Mathematics and General Aptitude

For CS & IT Students

Dr. Saktipada Nanda M.Sc., Ph. D.

Professor, Institute of Engineering and Management, Kolkata

Sibashis Nanda B.E.

IBM Certified Database Administrator
Project+ Certified Professional
Founder & CEO, Learnikx Education

A MindProBooks Initiative
www.mindprobooks.com

GATE Mathematics and General Aptitude For CS & IT Students

Dr. Saktipada Nanda
Sibashis Nanda

Second Printing January 2020

Trademarks

All terms mentioned in this book that are known to be trademarks or service marks have been appropriately capitalized. Mindprobooks.com cannot attest to the accuracy of this information. Use of a term in this book should not be regarded as affecting the validity of any trademark or servicemark.

Warning and Disclaimer

Every effort has been made to make this book as complete and as accurate as possible, but no warranty or fitness is implied. The information provided is on an "as is" basis. The authors and the publisher shall have neither liability nor responsibility to any person or entity with respect to any loss or damages arising from the information contained in this book or from the use of the electronic material accompanying it.

We Want to Hear from You!

Welcome to the second print of *GATE Mathematics and General Aptitude -CS & IT*. As the reader of this book, you are our most important critic and commentator. We value your opinion and want to know what we're doing right, what we could do better, what areas you'd like to see us publish in, and any other words of wisdom you're willing to pass our way. You can email and let us know what you did or didn't like about this book, as well as what we can do to make our books better.

Please note that we cannot help you with technical problems related to the topic of this book, and we might not be able to reply to every message.

When you write, please be sure to include this book's title and author as well as your name and contact information.

Email: *customerservice@mindprobooks.com*

[This page is intentionally left blank]

CONTENTS

- Each paper contains 65 questions of 100 marks and of 3 hours duration.

- In the 1 mark question, 1 mark is awarded for correct answer, and 1/3 mark is deducted for wrong answer.

- In the 2 mark questions, 2 marks are awarded for correct answer, and 2/3 mark is deducted for wrong answer.

- In the numerical type questions, there is no negative marking.

● Propositional and First Order Logic

1. *Proposition:* A declarative statement (or assertion) which is TRUE or False, but not both, is called a *proposition* (or, *statement*).
Sentence which are interrogative, imperative or exclamatory in nature are not propositions.

Statement	Proposition /NOT proposition	Truth value T(1)/F(0)
Calcutta is the Capital of Bihar	Proposition	F(0)
5 + 3 = 8	Proposition	T(1)
1 + 3 = 2	Proposition	F(0)
x + y = z	NOT proposition	------
What is your name?	NOT proposition	------

2. *Truth Table:* It is a table that displays the relationship between the truth values of sub propositions and that of the compound proposition constructed.

3. *Tautology and Contradiction*:
A compound proposition $P(p_1, p_2, ---, p_n)$ where $p_1, p_2, ---, p_n$ are elementary propositions is called a *Tautology*, if it is true for every truth assignment for $p_1, p_2, ---, p_n$.
A compound proposition $P(p_1, p_2, ---, p_n)$ where $p_1, p_2, ---, p_n$ are elementary propositions is called a *Contradiction*, if it is false for every

truth assignment for $p_1, p_2, ---, p_n$.

4. *Basic logical operators*: The basic logical operators (connectives) are

(i) Conjunction (AND) (ii) Disjunction (OR) and (iii) Negation (NOT)

(i) Conjunction: When $p \& q$ are any two propositions, the proposition

''p and q' denoted by $p \wedge q$ is called *conjunction* of $p \& q$.

It is TRUE (1) if $p \& q$ are both TRUE and is FALSE (0) otherwise.

(ii) Disjunction: When $p \& q$ are any two propositions, the proposition

denoted by $p \vee q$ is called *disjunction* of $p \& q$. It is FALSE (0) if $p \& q$

are both FALSE and is TRUE otherwise. $p \wedge q \vee r$ means $(p \wedge q) \vee r$

(iii) Negation: Given any proposition p, another proposition is formed by

writing the opposite truth values of p, is called the *Negation* of p and is

denoted by $\neg p$. It is TRUE if p is FALSE and FALSE if is TRUE.

$\neg p \wedge q$ means $(\neg p) \wedge q$. Also. $\neg(\neg p) \equiv p$.

Table for Truth values involving connectives on two propositions:

p	q	$p \wedge q$	$p \vee q$	$\neg p$
T	T	T	T	F
T	F	F	T	F
F	T	F	T	T
F	F	F	F	T

5. *Conditional and Bi-conditional Propositions*:

If $p \& q$ are any two propositions, the compound proposition 'if p then

q ' denoted by $p \rightarrow q$ is called *Conditional proposition*. It is FALSE (0)

when p is TRUE and q is FALSE (0) and TRUE (1) otherwise.

If $p \& q$ are any two propositions, the compound proposition 'p iff q' denoted by $p \leftrightarrow q$ is called *Bi-conditional proposition*. It is TRUE (1) if p and q have the same truth values and is FALSE (0) otherwise.

Table for Truth values involving conditional & Bi conditional operators:

p	q	$p \rightarrow q$	$p \leftrightarrow q$
T	T	T	T
T	F	F	F
F	T	T	F
F	F	T	T

6. *Contrapositive*: Let R be the proposition $p \rightarrow q$.

Its *converse* is $q \rightarrow p$ and its *contrapositive* is $\neg q \rightarrow \neg p$.

7. *Order for the use of logical operators*:

i) The negation operator has precedence over all logical operators.

ii) The Conjunction operator has precedence over Disjunction operator.

iii) The Conditional and Bi conditional operators have lower precedence over other operators. Of the two, the Conditional operator has precedence over Bi conditional operator.

8. *D-Morgan's Rules:*

(i) $\neg(p \wedge q) = \neg p \vee \neg q$ (ii) $\neg(p \vee q) = \neg p \wedge \neg q$

9. *Equivalences:*

(i) $p \rightarrow q \equiv \neg p \vee q$ (ii) $p \rightarrow q \equiv \neg q \rightarrow \neg p$

(iii) $p \leftrightarrow q \equiv (p \rightarrow q) \wedge (q \rightarrow p)$ (iv) $p \leftrightarrow q \equiv (\neg p \wedge \neg q) \vee (p \wedge q)$

10. *Distributive Laws*:

i) $p \vee (q \wedge r) = (p \vee q) \wedge (p \vee r)$ ii) $p \wedge (q \vee r) = (p \wedge q) \vee (p \wedge r)$

11. Identity Laws:

i) $i) p \vee T = p, ii) p \vee T = T, iii) p \wedge T = p, iv) p \wedge F = F$

12. *Complement Laws*:

i) $p \vee \neg p = T, ii) p \wedge \neg p = F, iii) \neg T = F, iv) \neg F = T$

13. *Fundamental principle of logical reasoning* **(Law of syllogism)**:

If p implies q and q implies r, then p implies r.

Symbolically, $p \rightarrow q, q \rightarrow r$ *then* $p \rightarrow r$.

14. *Quantifiers*: Let $p(x)$ be a propositional function defined on a set A.

(i) Universal Quantifier: "For every x in A, $p(x)$ is a true statement"-

Symbolically, $\forall x p(x)$ or $(\forall x \in A) p(x)$

(ii) Existential Quantifier: "For some x in A, $p(x)$ is a true statement"-

Symbolically, $\exists x p(x)$ or $(\exists x \in A) p(x)$

$\neg \forall x p(x) \equiv \exists x \neg p(x)$.

15. *Quantifier with two variables*:

Let $p(x, y)$ be a propositional function defined on a set $A = B \times B = B^2$.

"For every x in A, there exists a y such that $p(x, y)$ is a true

statement"- Symbolically, $\forall x \exists y, p(x, y)$ or $(\forall x \in A) p(x)$

$\neg(\forall x \exists y, p(x, y)) = \exists x \forall y, p(x, y)$

WORKED OUT EXAMPLES

Questions of one mark:

Example 1. If the truth values of the propositions $p \& q$ are T and F
respectively, then the truth value of $p \rightarrow q$ is

(A)   T   (B)   F   (C) both F and T   (D) none of these

4

Answer: (B)
By definition 4, the truth value is F.

Example 2. $\neg(p \to q)$ is equivalent to

(A) $p \lor \neg q$ (B) $\neg p \lor q$ (C) $p \land \neg q$ (D) $\neg p \land q$

Answer: (C)

$\neg(p \to q) = \neg(\neg p \lor q) = \neg(\neg p) \land \neg q$ [By DE-Morgan's Rule]
$= p \land \neg q$

Example 3. Let $p \& q$ are propositions, the expression $p \land (q \land \neg q)$ is

(A) a tautology   (B) a contradiction
(C) always TRUE when p is FALSE   (D) always TRUE when q is TRUE.

Answer: (B)

$p \land (q \land \neg q) \equiv p \land F \equiv F$ - which implies that it is a contradiction.

Example 4. The proposition $p \land (\neg p \lor q)$ is

(A)   a tautology   (B)   a contradiction
(C)   logically equivalent to q (D)   logically equivalent to $p \land q$

Answer: (D)

$p \land (\neg p \lor q) = (p \land \neg p) \lor (p \land q) = F \lor (p \land q) = p \land q$

Example 5. Consider the following logical inferences.

I_1 :   If it rains then the cricket match will not be played.
          The cricket match was played.
          Inference: There was no rain.

I_2 :   If it rains then the cricket match will not be played.
          It did not rain.

Inference: The cricket match was played.
Which of the following is **TRUE?**

(A) Both I_1 and I_2 are correct inferences

(B) I_1 is correct but I_2 is not a correct inference

(C) I_1 is not correct but I_2 is a correct inference

(D) Both I_1 and I_2 are not correct inferences   [CS 2012]

Answer: (B)

p : It rains q : The cricket match will not be played

$I_1 : p \to q$ which implies $\neg q \to \neg p$ that is if the cricket match was played there was no rain. I_1 is a correct inference.

I_2 : The statements are $p \to q$ and $\neg p$ lead to $\neg q$.

p	q	$p \to q$	$\neg p$	$\neg q$
T	T	T	F	F
T	F	F	F	T
F *	T	T	T	F
F *	F	T	T	T

$p \to q$ and $\neg p$ are both true in third and fourth rows but $\neg q$ is sometimes true and false. So, I_2 is not a correct inference.

Example 6. Let the propositions p be "Food is good", q be "Service is good" and r be "Restaurant is 5-star. The simple verbal sentence that describes the statement " It is not true that 5-star rating means good food and good service" in symbolic notation is

(A) $\neg(r \wedge (p \wedge q))$ (B) $r \to \neg(p \wedge q)$

(C) $\neg(r \to (p \wedge q))$ (D) $\neg(r \vee (p \wedge q))$

Answer: (C)

$p \wedge q$ means "Food is good", and "Service is good".

$\neg(p \wedge q)$ means -it is not true that "Food is good", and "Service is good".

5 star rating does not imply good food and good service is $r \to \neg(p \wedge q)$

Example 7. Let the propositions p be "It is cold" and q be "It is raining". The simple verbal sentence that describes the statement $q \vee \neg p$ is

(A) It is cold or it is not raining   (B) It is cold or it is raining

(C) It is not raining or it is not cold (D) It is raining or it is not cold

Answer: (D)

Example 8.

I. $\neg \forall x (P(x))$ II. $\neg \exists x (P(x))$ III. $\neg \exists \forall x (\neg P(x))$ IV. $\exists x (\neg P(x))$

Which two of the above are equivalent?

(A) I and III (B) I and IV (C) II and III (D) II and IV   [CS 2009]

Answer: (B)

By Negation property of quantifiers [Rule 14], $\neg \forall x p(x) \equiv \exists x \neg p(x)$.

Example 9. Let p, q, r denote the statements "It is raining", "It is cold", and "It is pleasant", respectively. Then the statement "It is not raining and it is pleasant, and it is not pleasant only if it is raining and it is cold" is represented by

(A) $(\neg p \wedge r) \wedge (\neg r \rightarrow (p \wedge q))$ (B) $(\neg p \wedge r) \wedge ((p \wedge q) \rightarrow \neg r)$
(C) $(\neg p \wedge r) \vee ((p \wedge q) \rightarrow \neg r)$ (D) $(\neg p \wedge r) \vee (r \rightarrow (p \wedge q))$ [CS/IT 2017]

Answer: (A)

p: "It is raining" q: "It is cold" r: "It is pleasant"

The correct representation for the statement "It is not raining and it is pleasant, and it is not pleasant only if it is raining and it is cold" is

$(\neg p \wedge r) \wedge (\neg r \rightarrow (p \wedge q))$

Example 10. Let the propositions p be "I study hard" and q be "I shall succeed". The contrapositive of the statement "If I study hard, I shall

7

succeed" is

(A) If I don't study hard, I will not succeed

(B) I study hard, I shall succeed

(C) If I don't succeed, I don't study hard (D) If I succeed, I study hard

Answer: (C)

The statement "If I study hard, I shall succeed" is $p \rightarrow q$.

The contrapositive of the statement is $\neg q \rightarrow \neg p$.

It is equivalent to "If I don't succeed, I don't study hard".

Example 11. Let $P(x)$ is the statement "x is wealthy", the set of all men is the domain. $\forall x(\neg P(x))$ states that

(A) Every man is wealthy   (B)   No man is wealthy

(C) At least one man is wealthy   (D) Some men are wealthy

Answer: (B)

Example 12. What is the correct translation of the following statement into mathematical logic?

"Some real numbers are rational"

(A) $\exists x(real(x) \vee rational(x))$ (B) $\forall x(real(x) \rightarrow rational(x))$

(C) $\exists x(real(x) \wedge rational(x))$ (D) $\exists x(rational(x) \leftarrow real(x))$ [CS 2012]

Answer: (C)

We can write the statement as - there exist some real number that are rational. So, quantifier $\exists$ is to be used and x will be real as well as rational.

Example 13. Suppose the predicate $F(x, y, t)$ is used to represent the person x can fool person y at time t. Which one of the statements

8

below expresses best the meaning of the formula $\forall x \exists y \exists t(F(x, y, t))$?

(A) Everyone can fool some person at some time

(B) No one can fool everyone all the time

(C) Everyone can fool some person all the time

(D) No one can fool some person at some time   [CS/IT 2010]

Answer: (B)

$\forall x \exists y \exists t(F(x, y, t)) \equiv \forall x \neg(\forall y \forall t F(x, y, t))$ which is same as (B).

Example 14. Let p, q and r are propositions and the expression

$(p \rightarrow q) \rightarrow r$ is a contradiction. Then the expression $(r \rightarrow p) \rightarrow q$ is

(A) a tautology   (B) a contradiction

(C) always TRUE when p is FALSE   (D) always TRUE when q is TRUE

Answer: (D)   [CS/IT 2017]

$(p \rightarrow q) \rightarrow r$ Is a contradiction, so r is false.

$(r \rightarrow p) \rightarrow q = (\neg r \lor p) \rightarrow q = \neg(\neg r \lor p) \lor q = (r \land \neg p) \lor q$

$= FALSE \lor q = q$ So, $(p \rightarrow q) \rightarrow r = q$ which is true if q is true.

Example 15. The statement $(\neg p) \Rightarrow (\neg q)$ is logically equivalent to which

of the statements below?

I. $p \Rightarrow q$ II. $q \Rightarrow p$ III. $\neg q \lor p$ IV. $\neg p \lor q$

(A) I only (B) I and IV only (C) II only (D) II and III only [CS/IT 2017]

Answer: (D)

By formula 7.ii), $(\neg p) \Rightarrow (\neg q)$ is equivalent to

$\neg(\neg q) \Rightarrow \neg(\neg p) \equiv q \Rightarrow p$ -the statement II.

By formula 7.i), $(\neg p) \Rightarrow (\neg q)$ is equivalent to

$\neg(\neg p) \vee \neg q \equiv p \vee \neg q$ - the statement III.

Example 16. Let $p \& q$ are propositions then the expression

$\neg(p \vee q) \vee (\neg p \wedge q)$ is equivalent to

(A) p (B) q (C) $\neg p$ (D) $\neg q$

Answer: (C)

$\neg(p \vee q) \vee (\neg p \wedge q) \equiv (\neg p \wedge \neg q) \vee (\neg p \wedge q)$ [De Morgan's

law]$\equiv \neg p \wedge (\neg q \vee q)$ [Associative law] $\equiv \neg p \wedge T = \neg p$

Example 17. Let $p \& q$ are propositions then the expression

$(q \rightarrow p) \wedge (\neg p \wedge q)$ is

(A) a tautology (B) a contradiction

(C) always TRUE when p is FALSE (D) always TRUE when q is TRUE.

Answer: (B)

$(q \rightarrow p) \wedge (\neg p \wedge q) = (\neg q \vee p) \wedge (\neg p \wedge q) = (\neg p \wedge q) \wedge (\neg q \vee p)$

$= (\neg p \wedge q \wedge \neg q) \vee (\neg p \wedge q \wedge p)$ [By Distributive Law]

$= (\neg p \wedge F) \vee (\neg p \wedge p \wedge q) = F \vee (F \wedge q) = F \vee F = F$

- which implies that it is a contradiction.

Example 18. Let $p \& q$ are propositions then the expression

$(p \wedge q) \rightarrow (p \vee q)$ is

(A) a tautology (B) a contradiction

(C) always TRUE when p is FALSE (D) always TRUE when q is TRUE

Answer: (A)

Let $p \wedge q = a$, $p \vee q = b$ then

$$(p \wedge q) \rightarrow (p \vee q) \equiv a \rightarrow b \equiv \neg a \vee b \equiv (\neg(p \wedge q) \vee (p \vee q))$$

$$\equiv (\neg p \vee \neg q) \vee (q \vee p) \equiv \neg p \vee (\neg q \vee q) \vee p \equiv \neg p \vee T \vee p = T$$

Example 19. Consider the following statements:

P. Good mobile phones are not cheap

Q. Cheap mobile phones are not good

L. P implies Q   M. Q implies P   N. P is equivalent to Q

Which one of the following about L, M, N is CORRECT?

(A)   Only L is TRUE   (B)   Only M is TRUE

(C)   Only N is TRUE   (D)   L, M and N is TRUE [CS/IT 2014]

Answer: (D)

Let p: Items are good q : Items are good

P. p then NOT q q, $p \rightarrow \neg q \equiv \neg p \vee \neg q$

Q. q then NOT p, $q \rightarrow \neg p \equiv \neg q \vee \neg p \equiv \neg p \vee \neg q$

So, L, M and N are TRUE.

Example 20. If $P(x) \equiv "x^2 < 10"$, where the universe consists of the

positive integers 1, 2, 3 and 4, then the truth value of $\forall x P(x)$ is

(A) T (B)   F   (C) not determined (D) non-existent

Answer: (B)

$$\forall x P(x) = P(1) \wedge P(2) \wedge P(3) \wedge P(4).$$

The truth value of $\forall x P(x)$ is $T \wedge T \wedge T \wedge F \equiv F$

Example 21. Let $P(x, y) : x$ is taller than y. If x is taller than y, then y is

not taller than x. This assertion can be symbolically represented as

(A) $\forall x \forall y (P(x, y) \rightarrow P(y, x))$ (B) $\forall x \forall y (P(x, y) \rightarrow \neg P(y, x))$

(C) $\forall x \forall y (P(x, y) \leftrightarrow \neg P(y, x))$ (D) $\forall x \forall y (P(x, y) \leftrightarrow P(y, x))$

Answer: (B)

$P(x, y) : x$ is taller than y.   so, $P(y, x) : y$ is taller than x.

$\neg P(y, x) : y$ is not taller than x. So, $P(x, y) \rightarrow \neg P(y, x)$ for all x & y.

Example 22. Let p & q are two propositions

p : It snows q : they drive the car

The negation of the statement "If it snows, then they do not drive the car" in symbolic form is

(A) $p \wedge q$ (B) $p \vee q$ (C) $\neg p \wedge q$ (D) $p \vee \neg q$

Answer: (A)

The statement in symbolic form is $p \rightarrow \neg q = \neg p \vee \neg q$

Its negation is $\neg(\neg p \vee \neg q) = \neg(\neg p) \wedge \neg(\neg q) = p \wedge q$

Example 23. Predicate $glitters(x)$ is true if x glitters and predicate $gold(x)$ is true if x is gold. Which one of the following logical formulae represents the above statement?
(A) $\forall x : glitters(x) \Rightarrow \neg gold(x)$ (B) $\forall x : gold(x) \Rightarrow glitter(x)$
(C) $\exists x : gold(x) \wedge \neg glitter(x)$ (D) $\exists x : glitters(x) \wedge \neg gold(x)$

[CS/IT 2014]

Answer: (D)
'All that glitters is gold' in symbolic representation:
$\forall x : glitters(x) \Rightarrow gold(x)$
Not all that glitters is gold' in symbolic representation:.
$\neg(\forall x : glitters(x) \Rightarrow gold(x)) \equiv \neg(\forall x : \neg glitter(x) \vee gold(x))$
$[\text{As } p \Rightarrow q \equiv \neg p \vee q)]$
$\equiv \exists x : \neg(\neg glitters(x)) \wedge \neg gold(x)$ [By De-Morgan's Rule]
$\equiv \exists x : glitters(x) \wedge \neg gold(x)$

Example 24. Let p & q are propositions then the expression
$(\neg p \vee q) \wedge (p \wedge (p \wedge q))$ is equivalent to
(A) $p \wedge q$ (B) $p \vee q$ (C) $p \rightarrow q$ (D) $q \rightarrow p$

Answer: (A)

$$(\neg p \vee q) \wedge (p \wedge (p \wedge q)) = (\neg p \vee q) \wedge ((p \wedge p) \wedge q)$$
$$(\neg p \vee q) \wedge (p \wedge q)$$
$$= (p \wedge q) \wedge (\neg p \vee q) \; ((p \wedge q) \wedge \neg p) \vee ((p \wedge q) \wedge q)$$
[By Distributive Law]
$$= (\neg p \wedge (p \wedge q)) \vee (p \wedge (q \wedge q)) = ((\neg p \wedge p) \wedge q) \vee (p \wedge q)$$
$$= (F \wedge q) \vee (p \wedge q) = F \vee (p \wedge q) = p \wedge q$$

Example 25. Let $p \& q$ are propositions then the expression

$\neg q \wedge (p \rightarrow q) \rightarrow \neg p$ is

(A) a tautology   (B) a contradiction

(C) always TRUE when p is FALSE   (D) always TRUE when q is TRUE

p	q	$b = \neg p$	$\neg q$	$p \rightarrow q$	$a = \neg q \wedge (p \rightarrow q)$	$a \rightarrow b$
T	T	F	F	T	F	T
T	F	F	T	F	F	T
F	T	T	F	T	F	T
F	F	T	T	T	T	T

- which implies that the compound proposition is a tautology.   Answer:

(A)

Questions of two marks:

Example 26. Which one of the following is NOT equivalent to $p \leftrightarrow q$?

(A) $(\neg p \vee q) \wedge (p \vee \neg q)$ (B) $(\neg p \vee q) \wedge (p \rightarrow q)$

(C) $(\neg p \wedge q) \vee (p \wedge \neg q)$ (D) $(\neg p \wedge \neg q) \vee (p \wedge q)$ [CS/IT 2015]

Answer: (C)

$$p \leftrightarrow q \equiv (p \rightarrow q) \wedge (q \rightarrow p) \equiv (\neg p \vee q) \wedge (p \vee \neg q) \rightarrow (A)$$

13

$$p \leftrightarrow q \equiv (p \rightarrow q) \wedge (q \rightarrow p) \equiv (q \rightarrow p) \wedge (p \rightarrow q)$$
$$\equiv (\neg p \vee q) \wedge (p \rightarrow q) \rightarrow \text{(B)}$$
$$p \leftrightarrow q \equiv (\neg p \wedge \neg q) \vee (p \wedge q) \rightarrow \text{(D)} \quad \text{[by 7.(iii)] So, (C) is not equivalent.}$$

Example 27. Let $p \& q$ are propositions then the expression

$\neg q \wedge (p \rightarrow q) \rightarrow \neg p$ is

(A) a tautology   (B) a contradiction

(C) always TRUE when p is FALSE   (D) always TRUE when q is TRUE

Answer: (A)

p	q	$\neg p = b$	$\neg q$	$p \rightarrow q$	$\neg q \wedge (p \rightarrow q) = a$	$a \rightarrow b$
T	T	F	F	T	F	T
T	F	F	T	F	F	T
F	T	T	F	T	F	T
F	F	T	T	T	T	T

So, it is a tautology.

Example 28. There are two shopping malls next to each other, one with sign board as 'Good items are not cheap' and second with sign board as 'Cheap items are not good'. The two statements are

(A) logically independent   (B) logically same

(C) logically different   (D) inconclusive

Answer: (B)

p : Items are good q : Items are cheap

First shopping mall board means 'if p then NOT q' that is $p \to \neg q$

Second shopping mall board means 'if q then NOT p' that is $q \to \neg p$

p	q	$\neg p$	$\neg q$	$p \to \neg q$	$q \to \neg p$
T	T	F	F	F	F
T	F	F	T	T	T
F	T	T	F	T	T
F	F	T	T	T	T

The truth values of the two propositions $p \to \neg q$ and $q \to \neg p$ are same. The two statements are same.

Example 29. Let p, q, r be propositions and the expression $(p \to q) \to r$ is a contradiction. Then the proposition $(r \to p) \to q$ is

(A) a tautology

(B) a contradiction

(C) always TRUE when p is FALSE

(D) always TRUE when q is TRUE   [CS/IT 2017]

Answer: (D)

As $(p \to q) \to r$ is FALSE, then $(p \to q)$ is TRUE and r is FALSE.

The possible cases are

(i) p TRUE, q TRUE, r FALSE

(ii) p FALSE, q TRUE, r FALSE

(iii) p FALSE, q False, r FALSE

For cases (i) and (ii) $(r \to p) \to q$ is TRUE, so it is not a contradiction, so (B) is not possible.

For (iii) $(r \to p) \to q$ is FALSE so, is is not a tautology so, (A) is not possible. For (iii) p is FALSE and $(r \to p) \to q$ is FALSE so (C) is NOT possible. Only possible is (D).

Example 30. Let $p \& q$ are propositions then the expression $(\neg p \to \neg q) \to (q \to p)$ is equivalent to

(A) p (B) q (C)   F   (D)   T

Answer: (D)

p	q	$\neg p$	$\neg q$	$\neg p \to \neg q = a$	$q \to p = b$	$a \to b$
T	T	F	F	T	T	T
T	F	F	T	F	T	T
F	T	T	F	T	F	T
F	F	T	T	T	T	T

So, $(\neg p \to \neg q) \to (q \to p)$ is equivalent to T.

Example 31. From the premises $H_1 : \neg p$ and $H_2 : p \vee q$, the valid conclusion should be

(A) $\neg p \wedge q$ (B) $p \wedge q$ (C) $p \leftrightarrow q$ (D) None of these

Answer: (A)

p	q	$H_1 : \neg p$	$H_2 : p \vee q$	$H_1 \wedge H_2$	$\neg p \wedge q$	$p \wedge q$	$p \leftrightarrow q$
T	T	F	T	F	F	T	T
T	F	F	T	F	F	F	F

F	T	T	T	T *	T *	F	F
F	F	T	F	F	F	F	T

$H_1 \wedge H_2$ is TRUE in the third row where $\neg p \wedge q$ is also TRUE. So, the valid conclusion is $C : \neg p \wedge q$

Example 32. In a room there are two types of people, namely Type 1 and Type 2. Type 1 people always tell the truth and Type 2 people always lie. You give a fair coin to a person in that room, without knowing which type he is from and tell him to toss it and hide the result from you till you ask for it. Upon asking, the person replies the following

"The result of the toss is head if and only if I am telling truth."

Which of the following options is correct?

(A) The result is head   (B) The result is tail

(C) If the person is of Type 2, then the result is tail

(D) If the person is of Type 1, then the result is tail   [CS/IT 2015]

Answer: (A)

Let r : person always tells truth t : result of the toss is head

The compound proposition is $r \leftrightarrow t$

r	t	$r \leftrightarrow t$
T	T	T
T	F	F
F	T	F
F	F	T

(a) The person always tells TRUTH (Type 1): r is true and $r \leftrightarrow t$ is true which occurs in the first row where t is also true i.e., the result is head.

(b) The person always LIES (Type 2): r is false and $r \leftrightarrow t$ is false which occurs in the third row where t is also true i.e., the result is head.

Example 33. What is the logical translation of the following statement?

"None of my friends are perfect"

(A) $\exists x\big(F(x) \wedge \neg P(x)\big)$ (B) $\exists x\big(\neg F(x) \wedge P(x)\big)$

(C) $\exists x\big(\neg F(x) \wedge \neg P(x)\big)$ (D) $\neg \exists x\big(F(x) \wedge P(x)\big)$ [CS/IT 2013]

Answer: (D)

Let $F(x): x$ is my friend and $P(x): x$ is perfect.

(A) There exist some friends who are not perfect - False

(B) There are some people who are not my friend but perfect- False

(C) There are some people who are not my friend and not perfect- False

(D) There are no person who is my friend and perfect - TRUE.

Example 34. Let p, q and r be the propositions and the expression $(p \to q) \to r$ be a contradiction. Then the expression $(r \to p) \to q$ is

(A) a tautology (B) a contradiction   (C) always TRUE when p is FALSE

(D) always TRUE when q is FALSE   [CS/IT 2017]

Answer: (D)

$(r \to p) \to q$ is TRUE if q is TRUE irrespective of the truth value of $r \to p$.

Example 35. Let p,q,r,s represent the following propositions:

$p: x \in \{8,9,10,11,12\}$ $q: x$ is a composite number

$r: x$ is a perfect square $s: x$ is a prime number

The integer $x \geq 2$ which satisfies

$\neg((p \to q) \wedge (\neg r \vee \neg s))$ is ____.   [CS/IT 2016]

Answer: 11

$$\neg(p \to q) = \neg(\neg p \vee q) = p \wedge \neg q = 11$$
$$\neg r \vee \neg s = (8,10,11,12) \vee (8,9,10,12) = (8.9.10.11.12)$$

So, $\neg((p \to q) \wedge (\neg r \vee \neg s)) = 11$

Example 36. Which one of the following is NOT logically equivalent to $\neg \exists x(\forall y(\alpha) \wedge \exists \forall z(\beta))$?

(A) $\forall x(\exists z(\neg \beta) \to \forall y(\alpha))$ (B) $\forall x(\forall z(\beta) \to \exists y(\neg \alpha))$

(C) $\forall x(\forall y(\alpha) \to \exists z(\neg \beta))$ (D) $\forall x(\exists y(\beta) \to \exists z(\alpha))$ [CS/IT 2013]

Answer: (A)

$\neg \exists x(\forall y(\alpha) \wedge \exists \forall z(\beta))$

$\equiv \forall x(\neg \forall y(\alpha) \vee z(\beta))$ [Rule (i) and (ii)]

$\equiv \forall x(\forall y(\alpha) \to \neg \forall z(\beta))$ [Rule (iii)]

$\equiv \forall x(\forall y(\alpha) \to \exists z(\neg \beta))$ [Rule (i)] which is Option (C).

Similarly, it can be shown that the given expression is equivalent to Options (B) and (D)

Example 37. Which one of the following options is CORRECT given three positive integers x, y and z and a predicate $P(x)$?

$$P(x) = \neg(x = 1) \wedge \forall y(\exists z(x = y * z) \Rightarrow ((y = x) \vee (y = 1)))$$

(A) $P(x)$ being true means that x is a prime number

(B) $P(x)$ being true means that x is a number other than1

(C) $P(x)$ is always true irrespective of the value of x

(D) $P(x)$ being true means that x has exactly two factors other than x

Answer: (A)   [CS 2011]

The logical operators are used in the order $\neg, \wedge, \vee, \rightarrow, \leftrightarrow$

$P(x)$ is true for

$(x \neq 1)$ AND (For all y if there exists a z such that $x = y * z$)

That is $(x \neq 1)$ AND (y must be x ($z = 1$) OR (y must be 1 ($z = x$))

So, x has only two factors 1 and x itself. That is x is a prime number.

Example 38. Consider the first-order logic sentence: $F : \forall x(\exists yR(x, y))$

Assuming non-empty logical domains, which of the sentences below are implied by F?

I. $\exists y(\exists x\, R(x, y))$ II. $\exists y(\forall x\, R(x, y))$

III. $\forall y(\exists x\, R(x, y))$ IV. $\exists x(\forall y\, R(x, y))$

(A) IV only (B) I and IV only (C) II only (D) II and III only   [CS/IT 2017]

Answer: (B)

$F : \forall x(\exists yR(x, y)) \rightarrow \exists x(\exists yR(x, y)) \rightarrow \exists y(\exists x\, R(x, y)) \Rightarrow I$ is true.

As $\exists y(\forall x\, R(x, y)) \rightarrow \forall x(\exists yR(x, y)) \Rightarrow II$ is not true.

As $\exists y$ does not imply $\forall y$ so, III is not true.

$F : \forall x(\exists yR(x, y)) \rightarrow \exists x(\forall yR(x, y)) \Rightarrow IV$ is true.

EXERCISE

Questions of one mark:

1. $\neg(p \rightarrow q)$ is equivalent to

(A) $p \vee \neg q$ (B) $\neg p \vee q$ (C) $p \wedge \neg q$ (D) $\neg p \wedge q$

20

2. Let the propositions p be "Food is good", q be "Service is good" and r be "Restaurant is 5-star. The simple verbal sentence that describes the statement " It is not the case that both food is good and rating is five star" in symbolic notation is

(A) $\neg(p \wedge r)$ (B) $\neg(p \vee r)$ (C) $\neg(p \rightarrow r)$ (D) $\neg(r \rightarrow p)$

3. Let the propositions p be "Food is good", q be "Service is good" and r be "Restaurant is 5-star". Then the statement − "If both the food and service are good then the rating will be 5-star" - in symbolic notation is

(A) $(p \wedge q) \rightarrow r$ (B) $r \rightarrow (p \wedge q)$

(C) $(\neg r \rightarrow (p \wedge q))$ (D) $\neg(r \vee (p \wedge q))$

4. Let $p \& q$ are two propositions

p : He swims q : Water is warm

The negation of the statement "He swims if and only if the water is warm" in symbolic form is

(A) $p \leftrightarrow \neg q$ (B) $\neg p \leftrightarrow q$ (C) $\neg p \leftrightarrow \neg q$ (D) $\neg q \leftrightarrow p$

Questions of two marks:

5. Which one of the following well formed formula is tautology?

(A) $\forall x \exists y \, R(x, y) \leftrightarrow \exists y \forall x \, R(x, y)$

(B) $(\forall x [\exists y \, R(x, y) \rightarrow S(x, y)]) \rightarrow \forall x \exists y \, S(x, y)$

(C) $[(\forall x \, y \, (p(x, y) \rightarrow R(x, y))] \rightarrow [\forall x \exists p \, (x, y) \vee R(x, y)]$

(D) $\forall x \forall y \, p(x, y) \rightarrow \forall y \forall y \, p(y, x)$ [CS/IT 2015]

Answers:

1. (C) 2. (A) 3. (A) 4. (B) 5. (C)

Chapter - 2
ALGEBRAIC SYSTEMS

1. *Set:* A well-defined collection of distinct objects is called a set. Each object is called an *element* or a *member* of the set. By the term *'well defined'* we mean that there is a rule (or a set of rules) by which we can determine whether a given object is a member of the collection or not. Generally, sets are denoted by the capital letters A, B, ---- and the elements by small letters a, b, -----.

2. *De- Morgan's Laws:* For any two sets A and B,

(a)   (A U B)' = A' ∩ B'   (b)   (A ∩ B)' = A' U B'

3. *Binary Relation on a Set:* A *Binary Relation* R on a non-empty set A is the subset of the Cartesian product A × A.
If (a, b) be an element of A × A and $(a, b) \in R$, then a is said to be related to b by the relation R and is expressed as $a \, R \, b$. A relation R on a non-empty set A is said to be
(i) *Reflexive* if $a \, R \, a$ holds for all $a \in$ A.
(ii) *Symmetric* if $a \, R \, b \Rightarrow b \, R \, a$ for any two elements $a, b \in$ A.
(iii) *Transitive* if $a \, R \, b$ and $b \, R \, c \Rightarrow a \, R \, c$ for any three elements $a, b, c \in$ A.
A *binary relation* R on a non-empty set A is called an *Equivalence Relation* (or, RST relation) if R is *Reflexive, symmetric and Transitive*.

4. *Mapping (or, Function):* Let A and B be two non-empty sets. A relation f between A and B is said to be a *mapping (or, a function)* from A to B if for each $x \in$ A there is related a unique element $y = f(x)$ in B. It is written as $f : A \to B$
$y = f(x)$ is called the image of x and x is called the pre-image of $y = f(x)$.

A mapping $f : A \rightarrow B$ is said to be

(i) *Injective (or, one-one)*
if $f(a) = f(b) \Rightarrow a = b$ & $f(a) \neq f(b) \Rightarrow a \neq b$.

(ii) *Surjective (or, onto)* if every element of B is the image of at least one element of A i.e., f (A) = B.

(iii) *Into* if at least one element of B has no pre-image in A i.e., $f(A) \subset B$

(iv) *Bijective* if f is both injective and surjective.

5. *Binary Operation: Binary Operation* in a non empty set is a rule which assigns exactly one element of S for every ordered pair of elements of S. It is denoted by the symbol $*$ or $\circ$.

So, $a, b \in$ S implies a unique element $c = a \circ b (or,\ a * b) \in$ S.

6. *Algebraic System:* A non-empty set S equipped with one or more binary operations on S is called an *Algebraic System*. It is symbolically represented by $(S, \circ)$ or $(S, \circ, *)$.

7. *Groupoid:* An Algebraic Structure $(S, \circ)$ is called a *Groupoid* if S is closed with the composition $\circ$ .i.e., $\forall a, b \in$ S $\Rightarrow a \circ b \in$ S(Closure Property).

8. *Semi-group:* An Algebraic Structure $(S, \circ)$ is called a *Semi-group* under the operation $\circ$ if the following two axioms (properties) hold:

(i) Closure Property: $\forall a, b \in$ S $\Rightarrow a \circ b \in$ S

(ii) Associative Property: $\forall a, b, c \in$ S, $a \circ (b \circ c) = (a \circ b) \circ c$

So, a *Groupoid* $(S, \circ)$ with the Associative Property (ii) is a *Semi-group*.

9. *Monoid:* An Algebraic Structure $(S, \circ)$ is called a *Monoid* under the operation $\circ$ if the following three axioms (properties) hold:

(i) Closure Property: $\forall a, b \in$ S $\Rightarrow a \circ b \in$ S

(ii) Associative Property: $\forall a, b, c \in$ S, $a \circ (b \circ c) = (a \circ b) \circ c$

(iii) Existence of Identity Element: There exists an Identity Element *e* in S such that $a \circ e = e \circ a = a, \forall a \in$ S.

So, a *Semi-group* $(S, \circ)$ with the Identity Element *e* is a *Monoid*.

10. *Group:* An Algebraic Structure $(G, \circ)$ is called a *Group* under the operation $\circ$ if the following four axioms (properties) hold:

(i) Closure Property: $\forall\, a,b \in S \Rightarrow a \circ b \in S$

(ii) Associative Property: $\forall\, a,b,c \in S,\ a \circ (b \circ c) = (a \circ b) \circ c$

(iii) Existence of Identity Element: There exists an *Identity Element e* in S such that $a \circ e = e \circ a = a, \forall\, a \in S.$

(iv) Existence of Inverse Element: For each element a in G, there exists an inverse element a^{-1} in G such that $a \circ a^{-1} = a^{-1} \circ a = e$ (Identity Element).

So, a Monoid $(G, \circ)$ with the Inverse of every element of G is a *Group*.

Note: The Identity element e and the inverse of an element a (i.e., a^{-1}) are both unique.

11. *Abelian Group or Commutative Group:* A Group $(G, \circ)$ is called *Abelian Group* or *Commutative Group* if $\circ$ is commutative i.e., $a \circ b = b \circ a \,\forall a, b \in G$

An Algebraic Structure $(G, \circ)$ is called an *Abelian Group* or a *Commutative Group* under the operation $\circ$ if the following five axioms (properties) hold:

(i) Closure Property: $\forall\, a,b \in G \Rightarrow a \circ b \in G$

(ii) Associative Property: $\forall\, a,b,c \in G,\ a \circ (b \circ c) = (a \circ b) \circ c$

(iii) Existence of Identity Element: There exists an Identity Element e in G such that $a \circ e = e \circ a = a, \forall\, a \in G.$

(iv) Existence of Inverse Element: For each element a in G, there exists an inverse element a^{-1} in G such that $a \circ a^{-1} = a^{-1} \circ a = e$ (Identity Element).

(v) Commutative Property: $a \circ b = b \circ a \,\forall a, b \in G$

12. *Properties of a Group:*

(i) The identity element (e) in a group is unique.

(ii) The inverse of any element a in a group is unique.

(iii) $(a^{-1})^{-1} = a$ \qquad\qquad (iv) $(a \circ b)^{-1} = (b^{-1} \circ a^{-1})$

(v) In a group $(G, \circ)$, $\forall a, b \in G$, the equations $a \circ x = b$ & $y \circ a = b$ have unique solutions in G. The solutions are respectively

$x = a^{-1} \circ b \ \& \ y = b \circ a^{-1}$

13. *Finite Group and its Order:* If a Group (G, $\circ$) consists of a finite number of distinct elements, then the Group is a *Finite Group*. Otherwise is an *Infinite Group*. The number of distinct elements of a Finite Group is called its *Order* and is denoted by O(G) or |G|.

14. *Order of an element of a Group:* In the group (G, $\circ$), let a be an element. The *order of the element* a is the least positive integer n such that $a^n = e$ (the identity element of G).

The order of a is denoted by O(a). If no such positive integer n exists such that $a^n = e$, then O(a) = 0 or infinity.

15. *Properties of order of an element of a Group:* Let a be an element of a Group (G, .), then

(i) O(a) = O(a^{-1})

(ii) If O(a) = n and $a^m = e$ then n is a divisor of m.

(iii) If O(a) = n then a, a^2,----, a^n (=e) are distinct elements of G.

(iv) If O(a) = n and p is prime to n then O(a^p) = n

(v) If O(a) is infinite and p is a positive integer then O(a^p) is also infinite.

16. *Cyclic Group:* If every element of a Group (G, .) can be expressed in the form a^n where $a \in$ G and n is a positive integer, then the Group is Cyclic. a is called the generator of the group.

17. *Properties of Cyclic Group:*

(i) If a be a generator of a group then a^{-1} is also a generator of the group.

(ii) A cyclic group of prime order p has exactly (p - 1) generators
a, a^2,--,a^{p-1}

(iii) A cyclic group having infinite number of elements has exactly two generators a and a^{-1}.

(iv) Every group of order 3 (in general, of prime order) is cyclic.

18. *Subgroup*: Let (G, ∘) ba group and H is a non-empty subset of G. If H itself forms a group with respect to the same composition ∘ as in G, then H is called a subgroup of G.

19. *Lagrange's theorem*: The order of every subgroup of a finite group is a divisor of the order of the Group.

20. *Properties of Subgroup*:
(i) A non-empty subset H of a group (G, ∘) is a subgroup of G if and only if $a,b \in H \Rightarrow a^{-1} \circ b \in H$ or $a,b \in H \Rightarrow a \circ b^{-1} \in H$.
(ii) The Identity element e of (G, ∘) is the Identity element of (H, ∘).
(iii) The inverse of $a (a \in H)$ in (H, ∘) is the same as the inverse of a in (G, ∘).
(iv) The *intersection* of two subgroups of a group is a subgroup of the group.
(v) The *union* of two subgroups of a group is not necessarily a subgroup of the group.

21. *Normal Subgroup*: A subgroup H of a Group G is said to be a nomal subgroup Of G if and only if aH = Ha, $\forall$ a ∈ G.

Question of one mark:

Example 1. The cardinality of the power set of
{0, 1, 2, 3, --------, 10} is ___.   [CS/IT 2015]
Answer: 2048
Here, $n = 11$ The cardinality of the power set of the given set
$= 2^n = 2^{11} = 2048$

Example 2. For a set A, the power set of A is denoted by 2^A. If
$A = \{5, \{6\}, \{7\}\}$ which of the following options are TRUE?
I. $\phi \in 2^A$ II. $\phi \subseteq 2^A$ III. $\{5, \{6\}\} \in 2^A$ IV. $\{5, \{6\}\} \subseteq 2^A$
(A)   I and III only   (B)   II and III only
(C)   I, II and III only   (D)   I, II and IV only [CS/IT 2015]
Answer: (C)
Number of elements of 2^A is $2^3 = 8$

$2^A \equiv \{\phi, 5, \{6\}, \{7\}, \{5, \{6\}, \{5, \{7\}, \{\{6\}, \{7\}, \{5, \{6\}, \{7\}\}$

I is true as null set is in 2^A II is true as null set is a subset of every set

III is true as $\{5,\{6\}\}$ is in 2^A

IV is false as $\{5,\{6\}\}$ is not a subset of 2^A but $\{\{5,\{6\}\}\}$ is a subset of 2^A.

Example 3. If $A = \{1.2,3\}, B = \{x, y, z\} \& f = \{(1, z), (2, y), (3, x)\}$

then f is

(A) *one-to-one* but *not onto*   (B) *onto* but *not one-to-one*

(C) neither *one-to-one* nor *onto*   (D) both *one-to-one* and *onto*

Answer: (D)

As $f(1), f(2), f(3)$ are all different, f is *one-to-one*.

The range of f is $B \Rightarrow f$ is *onto*. So, f is both *one-to-one* and *onto*.

Example 4. Consider the function $f : N \to N$ (where N is the set of

natural numbers including zero) is given by $f(j) = j^2 + 2$, then f is

(A) *one-to-one* but *not onto*   (B) *onto* but *not one-to-one*

(C) neither *one-to-one* nor *onto*   (D) both *one-to-one* and *onto*

Answer: (A)

$f(j) = f(i) \Rightarrow j^2 + 2 = i^2 + 2 \Rightarrow j^2 = i^2 \Rightarrow j = i$

[As $i \& j$ are both natural numbers]

Also, $f(j) \neq f(i) \Rightarrow j \neq i$ So, f is *one-to-one*.

As $f(j) = j^2 + 2$, the range of f is the set containing elements ≥ 2.

So, the elements 0 & 1 of the range set N are not images of any element

of the domain set N. There exists *no* j such that

$f(j) = 0$ or $f(j) = 1 \Rightarrow f$ is not *onto*.

Example 5. Let G be a finite group of 84 elements. The size of a largest

possible proper subgroup of G is ____.   [CS/IT 2018]

Answer: 42

$|G| = 84$.

By Lagrange's theorem, the order of a subgroup of G should be a divisor of 84. As the subgroup should be a proper subgroup, so it cannot have the same size as G. The largest divisor of 84 (other than 84) is 42.

Example 6. Let G be a group with 15 elements. Let L be a subgroup of G. It is known that L ≠ G and the size of L is at least 4.
The size of L is   [CS/IT 2014]
Answer: 5
$|G| = 15$.
By Lagrange's theorem, the number of elements of L should be a divisor of 15 either 3 or 5 or 15. As L ≠ G, it can not be 15.
As the size of L is at least 4, so the size of L is 5.

Example 7. The set G of all ordered pairs of real numbers (a,b), $a \neq 0$ with composition $\circ$ defined by $(a,b)\circ(c,d)=(ac, bc+d)$ forms
(A) a group   (B) a semi-group but not a monoid
(C) a monoid but not a group   (D) none of these
Answer (A)
Closure and associative properties hold in G w.r.t. the composition '$\circ$'.
(1, 0) is the identity element of G. The inverse of (a,b), $a \neq 0$
is $\left(\dfrac{1}{a}, -\dfrac{b}{a}\right) \in G$ So, G is a group w.r.t. the composition '$\circ$'.

Example 8. Statement: Every group of order m is commutative and every group of order n is cyclic.Then $(m,n) =$
(A)   (3, 4)   (B)   (4, 3)   (C)   (4, 5)   (D)   (5, 4)
Answer: (B)

Example 9. A group G is abelian if and only if $(ab)^{-1} =$
(A) ab (B) $b^{-1}a^{-1}$ (C) $a^{-1}b^{-1}$ (D) e
Answer (C)
As G is abelian, $ab = ba \Rightarrow (ab)^{-1} = (ba)^{-1} = a^{-1}b^{-1}$

28

Example 10. In the set of all positive rational numbers Q^+, the binary

relation $*$ is defined by $a * b = \dfrac{ab}{2}, \forall a, b \in Q^+$. The identity element of

Q^+ is

(A) 0   (B) 1   (C) 2   (D) does not exist

Answer (C)

Let e be the identity element of Q^+,

Then for every $\forall a \in Q^+$, $e * a = a \Rightarrow \dfrac{ea}{2} = a \Rightarrow e = 2$

Example 11. In the set of all positive rational numbers Q^+, the binary

relation $*$ is defined by $a * b = \dfrac{ab}{2}, \forall a, b \in Q^+$. The inverse of an

element $a \in Q^+$ is

(A) $\dfrac{1}{a}$ (B) $\dfrac{2}{a}$ (C) $\dfrac{4}{a}$ (D) does not exist

Answer (C)

From the previous example, $e = 2$ Let a' be the inverse of $a \in Q^+$,

then $a * a' = e \Rightarrow \dfrac{aa'}{2} = 2 \Rightarrow a' = \dfrac{4}{a}$

Example 12. In the set of real numbers $R - \{-1\}$, the binary relation $*$ is

defined by $a * b = a + b + ab, \forall a, b \in Q^+$. The identity element:

(A) 0   (B) 1   (C) 2   (D) does not exist

Answer (A)

Let e be the identity element of $R - \{-1\}$, then for every

$\forall a \in R - \{-1\}$, $e * a = a \Rightarrow e + a + ea = a \Rightarrow e(1 + a) = 0 \Rightarrow e = 0$

[as $a \neq -1, 1 + a \neq 0$]

Example 13. In the set of real numbers $R - \{-1\}$, the binary relation $*$ is

defined by $a * b = a + b + ab, \forall a, b \in Q^+$. The inverse of an element

$a \in R - \{-1\}$ is

(A) $\dfrac{1}{a}$ (B) $\dfrac{a}{1+a}$ (C) $-\dfrac{a}{1+a}$ (D)   does not exist

Answer (C)

From the previous example, $e = 0$ Let a' be the inverse of then
$$a * a' = e \Rightarrow a + a' + a.a' = 0 \Rightarrow a + a'(1+a) = 0$$

Or, $a'(1+a) = -a \Rightarrow a' = -\dfrac{a}{1+a}$ [as $a \neq -1 \Rightarrow 1+a \neq 0$]

Example 14. If the binary operation $\circ$ be defined on Z (the set of all integers) by $a \circ b = a + b + 1, a, b \in Z,$ The identity element of Z is

(A)   0   (B)   1   (C)   2   (D)   - 1

Answer (D)

Let e be the identity element of Z. Then for every
$$\forall a \in Z, e * a = a \Rightarrow e + a + 1 = a \Rightarrow e = -1$$

Example 15. If the binary operation $\circ$ be defined on Z (the set of all integers) by $a \circ b = a + b + 1, a, b \in Z,$ then the inverse of a with respect to $\circ$ is

(A) $-(1+a)$ (B) $-(1-a)$ (C) $-(2+a)$ (D) $-(2-a)$

Answer (C)

From the previous example, $e = -1$ Let a' be the inverse of $a \in Z,$
then $a * a' = e \Rightarrow a + a' + 1 = -1 \Rightarrow a' = -(2+a)$

Example 16. If S and T are two subgroups of a group G, then which of the following is a subgroup of G?

(A) $S \cup T$ (B) $S \cap T$ (C)   S - T   (D) $G - S$

Answer (B)

It follows from properties (iv) and (v) of Subgroup.

Question of two marks:

Example 17. Let X & Y be finite sets $f : X \rightarrow Y$ be a function. Which one of the following statements is TRUE?
(A) For any two subsets A and B of X, $|f(A \cup B)| = |f(A)| + |f(B)|$ (B) For any two subsets A and B of X, $f(A \cap B) = f(A) \cap f(B)$
(C) For any two subsets A and B of X,
$|f(A \cap B)| = \min\{|f(A)|, |f(B)|\}$
(D) For any two subsets S & T of Y, $f^{-1}(S \cap T) = f^{-1}(S) \cap f^{-1}(T)$

[CS/IT 2014]

Answer: (D)
Let X ={a, b, c} and Y = {1}.
The function maps each element of X to 1 that is f(a) = 1, f(b) = 1, f(c) = 1
A = {a, b}, B = {b, c}, A $\cup$ B = {a, b, c}, |f(A $\cup$ B)| =|f {a, b, c}|= 3 and
|f(A)| + |f(B)| = 2 + 2 = 4 $\Rightarrow$ (A) is not correct.
A $\cap$ B = {b}, f(A $\cap$ B) = f{b} = {1} and f(A) $\cap$ f(B) = {1, 1} $\cap$ {1, 1} = {1, 1}
So, f(A $\cap$ B) $\neq$ f(A) $\cap$ f(B) $\Rightarrow$ (B) is not correct.
|f(A $\cap$ B)| = 1 and min{|f(A)|, |f(B)|} = min{2, 2} = 2
So, |f(A $\cap$ B)| $\neq$ min{|f(A)|, |f(B)|} $\Rightarrow$ (C) is not correct.
For a function, a value can be mapped only to one function(D) is correct.

Example 18. There are two elements x, y in a group {G, *} such that every element of the group can be written as a product of some numbers of $x's$ and $y's$ in some order. It is known that
$$x * x = y * y = x * y * x * y = y * x * y * x = e,$$
where e is the identity element. The maximum number of elements in such a group is ___ .   [CS/IT 2014]
Answer: 4
$x * x = e \Rightarrow x$ is its own inverse.
Similarly, $y * y = e \Rightarrow y$ is its own inverse.

31

$(x * y) * (x * y) = e \Rightarrow x * y$ is its own inverse.

Similarly, $(y * x) * (y * x) = e \Rightarrow y * x$ is its own inverse.

Now, $(x * y) * (y * x) = x * (y * y) * x = x * e * x = x * x = e$

It implies that $x * y$ is inverse of $y * x$.

As the inverse of an element of a group is unique and it is shown that $x * y$ is inverse of $x * y$ so, $x * y = y * x$.

So, G contains only the elements $\{e, x, y, x * y = y * x\}$ that is the maximum number of elements of G is 4.

[The maximum number is 4 because $x \, \& \, y$ may be equal also.]

Example 19. The set P(X) of all subsets of a non-empty X, under the composition $\circ$ defined by the relation

$A \circ B = A \cup B, A, B \in P(X)$ constitute

(A) a group   (B) a semi-group but not a monoid

(C) a monoid but not a group   (D) none of these

Answer (C)

(i) Let $A, B \in P(X)$, then $A \subseteq X \, \& \, B \subseteq X$

Now, $A \circ B = A \cup B \subseteq (X \cup X) = X \Rightarrow A \cup B \in P(X)$ which proves closure property.

(ii) $A \cup (B \cup C) = (A \cup B) \cup C \Rightarrow A \circ (B \circ C) = (A \circ B) \circ C$ which proves associative property w.r.t. the composition '$\circ$'.

(iii) The empty set ϕ is a subset of X. So, $\phi \in P(X)$

Now, $A \cup \phi = \phi \cup A = A, \forall A \in P(X) \Rightarrow A \circ \phi = \phi \circ A = A$

So, ϕ is the identity element of P(X).

(iv) Let $A \subseteq X \, \& \, A \neq \phi$ then for every element $A' \in P(X)$ we don't have $A \circ A' = A' \circ A = \phi$. So, the inverse of every element of P(X) does not exist.

So, G is a monoid but not a group w.r.t. the composition '$\circ$'.

Example 20. If * is the operation defined on S = Q × Q, the set of ordered pair of rational numbers and given by $(a,b)*(x,y) = (ax, ay+b)$. If $a \neq 0$, the inverse of (a,b) is

(A) $\left(\dfrac{1}{a}, \dfrac{b}{a}\right)$ (B) $\left(-\dfrac{1}{a}, \dfrac{b}{a}\right)$ (C) $\left(\dfrac{1}{a}, -\dfrac{b}{a}\right)$ (D) $\left(-\dfrac{1}{a}, -\dfrac{b}{a}\right)$

Answer: (C)

If (e, f) is the identity element of S, then $\forall (a,b) \in S$,

$(a,b)*(e,f) = (a,b) \Rightarrow (ae, af+b) = (a,b)$

$\Rightarrow ae = a \ \& \ af+b = b \Rightarrow e = 1 \ \& \ f = 0$ as $a \neq 0$. Identity $= (1,0)$

If (c,d) be the inverse of (a,b), then

$(a,b)*(c,d) = (1,0) \Rightarrow ac = 1 \ \& \ ad+b = 0 \Rightarrow c = \dfrac{1}{a} \ \& \ d = -\dfrac{b}{a}$

As $a \neq 0$, the inverse of (a,b) is $\left(\dfrac{1}{a}, -\dfrac{b}{a}\right)$

Example 21. The Identity element in the Groupoid $(M, \cdot)$ where

$M = \left\{ \begin{pmatrix} a & a \\ a & a \end{pmatrix}, a \text{ is a non}-\text{zero real number} \right\}$ is

(A) $\begin{pmatrix} 1 & 1 \\ 1 & 1 \end{pmatrix}$ (B) $\begin{pmatrix} 1/2 & 1/2 \\ 1/2 & 1/2 \end{pmatrix}$ (C) $\begin{pmatrix} 1/\sqrt{2} & 1/\sqrt{2} \\ 1/\sqrt{2} & 1/\sqrt{2} \end{pmatrix}$ (D) $\begin{pmatrix} \sqrt{2} & \sqrt{2} \\ \sqrt{2} & \sqrt{2} \end{pmatrix}$

Answer: (B)

Let $A = \begin{pmatrix} a & a \\ a & a \end{pmatrix}$, $a \neq 0$ be an element of $(M, \cdot)$ and the identity element

$E = \begin{pmatrix} e & e \\ e & e \end{pmatrix}$, then $A.E = A \Rightarrow \begin{pmatrix} a & a \\ a & a \end{pmatrix} \cdot \begin{pmatrix} e & e \\ e & e \end{pmatrix} = \begin{pmatrix} a & a \\ a & a \end{pmatrix}$

Or, $\begin{pmatrix} 2ae & 2ae \\ 2ae & 2ae \end{pmatrix} = \begin{pmatrix} a & a \\ a & a \end{pmatrix} \Rightarrow 2ae = a \Rightarrow e = \dfrac{1}{2}$ as $a \neq 0$.

So, $E = \begin{pmatrix} 1/2 & 1/2 \\ 1/2 & 1/2 \end{pmatrix}$

Example 22 . The inverse of an element $B = \begin{pmatrix} b & b \\ b & b \end{pmatrix}$, $b \neq 0$ in the

Groupoid $(M,\cdot)$ where $M = \left\{ \begin{pmatrix} a & a \\ a & a \end{pmatrix}, a \text{ is a non} - \text{zero real number} \right\}$ is

(A) $\begin{pmatrix} 1/b & 1/b \\ 1/b & 1/b \end{pmatrix}$ (B) $\begin{pmatrix} 1/2b & 1/2b \\ 1/2b & 1/2b \end{pmatrix}$ (C) $\begin{pmatrix} 1/4b & 1/4b \\ 1/4b & 1/4b \end{pmatrix}$ (D) Does not exist

Answer: (C)

Let $C = \begin{pmatrix} c & c \\ c & c \end{pmatrix}$, $c \neq 0$ be inverse of $B = \begin{pmatrix} b & b \\ b & b \end{pmatrix}$, $b \neq 0$

Then $C.B = E \Rightarrow \begin{pmatrix} c & c \\ c & c \end{pmatrix} \cdot \begin{pmatrix} b & b \\ b & b \end{pmatrix} = \begin{pmatrix} 1/2 & 1/2 \\ 1/2 & 1/2 \end{pmatrix}$

Or, $\begin{pmatrix} 2bc & 2bc \\ 2bc & 2bc \end{pmatrix} = \begin{pmatrix} 1/2 & 1/2 \\ 1/2 & 1/2 \end{pmatrix} \Rightarrow 2bc = \frac{1}{2} \Rightarrow c = \frac{1}{4b}$ as $b \neq 0$

So, $B^{-1} = \begin{pmatrix} 1/4b & 1/4b \\ 1/4b & 1/4b \end{pmatrix}$

Example 23. Under matrix multiplication the set of matrices

$A_\theta = \begin{pmatrix} \cos\theta & -\sin\theta \\ \sin\theta & \cos\theta \end{pmatrix}$ forms

(A) a commutative group   (B) a group but not a commutative group
(C) a semi-group but not a group   (D) none of these
Answer (A)

Let $G = \{A_\theta, \theta \in R\}$ and $A_\alpha, A_\beta \in G$ then

$A_\alpha.A_\beta =$

$\begin{pmatrix} \cos\alpha & -\sin\alpha \\ \sin\alpha & \cos\alpha \end{pmatrix}\begin{pmatrix} \cos\beta & -\sin\beta \\ \sin\beta & \cos\beta \end{pmatrix} = \begin{pmatrix} \cos(\alpha+\beta) & -\sin(\alpha+\beta) \\ \sin(\alpha+\beta) & \cos(\alpha+\beta) \end{pmatrix} \in G$

Matrix multiplication is associative. $I = \begin{pmatrix} 1 & 0 \\ 0 & 1 \end{pmatrix} = \begin{pmatrix} \cos 0 & -\sin 0 \\ \sin 0 & \cos 0 \end{pmatrix}$

is the identity element of G.

$A_\alpha^{-1} = \begin{pmatrix} \cos\alpha & \sin\alpha \\ -\sin\alpha & \cos\alpha \end{pmatrix} = \begin{pmatrix} \cos(-\alpha) & -\sin(-\alpha) \\ \sin(-\alpha) & \cos(-\alpha) \end{pmatrix} \in G$

is the inverse of $A_\alpha = \begin{pmatrix} \cos\alpha & -\sin\alpha \\ \sin\alpha & \cos\alpha \end{pmatrix}$

$A_\alpha.A_\beta = \begin{pmatrix} \cos(\alpha+\beta) & -\sin(\alpha+\beta) \\ \sin(\alpha+\beta) & \cos(\alpha+\beta) \end{pmatrix} = A_{\beta+\alpha} = A_\beta.A_\alpha.$

So, $G = \{A_\theta, \theta \in R\}$ is a commutative group.

Example 24. If a be an element of a group $(G, \circ)$ and O(a) = 20, then the order of a^8 is

(A)   5   (B)   4   (C)   10   (D)   2

Answer (A)

Since O(a) = 20, a^{20} = e (identity element) -------(i)

Let O(a^8) = n, then $(a^8)^n$ = e   Or, a^{8n} = e ---------(ii)

where n is the least positive integer.

From (i) & (ii), 20 is a divisor of 8n i.e., 5 is a divisor of 2n   So, n = 5

or, O (a^8) = 5 [It follows from properties 10.(iv) and (v)]

Example 25. If x be an element of a multiplicative group G and O(x) = 30, then O(x^4) is

(A)   120   (B)   60   (C)   30   (D)   15

Answer (D)

O(x) = 30 $\Rightarrow$ x^{30} = e (identity element).

Let O(x^4) = n $\Rightarrow$ $(x^4)^n$ = e $\Rightarrow$ x^{4n} = e, where n is the least positive integer.

So, 30 is a divisor of 4n, i.e., 15 is a divisor of 2n. As n is the least positive integer , so n = 15 $\Rightarrow$ O(x^4) = 15

35

Example 26. Let R be the relation on the set of positive integers such that aRb if and only if a and b are distinct and have a common divisor other than 1. Which one of the following statements about R is true?

(A) R is symmetric and reflexive but not transitive

(B) R is reflexive but not symmetric and not transitive

(C) R is transitive but not reflexive and not symmetric

(D) R is symmetric and but not reflexive and not transitive [CS/IT 2015]

Answer: (D)

a and a are not distinct so aRa does not hold $\Rightarrow R$ is not reflexive.

If a and b are distinct and have a common divisor other than 1, then b and a are distinct and have a common divisor other than 1.

So, $aRb \Rightarrow bRa \Rightarrow R$ is symmetric.

$6 = 2 \times 3$ and $21 = 3 \times 7 \Rightarrow 6$ & 21 have a common divisor 3.

$21 = 3 \times 7$ and $35 = 7 \times 5 \Rightarrow 21$ & 35 have a common divisor 7.

But 6 & 35 have no common divisor.

aRb & bRc does not imply aRc. R is not transitive.

Example 27. Let R be the relation on the set of ordered pair of positive integers such that $((p,q),(r,s)) \in R$ if and only if $p - s = q - r$. Which one of the following is true about R?

(A) Both reflexive and symmetric (B) reflexive but not symmetric

(C) Not reflexive but symmetric   (D) Neither reflexive nor symmetric

[CS/IT 2015]

Answer: (C)

$((p,q),(p,q)) \notin R$ as $p - q \neq q - p \Rightarrow R$ is not reflexive.

If $((p,q),(r,s)) \in R$ then $p - s = q - r$.

Now, $((r,s),(p,q)) \in R$ if $r - q = s - p \Rightarrow p - s = q - r$ which is true.

So, $((p,q),(r,s)) \in R \Rightarrow ((r,s),(p,q)) \in R \Rightarrow R$ is symmetric.

Example 28. A binary relation R on $N \times N$ is defined as follows: $(a,b) R (c,d)$ if $a <= c$ or $b <= d$.

Consider the following propositions:

P. R is reflexive   Q. R is transitive

Which one of the following statements is TRUE?

(A) Both P and Q are true   (B) P is true and Q is false

(C) P is false and Q is true   (D)Both P and Q are false   [CS/IT 2016]

Answer: (B)

$(a,b)\,R\,(a,b)$ holds if $a <= a$ or $b <= b$ This is true for all $a\,\&\,b$.

So, R is reflexive.

We consider three pairs (2, 3), (3, 1) and (1, 1)

$(2,3)R(3,1)$ holds as $2 <= 3$ holds. $(3,1)R(1,1)$ holds as $1 <= 1$ holds.

But $(2,3)R(1,1)$ does not hold as $2 > 1$ and $3 > 1$.

So, $(2,3)R(3,1)$ and $(3,1)R(1,1)$ do not imply $(2,3)R(1,1)$ that is R is not transitive. So, R is not reflexive but not transitive.

Example 29. If R by the relation on N (the set of all positive integers) such that $(a,b) \in R$ if $a^2 + b$ is even.

Consider the following propositions:

L. R is reflexive   M. R is symmetric   N. R is transitive

Which one of the following statements is TRUE?

(A) L and M are true but N is false   (B) L and N are true but M is false

(C) M and N are true but L is false   (D) L, M, N are all true

Answer: (D)

$a^2 + a = a(a+1) =$ even as $a\,\&\,(a+1)$ are consecutive positive integers. So, $(a,a) \in R, \forall a \in N$ that is R is *reflexive*.

If $a^2 + b$ is even then $a\,\&\,b$ are both even or both odd. In either case, $b^2 + a$ is even. So, $aRb \Rightarrow bRa, \forall a,b \in N$ that is R is *symmetric*.

Let $a,b,c \in N$.

If aRb holds then either $a\,\&\,b$ are both even or both odd.

If bRc holds then either $b\,\&\,c$ are both even or both odd.

So, a,b,c are all even or all odd.

If a,b,c are all even then $a^2 + c$ is even. If a,b,c are all odd then $a^2 + c$ is also even. In either case, aRb, $bRc \Rightarrow aRc$

that is R is *transitive*.

Example 30. If R by the relation on N (the set of all positive integers) such that $(a,b) \in R$ if $3a + 4b = 7n$ for some integer n.

Consider the following propositions:

L. R is reflexive   M. R is symmetric   N. R is transitive

Which one of the following statements is TRUE?

(A) L and M are true but N is false   (B) L and N are true but M is false

(C) M and N are true but L is false   (D) L, M, N are all true

Answer: (D)

$\forall a \in N, 3a + 4a = 7a \Rightarrow (a, a) \in R \Rightarrow R$ is *reflexive*.

$\forall a, b \in N, aRb \Rightarrow 3a + 4b = 7n$

Now, $3b + 4a = (7a + 7b) - (3a + 4b) = 7(a + b) - 7n = 7(a + b - n)$

As $(a + b - n)$ is an integer, so, $aRb \Rightarrow bRa$ so, R is *symmetric*.

$\forall a, b, c \in R, aRb, bRc$ both hold. So, $3a + 7b = 7m$ & $3b + 7c = 7n$

Now, $3a + 4c = (7m - 4b) + (7n - 3b) = 7(m + n - b)$ can be written as $7n$ as m, n, b are all integers.

$\forall a, b, c \in R, aRb, bRc \Rightarrow aRc \Rightarrow R$ is *transitive*.

Example 31. The binary operator $\neq$ is defined by the following truth table:

p	q	$p \neq q$
0	0	0
1	0	1
0	1	1
1	1	0

Which one of the following is true about the binary operator $\neq$?

(A) Both commutative and associative

(B) Commutative but not associative

(C) Not commutative but associative

(D) Neither commutative nor associative   [CS/IT 2015]

Answer: (A)

$1 \neq 0 = 1 = 0 \neq 1 \Rightarrow$ Commutative

$1 \neq (1 \neq 0) = 1 \neq 1 = 0$ & $(1 \neq 1) \neq 0 = 0 \neq 0 = 0$

$\Rightarrow 1 \neq (1 \neq 0) = (1 \neq 1) \neq 0 \Rightarrow$ Associative.

Example 32. Consider the set of all function

$f : (0,1,2,---,1024) \rightarrow (0,1,2,---,1024)$ such that $f(f(i)) = i$, for all $0 \leq i \leq 1024$. Consider the following statements:

P. For each such function it must be the case that for every $i, f(i) = i$.

Q. For each such function it must be the case that for some $i, f(i) = i$.

R. Each such function must be onto.

Which one of the following is CORRECT?

(A)   P, Q and R are true   (B) Only Q and R are true

(C)   Only P and Q are true   (D) Only R are true   [CS/IT 2014]

Answer: (B)

If $f(i) = k$, then $f(f(i)) = f(f(k)) = i$.

Since the values of $i \& k$ will be equal in some case/s, the domain set and the co-domain set will intersect. So, Q is true.

but this is not the case for all values of i. So, P is not true.

As i ranges from 0 to 1024 so it may take 2015 different values. From the definition of a function, corresponding to each input there should be a unique output. The domain and the co domain are same. So, the function is *onto* which proves that R is true.

Example 33. The number of onto functions (surjective functions) from set X = {1, 2, 3, 4} to set Y = {a, b, c} is

(A)   36   (B)   64   (C)   81   (D)   72   [CS/IT 2015]

Answer: (A)

A function $f : X \rightarrow Y$ is called surjective if for all $y \in Y$, there exists $x \in X$ such that $f(x) = y$.

There are $^4c_2 = 6$ pairs {(1,2),(1,3),(1,4),(2,3,(2,4),(3,4)}

For every pair $(a,b) \in set$ {1,2,3,4} there are 3 ! = 6 onto functions.

Total number of onto functions = $6 \times 6 = 36$

Example 34. Let N be the set of natural numbers.

Consider the following sets :

P. Set of rational numbers (positive and negative)

Q. Set of functions from {0, 1} to N

R. Set of functions from N to {0, 1}

S. Set of finite subsets of N

Which of the sets above are countable?

(A) Q and S only   (B) P and S only

(C) P and R only   (D) P,Q and S only [CS/IT 2018]

Answer: (D)

Set of rational numbers is countable so, P is TRUE.

0 can be assigned in N ways, 1 can be assigned in N ways.

So, there are $N \times N$ functions and the set N is countable.

As the cross product of countable sets is countable so Q is TRUE.

As *finite* subsets of N contain finite number of natural numbers so, they are countable. S is TRUE.

Each of the numbers 0 and 1 represents a subset of N so, the set of such functions represent the power set of N which is uncountable by Cantor's theorem. [Power set of a countable infinite set is always uncountably infinite. So, R is FALSE.

Hence, P,Q & S are only countable.

Example 35. A function $f : N^+ \rightarrow N^+$ defined on the set of positive integers N^+, satisfied the following properties:

$f(n) = f(n/2)$, if n is even

$f(n) = f(n+5)$, if n is odd

Let $R = \{i / \exists\, j : f(j) = i\}$ be the set of distinct values that f takes.

The maximum possible size of R is ____.   [CS/IT 2016]

Answer: 2

Let $f(1) = x\, and\, f(5) = y$, then $f(2) = f(2/2) = f(1) = x$

$f(3) = f(3+5) = f(8) = f(8/2) = f(4) = f(4/2) = f(2) = x$

$f(4) = f(4/2) = f(2) = x$ and

$f(5) = y, f(10) = f(10/2) = f(5) = y$

So, when n is any positive integer $\neq$ (5 or any multiple of 5), $f(n) = x$

and n is any positive integer $=$ (5 or any multiple of 5), $f(n) = y$

So, f has two distinct values only.

Example 36. Let S denote the set of all functions $f : \{0,1\}^4 \rightarrow \{0,1\}$.

Denote by N the number of functions from S to the set $\{0,1\}$. The value of

$\log_2 \log_2 N$ is

(A)   12   (B)   13   (C)   15   (D)   16   [CS/IT 2014]

Answer: (D)

The given mapping S is is defined by $f : \{0,1\}^4 \rightarrow \{0,1\}$.

No. of functions from S is 2^{16} N is defined by $f : S \rightarrow (0,1)$

No. of functions from S to {0,1} should be 2^S.

$\log_2 \log_2 N = \log_2 S = \log_2 2^{16} = 16$

Example 37. If V_1 & V_2 are 4 dimensional sub spaces of a 6 dimensional

vector space V, then the smallest possible dimension of $V_1 \cap V_2$

is _____   [CS/IT 2014]

Answer: 2

$\dim\left(V_1 \cup V_2\right) \leq 6$ and $\dim\left(V_1\right) = 4$, $\dim\left(V_2\right) = 4$ and $\dim\left(V_1 \cup V_2\right) = x$

$$\dim\left(V_1\right)+\dim\left(V_2\right)-\dim\left(V_1\cap V_2\right)=\dim\left(V_1\cup V_2\right)\leq6$$

Or, $4+4-x\leq6\Rightarrow-x\leq-2\Rightarrow x\geq2$

Example 38. Let u and v be two vectors in R_2 whose Euclidean norms satisfy $|u|=2|v|$. What is the value of α such that $w=u+\alpha v$ bisects the angle between u and v?

(A) 2   (B) 1/2   (C) 1   (D)   - 1/2   [CS/IT 2017]

Answer: (A)

$|u|=2|v|=|2v|$. So, the vectors u and $2v$ have the same magnitude in directions of u and v. Their sum $u+2v$ bisects the angle between u and v. So, $u+\alpha v=u+2v\Rightarrow\alpha=2$

EXERCISE

Question of one mark:

1. The number of relations from A = {a, b, c} to B = {1, 2} is

(A)   9   (B)   8   (C)   32   (D)   64

Question of two marks:

2.Consider the function $f:N\to N$ where *N* is the set of natural numbers including 0. $f(k)=k^2+2$. Then f is

(A) *one - one* but not *onto*   (B)   *onto* but not *one - one*
(C) both *one - one* and *onto*   (D)   neither *one - one* nor *onto*

3.Let A be the set of non-zero integers and let $\approx$ be a relation on A × A defined by (a, b) $\approx$ (c, d) whenever ad = bc.
Consider the following propositions:
L. R is reflexive   M. R is symmetric   N. R is transitive
Which one of the following statements is TRUE?
(A) L and M are true but N is false   (B) L and N are true but M is false
(C) M and N are true but L is false   (D) L, M, N are all true

42

4.How many onto (or, surjective)functions are there from an n-element to a 2-element set?

(A) 2^n (B) $2^n - 1$ (C) $2^n - 2$ (D) $2(2^n - 2)$ [CS 2012]

5.Let S = N × N and * be a binary relation on S defined by (a, b)*(c, d) = (a + c, b + d) then

(A) S is an abelian group   (B) S is an cyclic group

(C) S is a group but not abelian   (D) S is a semigroup

Answers:

1. (D)   2. (C)   3. (D)   4. (C)   5. (D)

43

1. *Planar graph*: A graph is a *Plane graph* if it can be drawn on a plane in such a way that any two of its edges either meet only at their end vertices or do not meet at all.

A graph which is isomorphic to a plane graph is *Planar graph.*

A graph which is not planar is a *non-planar graph.*

A graph with parallel edges and loops is called a *pseudo-graph.*

A graph whose each edge has direction from its initial vertex to its terminal vertex is called a *digraph (directed graph).*

A simple graph with exactly one edge between each pair of distinct vertices is called a *complete graph.*

A graph G is called a *connected graph* if there exists a path from one vertex to other vertex; otherwise the graph is *disconnected graph.*

2. *Order and size of a graph*:

The number of vertices (or nodes) in a graph $G = (V, E)$ is called the *order* of the graph G. So, order of G is $|V|$.

The number of edges (or arcs) in a graph $G = (V, E)$ is called the *size* of the graph G. So, size of G is $|E|$.

Results:

3. The number of vertices in a graph $G = (V, E)$ is called the order of the graph G. So, order of G is $|V|$.

4. *Kuratowski's Graphs*: A complete graph with five vertices ((K_5) is *Kuratowski's first graph.*

A regular connected graph with six vertices and nine edges $(K_{3,3})$ is *Kuratowski's second graph.*

5. *Euler's formula*: If a connected planar graph has n vertices, e edges and f regions, then $n - e + f = 2$

A planar graph is with n vertices, e edges, f regions and k number of connected components then $n - e + f - k = 1$

6. *Results:* In any simple connected planar graph with n vertices, e (> 2) edges and f regions

i) $\quad e \geq \dfrac{3}{2} f$ 　　　　　　ii) 　　$e \leq 3n - 6$

iii) A complete graph with 5 vertices (K_5) is non-planar.

iv) Any graph with 4 or lesser number of vertices is planar.

v) Any graph with 8 or lesser number of edges is planar.

7. *Dual of a graph:* In a graph G having n regions $F_1, F_2, \quad , F_n$ -

n points $v_1, v_2, ---, v_n$ are inserted one in each region. A new graph

G^* whose vertices are $v_1, v_2, ---, v_n$ and the line segments joining

these points are edges is called the dual of the graph G.

If the graph G is self dual (i.e. G and G^* are same or isomorphic) with n vertices and e edges, then $e = 2n - 2$.

8. *Relation between a Graph G and its Dual G^*:*

i) A pendant edge of G gives a loop in G^*

ii) A self loop in G gives a pendant edge in G^*

iii) Number of vertices of $G^* = $ Number of regions of G

iv) Number of edges of $G^* = $ Number of edges of G

v) Number of regions of $G^* = $ Number of vertices of G

vi) G^* is always connected even if G is disconnected

9. *Adjacency Matrix:* $G = (V, E)$ be a graph with vertex set V and edge set E. The adjacency matrix of G is an $n \times n$ matrix $A = \left[a_{ij} \right]_{n \times n}$ where

a_{ij} denotes the connectivity of the vertices.

i) For an *undirected graph with no parallel edges*

$$a_{ij} = \begin{cases} 1, & \text{if } v_i v_j \text{ is an edge of } G \\ 0, & \text{if there is no edge between } v_i \text{ \& } v_j \text{ in } G \end{cases}$$

A self loop at the vertex v_i corresponds to $a_{ii} = 1$.

ii) For a *directed graph with no parallel edges*

$$a_{ij} = \begin{cases} 1, & \textit{if an edge in G is directed from } v_i \textit{ to } v_j \\ 0, & \textit{if there is no edge between } v_i \textit{ \& } v_j \textit{ in } G \end{cases}$$

A self loop at the vertex v_i corresponds to $a_{ii} = 1$.

10. *Tree*: A connected graph with n vertices and $(n-1)$ edges is a tree. Leaves of a tree are pendant vertices.

11. *Spanning Tree*: A spanning tree T of a connected, undirected graph G is a sub graph containing all vertices of G and is a tree.

12. *Minimum Spanning Tree (MST)*: A spanning tree T of a weighted graph G is said to be a minimum spanning tree (MST) if the sum of weights of edges of T is minimum.

13. A tree with n vertices has $(n-1)$ edges.

14. *Chromatic number of a complete graph* (K_n):

The chromatic number of a complete graph (K_n) with n vertices is n.

The chromatic number of the graph $K_n - v_i$ obtained by deletion of a vertex v_i from K_n is $n-1$.

15. *Properties:*

i) If a simple graph G is k – vertex colour able but not $(k-1)$ vertex colour able then G is a k – chromatic graph and its chromatic number $\aleph(G)$ is k.

ii) If $\aleph(G) = k$ then k is the minimum number such that G is k-vertex colour able.

iii) If a graph G has n number of vertices, then $\aleph(G) \le n$

iv) If G is a single vertex graph or a null graph then G is 1-chromatic graph.

16. *Chromatic number of a Circuit* (C_n): The chromatic number of a circuit with n vertices (C_n) is

a)   2 if n is even   b) 3 if n is odd.

17. *Chromatic Polynomial*: Let $f(G,x)$ be the number of different colourings of a graph G with x or lesser number of colours. Then $f(G,x)$ is a polynomial of x which is called chromatic polynomial of G.

Results:

i) Chromatic polynomial for a complete graph with n vertices (K_n) is

$$f(K_n, x) = x(x-1)(x-2)------(x-n+1)$$

ii) The chromatic polynomial of a tree with n vertices is $x(x-1)^{n-1}$

Question of one mark:

Example 1. If K_n is a complete graph with n vertices, then all integral values of $n \geq 2$ for which K_n is to be planar is

(A)   2, 3   (B)   2, 4   (C)   3, 4   (D)   3, 5

Answer: (C)

Example 2. A simple graph has 6 nodes, two of degree 4 and four of degree 2. The number of edges of the graph

(A)   6   (B)   12   (C)   10   (D)   8

Answer: (D)

Total degree of the vertices = 2 × 4 + 4 × 2 =16

Sum of degree of all vertices in a graph = 2 × number of edges

$\Rightarrow$ 16 = 2 × number of edges $\Rightarrow$ Number of edges = 8

Example 3. In any simple connected planar graph with 20 vertices, each of degree 3, the number of regions does a representation of this planar graph divide the plane is

(A)   9   (B)   12   (C)   15   (D)   18

Answer: (B)

n = 20, degree of each vertex is 3, so $\sum_i d(v_i) = 3 \times 20 = 60$

But $\sum_i d(v_i) = 2e \Rightarrow 2e = 60 \Rightarrow e = 30$

If f be the number of regions, $n - e + f = 2 \Rightarrow f = 2 - 20 + 30 = 12$

Example 4. In a regular graph $G,$ the degree of each of its vertices is 4. If G determines 10 regions, then the number of vertices of G is
(A)   12   (B)   10   (C)   8   (D)   6
Answer: (C)
$n =$ number of vertices, $e =$ number of edges,
$f =$ number of regions = 10,   Sum of degrees of all vertices = $2e$
$n \times 4 = 2e \Rightarrow e = 2n$
By Euler's formula, $f = e - n + 2 \Rightarrow 10 = 2n - n + 2 \Rightarrow n = 8$

Example 5. The maximum number of vertices in a connected graph having 17 edges is
(A)   16   (B)   18   (C)   20   (D)   22
Answer: (B)
The minimum number of edges in a connected graph with n vertices is $n - 1$. So, the maximum number of vertices with $n - 1$ edges is n.
So, $n - 1 = 17 \Rightarrow n = 18$

Example 6. A regular planar graph G determines eight regions, degree of each vertex being 3. The number of vertices of G is
(A)   12   (B)   10   (C)   8   (D)   6
Answer: (A)
$$f = 8, 2e = \sum \deg ree(v_i) = 3n \Rightarrow e = \frac{3n}{2}$$

By Euler's formula, $f = e - n + 2 \Rightarrow 8 = \dfrac{3n}{2} - n + 2 \Rightarrow 6 = \dfrac{n}{2} \Rightarrow n = 12$

Example 7. Consider the following directed graph:

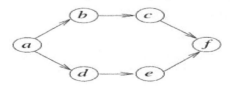

48

The number of different topological orderings of the vertices of the graph is

(A)   1   (B)   2   (C)   4   (D)   6   [CS/IT 2016]

Answer: (D)

The different topological orderings of the vertices of the graph are

a - b - c - d - e - f   a - d - e - b - c - f   a - b - d - c - e - f

a - d - b - c - e - f   a - d - b - e - c - f   a - d - b - e - c - f

Example 8. The condition $e \leq 3n - 6$

(A) is a necessary but not a sufficient condition for planarity

(B) is a sufficient but not a necessary condition for planarity

(C) is a necessary as well as a sufficient condition for planarity

(D) is neither a necessary nor a sufficient condition for planarity

Answer: (A)

Example 9. G and G^* are two graphs dual to each other. Which one of the following equals the 'Number of vertices of G ' ?

(A)   No. Of vertices of G^* (B)   No. Of edges of G^*

(C)   No. Of regions of G^* (D)   No. Of components of G^*

Answer: (C)

It follows from Property 8. v)

Example 10. G and G^* are two graphs dual to each other. Which one of the following equals the 'Number of edges of G' ?

(A)   No. of edges of G^* (B) No. of vertices of G^*

(C)   No. of regions determined by G^* (D) No. of components of G^*

Answer: (A)

It follows from Result 8. iv)

Example 11. If there is one and only path between every pair of vertices in a graph G then G is a

(A)   Kuratowski's First Graph   (B)   Kuratowski's Second Graph

(C)   Disconnected Graph   (D)   Tree

Answer: (D)

Example 12. Let T be a tree with 10 vertices. The sum of the degrees of all the vertices in T is _____.   [CS/IT 2017]

49

Answer: 18

Number of edges = e = v-1 = 10 − 1 = 9

Sum of degrees = 2e = 2 × 9= 18

Example 13. Consider a binary tree with 15 nodes then the minimum and maximum height of the tree is

Note: the height of a tree with single node is zero

(A) 4 and 5 respectively   (B) 4 and 14 respectively

(C) 3 and 14 respectively   (D) 3 and 15 respectively [CS/IT 2017]

Answer: (B)

Minimum height = [log (15)] = 3, Maximum height = 15 − 1 = 14

Example 14. The number of ways a tree on 5 vertices can be coloured with at most 4 colours is

(A)   324   (B)   125   (C)   256   (D)   625

Answer: (A)

Let G be a tree. The chromatic polynomial of G on 5 vertices is

$$f(G,x) = x(x-1)^4 \Rightarrow f(G,4) = 4(4-1)^4 = 324$$

Example 15. A binary tree T has 20 leaves. The number of nodes in T having two children is _____.

(A) 18 (B) 19 (C) 17 (D) Any number between 10 and 20 [CS/IT 2015]

Answer: (B)

$$(k-1)I + 1 = L ----(i)$$

where k = No. of children = 2, L = No. of leaves = 20 and

I = No. Internal nodes.   From $(i), (2-1).I + 1 = 20 \Rightarrow I = 19$

Example 16. Consider a binary tree T that has 200 leaf nodes. Then, the number of nodes in T that have exactly two children are __ [CS/IT 2015]

Answer: 199

Here, $L = 200, k = 2$ Following the formula, $(k-1)I + 1 = L, I = 199$

Example 17. The chromatic number of a circuit (cycle) with 107 edges is

(A)   4   (B)   3   (C)   2   (D)   1

Answer: (B)

A circuit with 107 edges will have 107 vertices.

Since 107 is odd, so $\aleph(C_{107}) = 3$

Example 18. The chromatic number of a circuit (cycle) with n vertices is
(A) 2 if n is even and 3 if n is odd   (B) 3 if n is even and 2 if n is odd
(C) 4 if n is even and 3 if n is odd   (D) 2 if n is even and 4 if n is odd
Answer: (A)

Example 19. The chromatic number of a complete graph with 15 number of vertices is
(A)   12   (B)   13   (C)   14   (D)   15
Answer: (D)

Example 20. The chromatic number of the following graph is [CS/IT 2018]
(A)   2   (B)   4   (C)   3   (D)   5
Answer: (B)

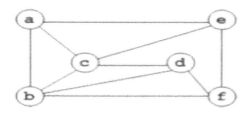

For a planar graph with odd length cycle, the chromatic number is less than or equal to 3 but can not be less than 3.
So, the chromatic number may be 3 or 4.
By using 3 colors red, blue and green we can color all the vertices.
So, chromatic number = 3

Example 21. Which of the following is/are correct in-order traversal sequence(s) of binary search tree(s)?
1.   3, 5, 7, 8, 15, 19, 25   2.   5, 8, 9, 12, 10, 15, 25

3.   2, 7, 10, 8, 14, 16, 25   4.   4, 6, 7, 9, 18, 20, 25

(A) 1 and 4 only   (B) 2 and 3 only
(C) 2 and 4 only   (D) 2 only   [CS/IT 2015]
Answer: (A)

The in-order traversal sequence(s) of binary search tree(s) should be in increasing order. Only 1 and 4 are in increasing order.

Example 22. The post- order traversal of a binary tree is 8, 9, 6, 7, 4, 5, 2, 3, 1. The in- order traversal of the same tree is 8, 6, 9, 4, 7, 2, 5, 1, 3. The height of a tree is the length of the longest path from the root to any leaf. The height of the binary tree above is _____.   [CS/IT 2018]

Answer: 4

A binary tree with post- order and in- order traversals are constructed:

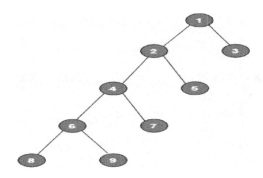

The height of the binary tree above is 4. Let G be a weighted connected undirected graph with distinct positive edge weights. If every edge weight is increased by the same value, then which of the following statements is/are **TRUE**?

P: Minimum spanning tree of G does not change

Q: Shortest path between any pair of vertices does not change

(A)P only (B) Q only (C)Neither P nor Q (D)Both P and Q [CS/IT 2016]

Answer: (A)

There may be different number of edges in different paths from S to T. So, the shortest path may change, Q is FALSE.

In Kruskal's algorithm, the edges are sorted first. So, if all the weights are increased, the order of edges would not change. So, P is TRUE.

Example 23. Consider a binary tree with 15 nodes then the minimum and maximum height of the tree is

Note: the height of a tree with single node is zero

(A) 4 and 5 respectively   (B) 4 and 14 respectively

(C) 3 and 14 respectively   (D) 3 and 15 respectively [CS/IT 2017]

Answer: (C)

Minimum height = [log (15)] = 3, Maximum height = 15 − 1 = 14

Example 24. Let T be a tree with 10 vertices. The sum of the degrees of all the vertices in T is _____.   [CS/IT 2017]

Answer: 18

Question of two marks:

Example 25. If G be a simple planar graph with less than 12 vertices. Then G has a vertex whose least degree is

(A)   6   (B)   3   (C)   8   (D)   4

Answer: (D)

Let G has n number of vertices $v_1, v_2, - - -, v_n$ and e edges.

SO, $n < 12$ If possible, $\deg ree(v_i) \geq 5$ for all i.

Then $\sum \deg(v_i) \geq 5n \Rightarrow 2e \geq 5n$ But for a planar graph

$e \leq 3n - 6 \Rightarrow 5n \leq 2(3n - 6) \Rightarrow 5n \leq 6n - 12 \Rightarrow n \geq 12$ which

contradicts the hypothesis that $n < 12$. So, $degree(v_i) < 4$

Example 26. T is a binary tree on n vertices and p is the number of pendant vertices in T. The number of vertices of degree 3 in T is

(A) $\dfrac{n-1}{2}$ (B) $\dfrac{n-2}{2}$ (C) $\dfrac{n-3}{2}$ (D) $\dfrac{n-4}{2}$

Answer: (C)

Number of edges $= n - 1 \Rightarrow$ total degree $= 2(n - 1)$

Number of total vertices $= n$, number of pendant vertices $= p$,

number of two-degree vertices $= 1$

So, number of three degree vertices $= n - p - 1$

Total degree $= 1 \times p + 1 \times 2 + 3(n - p - 1) = 3n - 2p - 1$

$\Rightarrow 3n - 2p - 1 = 2(n - 1) \Rightarrow p = \dfrac{n+1}{2}$

Number of 3-degree vertices $= n - p - 1 = n - 1 - \dfrac{n+1}{2} = \dfrac{n-3}{2}$

Example **27.** G is an undirected graph with n vertices and 24 edges such that each vertex of G has degree at least 3. Then the maximum possible value of n is _____.   [CS/IT 2017]

Answer: 16

$e = 25, \quad n \leq ?$

Each vertex has at least 3 degree and $2e = \Sigma degree$

So, $2e \geq 3n \Rightarrow n \leq \dfrac{2e}{3} \Rightarrow n \leq \dfrac{2 \times 25}{3} \Rightarrow n \leq 16.6$

The maximum possible value of n is 16

Example **28.** The pre-order traversal of a binary search tree is given by

12, 8, 6, 2, 7, 9, 10, 16, 15, 19, 17, 20.

Then the post-order traversal of this tree is

(A)   2, 6, 7, 8, 9, 10, 12, 15, 16, 17, 19, 20
(B)   2, 7, 6, 10, 9, 8, 15, 17, 20, 19, 16, 12
(C)   7, 2, 6, 8, 9, 10, 20, 17, 19, 15, 16, 12
(D)   7, 6, 2, 10, 9, 8, 15, 16, 17, 20, 19, 12   [CS/IT 2017]

Answer: (B)

A binary search tree is drawn from pre- order nodes.

In pre-order, the first node is the root. So, root is 12.

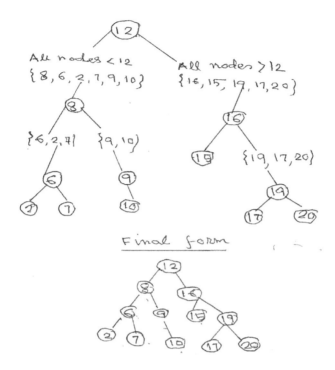

All nodes < 12
{8, 6, 2, 7, 9, 10}

All nodes > 12
{16, 15, 19, 17, 20}

{6, 2, 7} {9, 10}

{19, 17, 20}

Final form

Example 29. G is a non-directed graph with 12 edges. If it has six vertices each of degree 3 and rest have degree less than 3, then the minimum number of vertices of G is

(A)   3   (B)   6   (C)   9   (D)   12

Answer: (C)

Sum of degrees of all vertices = 2 × 12 = 24

Total degree $= 6 \times 3 + \sum_{i=1}^{3} d(V_i) = 24 \Rightarrow \sum_{i=1}^{3} d(V_i) = 6$

If n = number of vertices, then $2n \geq 6 \Rightarrow n \geq 3$

So, the minimum number of vertices of $G = 6 + 3 = 9$

Example 30. Consider the weighted undirected graph with 4 vertices, where the weight of $\{i, j\}$ is given by the entry W_{ij} in the matrix W.

$$W = \begin{bmatrix} 0 & 2 & 8 & 5 \\ 2 & 0 & 5 & 8 \\ 8 & 5 & 0 & x \\ 5 & 8 & x & 0 \end{bmatrix}$$ The largest possible integer value of x, for which

at least one shortest path between some pair of vertices will contain the edge with vertex x is _____.   [CS/IT 2016]

Answer: 12

Let the vertices be 0, 1, 2 and 3. x directly connects vertices 2 to 3. The shortest path (excluding x) from vertex 2 to vertex 3 is of weight 12 (2 - 1 - 0 - 3)

Example 31. Consider the following undirected graph G:

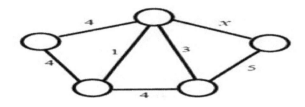

Choose a value for x that will maximize the number of minimum spanning trees (MWSTS) of G. The number of MWSTS of G for this value of x is

(A)   4   (B)   5   (C)   2   (D)   3 [CS/IT 2018]

Answer: (C)

The value of x that will maximize the number of MWSTS is 5 as it will have two more choices for corner vertices.

Following Kruskal's algorithm for MSTS

i) Edges with weights 1 and 3 are selected

ii) Bottom edge with weight 4 will not be considered as it will make cycle on MST.

iii) Both corner vertices have two-two choices to select the vertices.

So, these corner edges with weights 4 and 5 will lead to 2 × 2 = 4 MSTS.

Example 32. Let G be a connected planar graph with 10 vertices. If the number of edges on each face is three, then the number of edges in G is

(A)   24   (B)   20   (C)   32   (D)   64 [CS/IT 2015]

Answer: (A)

For a finite planar graph without edge intersection, $v - e + f = 2 - -(i)$

Here, $v = 10$, $2e = 3f \Rightarrow f = \dfrac{2e}{3}$ [Every edge is shared by 2 faces]

From (i), $10 - e + \dfrac{2e}{3} = 2 \Rightarrow e = 24$

Example 33. G = (V, E) be any connected undirected edge- weighted graph. The weights of the edges in E are positive and distinct.
Consider the following statements:
I. Minimum Spanning Tree of G is always unique
II. Shortest path between any two vertices of G is always unique.
Which of the following statements is/are necessarily true?
(A) I only (B) II only   (C) both I and II (D) neither I nor II   [CS/IT 2017]
Answer: (A)
As the graph G contains distinct weight edge then MST is always unique.
So, (I) is true.
The shortest path between any two vertices need be unique.
So, (II) is not necessarily true.

Example 34. The graph shown below has 8 edges with distinct integer edge weights. The minimum spanning tree (MST) is of weight 36 and contains the edges: {(A, C), (B, C), (B, E), (E, F), (D, F)}. The edge weights of only those edges which are in the MST are given in the figure are shown below. The minimum possible sum of weights of all 8 edges of this graph is

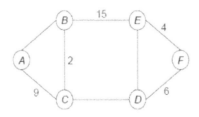

(A)   66   (B)   69   (C)   68   (D)   70   [CS/IT 2015]
Answer:(B)
In every cycle, the weight of an edge that is not part of MST must be greater than or equal to weights of other edges that are part of MST.

So, the minimum possible weight of ED is 7[> (6, 4)],
of CD is 16 [> (15, 2)], of AB is [10 > (9, 2)]
So, the minimum possible sum of weights of all 8 edges
= 10 + 15 + 4 + 6 + 16 + 7 + 2 + 9 = 69

Example 35. The height of a tree is the length of the longest root-to-leaf
path in it. The maximum and minimum number of nodes in a binary tree
of height 5 are

(A) 63 and 6, respectively   (B) 64 and 5, respectively

(C) 32 and 6, respectively   (D) 31 and 5, respectively   [CS/IT 2015]

Answer: (A)

Number of nodes is maximum for a perfect binary tree.

A perfect binary tree of height h has $2^{h+1} - 1$ nodes. Here $h = 5$

The maximum number of nodes $= 2^{5+1} - 1 = 63$

The number of nodes is minimum for a skewed binary tree.

If its height h number of nodes $= h + 1 = 5 + 1 = 6$

Number of edges = e = 10 − 1 = 9   Sum of degrees = 2e = 2 × 9 = 18

Example 36. Let G be a connected undirected graph of 100 vertices and
300 edges. The weight of a minimum spanning tree of G is 500. When
the weight of each edge of is G is increased by five, the weight of a
minimum spanning tree becomes _____.

(A)   1000   (B)   995   (C)   2000   (D)   1995   [CS/IT 2015]

Answer: (B)

As there are 100 vertices, the no. of edges in MST is 99.

If the weight of every edge is increased by 5, the increase in weight of
MST is 99 × 5 = 495

So, after increase the weight of MST becomes 500 + 495 = 995

Example 37. Let G be a graph with 100 vertices, with each vertex labeled
by a distinct permutation of numbers 1, 2, ---------, 100. There is an edge
between vertices u and v if and only if the label of u can be obtained by
swapping two adjacent numbers in the label of v. Let y denote the
degree of a vertex in $G,$ and z denote the number of connected
components in G. Then $y + 10z = $ _____

(A)   109   (B)   110   (C)   119   (D) None of these [CS/IT 2018]
Answer: (A)

There is an edge between the vertices u & v iff the label of u can be obtained by swapping two adjacent numbers in the label of v. Then the set of swapping numbers will be {(1, 2), (2 , 3), ----, (9,9)}.
There will be 99 such sets, that is the number of edges = 99
Each vertex will have 99 edges $\Rightarrow y = 99$

As the vertices are connected together, so the number of connected components formed $= 1 \Rightarrow z = 1$
So, $y + 10z = 99 + 10 \times 1 = 109$

***Example* 38.** If a graph G has at least one edge then the sum of the coefficients of its chromatic polynomial is _____.
Answer: 0
The chromatic polynomial of G is
$$f(G,x) = a_0 x^n + a_1 x^{n-1} + - - - + a_n - - - - - (i)$$
As G has at least one edge then at least two vertices of G are adjacent.
So, at least two colours are necessary for the colouring of
$G.\, \aleph(G) \geq 2 \Rightarrow f(G,1) = 0$

From (i) $f(G,1) = a_0 1^n + a_1 1^{n-1} + - - - + a_n = 0$

EXERCISE

Question of one mark:

1. A regular graph G determines 8 regions, degree of each vertex being 3. The number of vertices of G is
(A)   16   (B)   18   (C)   20   (D)   24

2. A connected planar graph has 5 edges and 3 vertices. The number of regions determined by the graph is
(A)   4   (B)   2   (C)   5   (D)   6

3. A non-directed graph G has 8 edges. If the degree of each vertex is 2, then the number of vertices of G is
(A)   4   (B)   6   (C)   8   (D)   10

4. For each graph G the constant term in its chromatic polynomial is _.

Question of two marks:

5. Let G be a complete undirected graph on 4 vertices, having 6 edges with weights being 1, 2, 3, 4, 5 and 6. The maximum possible weight that a minimum weight spanning tree of G can have is _____.   [CS/IT 2016]

6. G = (V, E) is an undirected simple graph in which edge has a distinct weight, and e is a particular edge of G. Which of the following statements about the minimum spanning trees (MSTs) of G is/are TRUE?
I. If e is the lightest edge of some cycle in G, then every MST of G includes e.
II. If e is the heaviest edge of some cycle in G, then every MST of G excludes e.
(A) I only (B) II only   (C) both I and II (D) neither I nor II   [CS/IT 2016]

7. While inserting the elements 71, 65, 84, 69, 67,83 in an empty binary search tree (BST) in the sequence shown. The element in the lowest level is
(A)   65   (B)   67   (C)   69   (D)   83 [CS/IT 2015]

8. Let δ denote the minimum degree of a vertex in a graph. For all planar graphs on n vertices with $\delta \geq 3$, which one of the following is TRUE?

(A) In any planar embedding, the number of faces is at least $\dfrac{n}{2} + 2$

(B) In any planar embedding, the number of faces is less than $\dfrac{n}{2} + 2$

(C) There is a planar embedding in which the number of faces is less than $\dfrac{n}{2} + 2$

(D) There is a planar embedding in which the number of faces is at most $\dfrac{n}{\delta + 1}$

[CS/IT 2014]

9. A graph is self-complementary if it is isomorphic to its complement. For all self-complementary graphs on n vertices, n is
(A) A multiple of 40   (B) Even
(C) Odd   (D) Congruent to 0 mod 4 or 1 mod 4 [CS/IT 2015]

10. In a connected graph, a bridge is an edge whose removal disconnects a graph. Which one of the following statements is true?
(A) A tree has no bridges (B) A bridge can not be part of a simple cycle
(C) Every edge of a clique with size ≥ 3 is a bridge (A clique is any complete sub graph of a graph)
(D) A graph with bridges can not have a cycle   [CS/IT 2015]

Answers:

1. (B) 2. (A) 3. (C) 4. 0 (zero) 5. (7) 6. (B) 7. (B) 8. (A) 9. (D) 10. (B)

- Counting Techniques
- Recurrence Relations
- Generating Functions
- *Counting Techniques*

It deals with the number of ways of arranging or choosing objects from a finite set according to some specified rule/rules. It deals with the problems of *Permutations* and *Combinations*.

1. Permutation: An ordered arrangement of r elements of a set containing n distinct elements is called r-permutation of n elements and is denoted by $P(n,r), r \leq n.$ $\quad P(n,r) = \dfrac{n!}{(n-r)!}, P(n,n) = n!$

2. Combination: An unordered selection of r elements of a set containing n distinct elements is called r-combination of n elements and is denoted by $C(n,r), r \leq n.$ $C(n,r) = \dfrac{n!}{r!(n-r)!}, C(n,n) = 1, \ C(n,r) = \dfrac{P(n,r)}{P(r,r)}$

3. Pascal,s identity: $C(n,r) + C(n,r-1) = C(n+1,r)$

rinciple of Inclusion and Exclusion:

i) If $A \& B$ are finite subsets of a finite universal set U, then

$|A \cup B| = |A| + |B| - |A \cap B|,$ where $|A|$ denotes the number of distinct elements (cardinality) of te set A etc.

ii) If $A, B \& C$ are finite subsets of a finite universal set U, then

$|A \cup B \cup C| = |A| + |B| + |C| - |A \cap B| - |B \cap C| - |A \cap C| + |A \cap B \cap C|$

where $|A|$ denotes the number of distinct elements (cardinality) of te set A etc.

4. If n pigeons are accommodated in m pigeonholes $(n > m)$, then one of the pigeonholes must contain at least $\left[\dfrac{n-1}{m}\right]+1$ pigeons where $[x]$ denotes the greatest integer $\le x$ (real).

Question of one mark:

Example 1. The number of bit strings of length 10 containing exactly four 1's is

(A)   250   (B)   210   (C)   120   (D)   48

Answer: (B)

10 positions may be filled by four 1's and six 0's in

$$\frac{10!}{4!6!}=\frac{7.8.9.10}{2.3.4}=210$$

Example 2. The number of bit strings of length 10 containing at most three 1's is

(A)   196   (B)   144   (C)   176   (D)   268

Answer: (C)

The cases consist of:

no 1and 10 zeroes, one 1 and 9 zeroes, two 1 and 8 zeroes, three 1 and 7 zeroes,

$$=\frac{10!}{0!10!}+\frac{10!}{1!9!}+\frac{10!}{2!8!}+\frac{10!}{3!7!}=1+10+45+120=176$$

Example 3. The number of bit strings of length 10 containing an equal number of 1's and 0's is

(A)   252   (B)   256   (C)   276   (D)   296

Answer: (A)

10 positions may be filled by five 1's and five 0's in

$$\frac{10!}{5!5!}=\frac{6.7.8.9.10}{2.3.4.5.6}=252$$

Example 4. How many different paths in the x-y plane are there from the point (1, 3) to (5, 6), if a path proceeds one step at a time by going one step to the right ® or one step upward (U)?

(A)   45   (B)   55   (C)   25   (D)   35

Answer: (D)

TO reach the point (5, 6) from (1, 3), one has to cover (5 - 1) = 4 steps to the right and (6 - 3) = 3 steps upwards. Total 7 steps of which 4 R & 3 U.

It can be done in $\dfrac{7!}{4!3!} = \dfrac{5.6.7}{3.2} = 35$

Example 5. In how many ways can five examinations be scheduled in a week so that no two examinations are scheduled on the same day considering Sunday as a holiday?

(A)   6   (B)   10   (C)   15   (D)   30

Answer: (A)

Excluding Sunday, 5 examinations are to be scheduled on 6 days so that there will be at most one examination in a day.

It can be done in $^6P_5 = \dfrac{6!}{5!} = 6$

Example 6. The number of divisors of 2100 is

(A)   42   (B)   36   (C)   78   (D)   72 [CS/IT 2015]

Answer: (B)

$2100 = 2^2 \times 5^2 \times 3^1 \times 7^1$

So, the number of divisors of 2100 is (2 + 1)(2 + 1)(1 + 1)(1 + 1) = 36

Example 7. The number of integer solutions of the equation $x_1 + x_2 + x_3 + x_4 = 32 \ (x_i > 0, 1 \le i \le 4)$ is

(A)   4495   (B)   4385   (C)   4275   (D)   4165

Answer: (A)

As the solutions are integer $x_i \ge 1, \ 1 \le i \le 4$

Put $u_i = x_i - 1$ so that $u_i \ge 0, \ 1 \le i \le 4$

The given equation becomes $u_1 + u_2 + u_3 + u_4 = 28$ for which the number of non-negative solution is C(4 + 28 - 1, 28) = C(31, 28)

$= \dfrac{31!}{28!3!} = \dfrac{31.30.29}{3.2} = 31.5.29 = 4495$

Example **8.** The number of non-negative integer solutions of the inequality $x_1 + x_2 + x_3 + x_4 + x_5 + x_6 < 10$ is

(A)   5035   (B)   5025   (C)   5015   (D)   5005

Answer: (D)

We convert the inequality to an equality by introducing an auxiliary variable $x_7 > 0$

So, the inequality becomes

$x_1 + x_2 + x_3 + x_4 + x_5 + x_6 + x_7 = 10, x_i \geq 0, i = 1,2,--,6 \& x_7 > 0$

Put $u_i = x_i, i = 1,2,--,6 \& u_7 = x_7 - 1,$

the above equation becomes

$u_1 + u_2 + u_3 + u_4 + u_5 + u_6 + u_7 = 10 - 1 = 9, u_i \geq 0$ for $i \leq i \leq 7$

The number of non-negative solution is C(7 + 9 - 1, 9) = C(15, 9)=5005

Example **9.**The number of 4 digit numbers having their digits in non-decreasing order (from left to right) constructed by using the digits belonging to the set {1, 2, 3} is _____.   [CS/IT 2015]

Answer: 15

The arrangements are

(1,1,1,1)   (1,1,1,2)   (1,1,1,3)   (1,1,2,3)   (1,1,2,2)   (1,1,3,3)
(1,2,2,2)   (1,2,2,3)   (1,2,3,3)   (1,3,3,3)
(2,2,2,2)   (2,2,2,3)   (2,2,3,3)   (2,3,3,3)
(3,3,3,3)

Total number of arrangements = 15

Example **10.** In a certain programming language, variable should be of length three and be made up of two letters followed by a digit or of length two made up of one letter followed by a digit. If repetition of letters is not allowed then the possible number of variables is

(A)   4440   (B)   5550   (C)   6760   (D)   7320

Answer: (C)

(a) Total variables of length three = 26 × 25 × 10 = 6500

(b) Total variables of length two = 26 × 10 = 260

Total variables of length three or two = 6500 + 260 = 6760

Coefficient of $x^{12} = \dfrac{3.(3+1)(3+2)}{6} = 10$

Example 11. A word that reads the same when read in forward or backward is called as 'palindrome'. The number of seven letter palindromes that may be formed from English alphabets is _____.

Answer: 26^4

Here, the first letter is same as the seventh, the second letter is same as the sixth, the third letter is same as the fifth, the fourth letter is fixed.

So, first four positions are to be filled with 26 letters. Answer is 26^4.

Example 12. The coefficient of x^{12} in $\left(x^3 + x^4 + x^5 + ---\right)^3$
is _   [CS/IT 2016]

Answer: 10

$$\left(x^3 + x^4 + x^5 + ---\right)^3 = x^9\left(1 + x + x^2 + ---\right)^3 = x^9\left\{(1-x)^{-1}\right\}^3$$

$$= x^9(1-x)^{-3} = x^9\left\{1 + 3x + \frac{3(3+1)}{2!}x^2 + \frac{3.(3+1)(3+2)}{3!}x^3 + --\right\}$$

Coefficient of x^{12} is $= \dfrac{3(3+1)(3+2)}{3!} = 10$

Question of two marks:

Example 13. $\displaystyle\sum_{x=1}^{99} \frac{1}{x(x+1)} =$ [CS/IT 2015]

Answer: 0.99

$$\sum_{x=1}^{99} \frac{1}{x(x+1)} = \sum_{x=1}^{99}\left(\frac{1}{x} - \frac{1}{x+1}\right)$$

$$= \left(1 - \frac{1}{2}\right) + \left(\frac{1}{2} - \frac{1}{3}\right) + --- + \left(\frac{1}{98} - \frac{1}{99}\right) + \left(\frac{1}{99} - \frac{1}{100}\right)$$

$$= 1 - \frac{1}{100} = \frac{99}{100} = 0.99$$

Example 14. There are 250 students in an Engineering college. Of them 188 have taken a course in FORTRAN, 100 in C, 35 in PYTHON. Further 88 have taken a course in both FORTRAN & C, 23 in C & PYTHON and 29 in FORTRAN & PYTHON. If 19 of them have taken all the three courses then the number of students not having any of the three programming languages is

(A)　　35　　(B)　　48　　(C)　　56　　(D)　　64

Answer: (B)

If F, C and P denote the sets of students taking the programming languages FORTRAN, C and PYTHON respectively, then

$$|F| = 188, |C| = 100, |P| = 35, |F \cap C| = 88, |C \cap P| = 23, |F \cap J| \ \&$$
$$|F \cap C \cap P| = 19$$

The number of students having at least one of the three programming languages is given by

$$|F \cup C \cup P| = |F| + |C| + |P| - |F \cap C| - |C \cap P| - |F \cap P| + |F \cap C \cap P|$$
$$= 188 + 100 + 35 - 88 - 23 - 29 + 19 = 342 - 140 = 202$$

The number of students NOT having any of the three programming languages = 250 - 202 = 48

Example 15. The number of integers between 1 and 250 (both inclusive) that are not divisible by any of the three integers 2, 3 and 5 is

(A)　　76　　(B)　　68　　(C)　　66　　(D)　　72

Answer: (C)

If A, B and C denote the sets of integers lying between 1 and 250 that are divisible by 2, 3 and 5 respectively, then

$$|A| = \left[\frac{250}{2}\right] = 125, |B| = \left[\frac{250}{3}\right] = 83, |C| = \left[\frac{250}{5}\right] = 50$$

$$|A \cap B| = \left[\frac{250}{6}\right] = 41, |B \cap C| = \left[\frac{250}{15}\right] = 16, |A \cap C| = \left[\frac{250}{10}\right] = 25$$

and $|A \cap B \cap C| = \left[\frac{250}{30}\right] = 8$ where $[x]$ is the greatest integer $\leq x$.

The number of integers between 1 and 250 (both inclusive) that are divisible by at least one of the integers 2, 3, 5 is given by

$$|A \cup B \cup C| = |A| + |B| + |C| - |A \cap B| - |B \cap C| - |A \cap C| + |A \cap B \cap C|$$
$$= 125 + 83 + 50 - 41 - 16 - 25 + 8 = 266 - 82 = 184$$

The number of integers between 1 and 250 (both inclusive) that are NOT divisible by any of the three integers 2, 3 and 5 is 250 - 184 = 66

Example 16. The number of integers between 1 and 500 (both are inclusive) that are divisible by 3 or 5 or 7 is ____. [CS/IT 2017]

Answers: 271

Let P, Q, R be the sets of numbers lying between 1 and 500 and divisible by 3, 5, 7 respectively.

$$n(P) = \left[\frac{500}{3}\right] = 166, n(Q) = \left[\frac{500}{5}\right] = 100, n(R) = \left[\frac{500}{7}\right] = 71$$

$$n(P \cap Q) = \left[\frac{500}{15}\right] = 33, n(Q \cap R) = \left[\frac{500}{35}\right] = 14, n(P \cap R) = \left[\frac{500}{21}\right] = 23$$

and $n(P \cap Q \cap R) = \left[\frac{500}{105}\right] = 4$

No. of numbers are divisible by 3 or 5 or 7 is

$n(P \cup Q \cup R)$
$= n(P) + n(Q) + n(R) - n(P \cap Q) - n(Q \cap R) - n(P \cap R) + n(P \cap Q \cap R)$
$= 166 + 100 + 71 - 33 - 14 - 23 + 4 = 271$

Example 17. A man hiked for 10 hours and covered a total distance of 45 km. It is known that he hiked 6 km in the first hour and only 3 km in the last hour. The least distance (in km) he must have hiked within a certain period of two consecutive hours is

(A)   6   (B)   7   (C)   8   (D)   9

Answer: (D)

He has hiked (36 - 6 - 3) = 27 km in 2nd to 9th hour, the consecutive two hour periods are (2 & 3), (4 & 5), (6 & 7), (8 & 9) - totally 4 time periods.

We take number of pigeons $= n = 36 \, km$ and

number of pigeonholes $= m = 4$ time periods.

Least distance hiked within a certain period of two consecutive hours
= Least number of pigeons accodommated in 4 pigeonholes

$$= \left[\frac{n-1}{m}\right] + 1 = \left[\frac{36-1}{4}\right] + 1 = [8.75] + 1 = 9 \, km.$$

● *Recurrence Relations or Difference Equations:*

An equation that expresses a_n, the general term of the sequence $\{a_n\}$ in terms of one or more of the evious terms of the sequence is called a *Recurrence Relation* (or, a *Difference Equation*) for $\{a_n\}$.

A linear recurrence relation of 2nd degree with constant coefficients can be written as $c_0 a_n + c_1 a_{n-1} + c_2 a_{n-2} = f(n)$.
The recurrence relation is homogeneous if $f(n) = 0$.

Question of one mark:

Formulation of Recurrence relations:
Example 18. Let a_n be the number of $n-$bit strings that do NOT contain two consecutive 1's. Which one of the following is the recurrence relation for a_n?

(A) $\quad a_n = a_{n-1} + 2a_{n-2}$ (B) $\quad a_n = a_{n-1} + a_{n-2}$

(C) $\quad a_n = 2a_{n-1} + a_{n-2}$ (D) $\quad a_n = 2a_{n-1} + 2a_{n-2}$ [CS/IT 2016]

Answer: (B)

It may be of the form 1 0 1. If $a_{n-2} = 1$, then $a_{n-1} = 0$ & $a_n = 1$

$\Rightarrow a_n = a_{n-1} + a_{n-2}$

Example 19. Which one of the following is the recurrence equation for the worst case time complexity of the Quicksort algorithm for sorting $n(\geq 2)$ numbers? In the options given below, c is a constant.
(A) $T(n) = 2T(n/2) + cn$ (B) $T(n) = T(n-1) + T(1) + cn$
(C) $T(n) = 2T(n-1) + cn$ (D) $T(n) = T(n/2) + cn$ [CS/IT 2015]
Answer: (B)
For worst case time,the pivot is in the central position and
i) sub-array in the left of pivot is of size $(n-1)$
ii) sub-array in the right of pivot is of size 0.

Example 20. The recurrence relation corresponding to the Fibonacci sequence is _____.
Answer: $F_{n+2} = F_{n+1} + F_n$
The Fibonacci numbers are 0, 1, 1, 2, 3, 5, 8, 13, $------$

$$F_0 = 0, F_1 = 1, F_{n+2} = F_{n+1} + F_n$$

Example 21. Let $a_n = \sum_{k=1}^{n} k^2$. The recurrence relation for a_n is ____

Answer: $a_n - a_{n-1} = n^2$

$a_n = \sum_{k=1}^{n} k^2 = 1^2 + 2^2 + - - + (n-1)^2 + n^2, a_{n-1} = 1^2 + 2^2 = - - - + (n-1)^2$

$\Rightarrow a_n - a_{n-1} = n^2$

Example 22. There are n guests in a hall. Each person shakes hand with everybody else exactly once. If a_n denotes the number of handshakes that occur recursively, then the recurrence relation for a_n is ____.

Answer: $a_n = a_{n-1} + (n-1); n \geq 2, a_1 = 0$

As no person handshakes with himself so $a_1 = 0, n \geq 2$. Consider one of the guests, say, G. The number of handshakes by the remaining $(n-1)$ guests is a_{n-1}. G shakes hand with $(n-1)$ guests, yielding $(n-1)$ additional handshakes. So, $a_n = a_{n-1} + (n-1); n \geq 2, a_1 = 0$

Example 23. It is given that white tiger population of Odissa was 30 at time $t = 0$ and 32 at time $t = 1$. The recurrence relation for growth rate of tiger is _____.

Answer: $t_n - 3t_{n-1} + 2t_{n-2} = 0$.

$t_n - t_{n-1} = 2(t_{n-1} - t_{n-2}) \Rightarrow t_n - 3t_{n-1} + 2t_{n-2} = 0$.

Example 24. Consider the recurrence function $T(n) = \begin{cases} T(n/2) + 1, n \geq 2 \\ 1 \qquad\qquad n = 1 \end{cases}$

Then $T(n)$ in terms of Θ notation is

(A) $\Theta(\log\log n)$ (B) $\Theta(\log n)$ (C) $\Theta(\sqrt{n})$ (D) $\Theta(n)$

Answer: (B)

Let us tabulate the values the values of of the recurrence on the first few powers of 2 as

70

n	$1 = 2^0$	$2 = 2^1$	$4 = 2^2$	$8 = 2^3$	$16 = 2^4$	$32 = 2^5$	-----
$T(n)$	1	1	2	3	4	5	

$T(n) = k$ where $n = 2^k \Rightarrow T(n) = \log_2 n =$

Now, $\sqrt[k]{n} = 2 \Rightarrow n = 2^k \Rightarrow k = \log_2 n$ So, $T(n) = \Theta(\log n)$

Example 25. Consider the recurrence function

$$T(n) = \begin{cases} 2T(\sqrt{n}) + 1, n > 2 \\ 2 \qquad\quad 0 < n \le 2 \end{cases}$$ Then $T(n)$ in terms of Θ notation is

(A) $\Theta(\log\log n)$ (B) $\Theta(\log n)$ (C) $\Theta(\sqrt{n})$ (D) $\Theta(n)$ [CS/IT 2017]

Answer: (B)

$T(n) = 2T(\sqrt[2]{n}) + 1 = ---- = 2^k T\left(\sqrt[k]{n}\right) + k$

Now, $\sqrt[k]{n} = 2 \Rightarrow n = 2^k \Rightarrow k = \log_2 n$ So, $T(n) = \Theta(\log n)$

Methods of solution of Recurrence Relation:

$c_0 a_n + c_1 a_{n-1} + c_2 a_{n-2} = f(n)$:

<u>*Question of two marks:*</u>

i) Iterative Method:

Example 26. The solution of recurrence relation

$a_n = a_{n-1} + (n-1), n \ge 2, a_1 = 0$ is $a_n =$

(A) $\dfrac{(n+1).n}{2}$ (B) $\dfrac{(n-1).n}{2}$ (C) $\dfrac{(n-1).(n+1)}{2}$ (D) n^2

Answer: (B)

$a_n = a_{n-1} + (n-1) \Rightarrow a_{n-1} = a_{n-2} + (n-2)$

$\Rightarrow a_n = a_{n-2} + (n-2) + (n-1)$

$\Rightarrow a_n = a_{n-3} + (n-3) + (n-2) + (n-1)$

$\Rightarrow a_n = a_1 + 1 + 2 + 3 + --- + (n-2) + (n-1)$

$= 0 + 1 + 2 + 3 + --- + (n-1)$

$$= \frac{(n-1)\{(n-1)+1\}}{2} = \frac{(n-1).n}{2}$$

Example 27. The solution of recurrence relation

$t_n = t_{n-1} + 4n, n \geq 1, t_0 = 0$ is $t_n =$

(A) $\frac{(n+1).n}{2}$ (B) $n(n+1)$ (C) $2n(n+!)$ (D) n^2

Answer: (C)

$t_n = t_{n-1} + 4n \Rightarrow t_{n-1} = t_{n-2} + 4(n-1)$

$\Rightarrow t_n = t_{n-2} + 4(n-1) + 4n \Rightarrow t_n = t_{n-3} + 4(n-2) + 4(n-1) + 4n$

$\Rightarrow t_n = t_0 + 4.1 + 4.2 + 4.3 + - - - + 4.n = 0 + 4(1 + 2 + 3 + - - - + n)$

$= 4.\frac{n(n+1)}{2} = 2n(n+1)$

Example 28. Consider the recurrence relation $a_1 = 8, a_n = 6n^2 + 2n + a_{n-1}$.

Let $a_{99} = K \times 10^4$. The value of K is ____.   [CS/IT 2016]

Answer: 198

$a_n - a_{n-1} = 6n^2 + 2n$

$n = 99, a_{99} - a_{98} = 6 \times 99^2 + 2 \times 99$

$n = 98, a_{98} - a_{97} = 6 \times 98^2 + 2 \times 98$ ----------------------

$n = 3, a_3 - a_2 = 6 \times 3^2 + 2 \times 3$

$n = 2, a_2 - a_1 = 6 \times 2^2 + 2 \times 2$

Adding,

$a_{99} - a_1 = 6(99^2 + 98^2 + - - - + 2^2) + 2(99 + 98 + - - - + 2)$

$$a_{99} = 6([99^2 + 98^2 + ---- + 2^2 + 1^2] - 1^2) + 2([99 + 98 + ---- + 2 + 1] - 1) + a_1$$

$$= 6 \cdot \frac{99.(99+1)(2.99+1)}{6} - 6 + 2 \cdot \frac{99.(99+1)}{2} - 2 + 8 = 1970100 + 9900$$

$$= 1980000 = 198 \times 10^4 = K \times 10^4 \text{ (given)} \Rightarrow K = 198$$

ii) Characteristics Roots Method:

Solution for the homogeneous part $\left(a_n^{(h)}\right)$:

Let $a_n = r^n$ is a solution of $c_0 a_n + c_1 a_{n-1} + c_2 a_{n-2} = 0$

The characteristic equation is $c_0 r^2 + c_1 r + c_2 = 0$

The roots of the characteristic equation are characteristics roots, say $r_1 \& r_2$.

Case i) If $r_1 \& r_2$ are real and unequal i.e., $r_1 \neq r_2$,

$a_n^{(h)} = c_1 r_1^n + c_2 r_2^n$ where $c_1 \& c_2$ are arbitrary constants.

Case ii) If $r_1 \& r_2$ are real and equal i.e., $r_1 = r_2 = r$ (say)

$a_n^{(h)} = (c_1 + c_2 n) r^n$ where $c_1 \& c_2$ are arbitrary constants.

solution for the homogeneous part Case iii) If $r_1 \& r_2$ are complex

conjugate, i.e., $r_1 = r(\cos\theta + i\sin\theta)$. $r_2 = r(\cos\theta - i\sin\theta)$

$a_n^{(h)} = r^n(c_1 \cos n\theta + c_2 \sin n\theta)$ where $c_1 \& c_2$ are arbitrary constants.

[$a_n^{(h)}$ *is the solution for the homogeneous part*]

Particular solutions $\left(a_n^{(p)}\right)$:

	$f(n)$	$a_n^{(p)}$
1	C (constant)	A (constant)
2	n	$A_0 n + A_1$

3	n^2	$A_0 n^2 + A_1 n + A_2$
4	r^n	Ar^n
5	$\sin \alpha n \; or \; \cos \alpha n$	$A \sin \alpha n + B \cos \alpha n$
6	$r^n \sin \alpha n \; or \; r^n \cos \alpha n$	$r^n (A \sin \alpha n + B \cos \alpha n)$

The general solution is $a_n = a_n^{(h)} + a_n^{(p)}$

Question of two marks:

Example 29. Consider the recurrence relation

$a_n = 4(a_{n-1} - a_{n-2})$ subject to the initial conditions $a_n = 1$ for

$n = 0 \, \& \, n = 1.$ The solution is given by $a_n =$

(A) $2^n - 2^{n-1}$ (B) $2^n - n2^{n-1}$ (C) $2n - n2^n$ (D) $2(n-1) - 2^n$

Answer: (B)

The characteristic equation is $r^2 - 4r + 4 = 0 \Rightarrow r = 2,2$

$a_n = (A + Bn).2^n$ $\qquad\qquad$ $a_n = 1$ for $n = 0 \, \& \, n = 1.$

So, $1 = A \, \& \, 1 = (A + B).2 \Rightarrow A = 1 \, \& \, B = -1/2$

The solution is $a_n = 2^n - n2^{n-1}$

Example 30. The solution of the recurrence relation

$t_n - 3t_{n-1} + 2t_{n-2} = 0, t_0 = 30, t_1 = 32$ is given by $t_n =$

(A) $32 + 2^n$ (B) $32 + 2^{n+1}$ (C) $28 + 2^{n+1}$ (D) $30 + 2^n$

Answer: (C)

The characteristic equation is $r^2 - 3r + 2 = 0 \Rightarrow r = 1,2$

The general solution is $t_n = A + B.2^n$

$t_0 = 30 \Rightarrow A + B = 30 \; \& \; t_1 = 32 \Rightarrow A + 2B = 32$

Solving, $A = 28 \; \& \; B = 2$ The solution is $t_n = 28 + 2^{n+1}$

Example 31. The solution of the recurrence relation $t_n = 3t_{n-1} + 4t_{n-2}$,

$t_n = 0$ if $n = 0$ and $t_n = 5$ if $n = 1$ is given by $t_n =$

(A) $4^n + (-1)^n$ (B) $4^n - (-1)^n$ (C) $4^n + 2(-1)^n$ (D) $4^n - 2(-1)^n$

Answer: (B)

The characteristic equation is $r^2 - 3r - 4 = 0 \Rightarrow r = 4, -1$

The general solution is $t_n = A.4^n + B.(-1)^n$

$t_0 = 0 \Rightarrow A + B = 0$ & $t_1 = 5 \Rightarrow 4A - B = 5$ Solving, $A = 1$ & $B = -1$

The solution is $t_n = 4^n - (-1)^n$

Example 32. The solution of the recurrence relation

$a_n = 2a_{n-1} + 2^n, a_0 = 2$ is given by $a_n =$ _____.

Answer: $(2 + n).2^n$

$a_n = 2a_{n-1} + 2^n \Rightarrow a_n - 2a_{n-1} = 2^n$ The characteristic equation is

$r - 2 = 0 \Rightarrow r = 2 \Rightarrow a_n^{(h)} = A.2^n$ & $a_n^{(p)} = B.n.2^n$

Taking, $a_n^{(p)} = a_n = B.n.2^n, a_{n-1} = B(n-1).2^{n-1}$

From the given relation,

$B.n.2^n = 2.B(n-1).2^{n-1} + 2^n \Rightarrow Bn = B(n-1) + 1 \Rightarrow B = 1$

The solution is $a_n = a_n^{(h)} + a_n^{(p)} = A.2^n + .n.2^n \,--(i)$

$a_0 = 2 \Rightarrow A = 2$ So, the solution is $a_n = (2 + n).2^n$

Example 33. Consider the recurrence relation $a_n = 6a_{n-1} - 9a_{n-2} + 3^n$

The general solution is given by $a_n = (A + Bn).3^n + f(n)$, then $f(n) =$

(A) $\dfrac{1}{6}n^2.3^n$ (B) $\dfrac{1}{4}n^2.3^n$ (C) $\dfrac{1}{3}n^2.3^n$ (D) $\dfrac{1}{2}n^2.3^n$

Answer: (D)

The characteristic equation is $r^2 - 6r + 9 = 0 \Rightarrow r = 3, 3$

$a_n^{(h)} = (A + Bn).3^n$ As $a_n^{(h)}$ contains 3^n & $n.3^n$ so, we take

$a_n^{(p)} = Cn^2.3^n$ Now, $a_n^{(p)} = Cn^2.3^n$ satisfies the recurrence relation

so $Cn^2.3^n = 6C(n-1)^2.3^{n-1} - 9C(n-2)^2.3^{n-2} + 3^n$

$\Rightarrow Cn^2.3^n = 2C(n-1)^2.3^n - C(n-2)^2.3^n + 3^n$

$\Rightarrow Cn^2 = 2C(n-1)^2 - C(n-2)^2 + 1$

Comparing the constant terms on both sides,

$0 = 2C - 4C + 1 \Rightarrow C = 1/2 \Rightarrow a_n^{(p)} = Cn^2.3^n = \dfrac{1}{2}n^2 3^n = f(n)$

Example 34. A particle is moving in the horizontal direction. The distance travels in each second is equal to two times the distance it travelled in the previous second. If a_r denotes the position of the particle in the r-th second and $a_0 = 3, a_3 = 10$ then $a_8 =$

(A) 258 (B) 130 (C) 514 (D) 66

Answer: (A)

$a_{r+2} - a_{r+1} = 2(a_{r+1} - a_r) \Rightarrow a_{r+2} - 3a_{r+1} + 2a_r = 0$

The characteristic equation is $m^2 - 3m + 2 = 0 \Rightarrow m = 1,2$

$a_r = A.2^r + B.1^r = A.2^r + B ----(i)$

$a_0 = A + B = 3$ & $a_3 = A + 8B = 10$ Solving, $A = 2, B = 1$

$(i) \rightarrow a_r = 2 + 2^n$, $n = 8, a_8 = 2 + 2^8 = 258$

ii) Generating Function Method

The generating function for the sequence of real numbers

i) $a_0, a_1, a_2, ---, a_n, ---- $ is

$$G(x) = a_0 + a_1x + a_2x^2 + ---- + a_nx^n + ---- = \sum_{n=0}^{\infty} a_nx^n$$

ii) 1, 1, 1, ------------ is

$$G(x) = 1 + x + x^2 + ---- + x^n + ---- = \sum_{n=0}^{\infty} x^n = \dfrac{1}{1-x}, -1 < x < 1$$

iii) 1, 2, 3, ------------ is

$$G(x) = 1 + 2x + 3x^2 + ---- + (n+1)x^n + ----$$

$$= \sum_{n=0}^{\infty}(n+1)x^{n} = \frac{1}{(1-x)^{2}}, \ -1 < x < 1$$

iv) 0, 1, 2, 3, -------- i.e., is of n is

$$G(x) = 0 + 1x + 2x^{2} + - - - + nx^{n} + - - -$$

$$= x\left(1 + 2x + 3x^{2} + - - -\right) = \frac{x}{(1-x)^{2}}, \ |x| < 1$$

v) $1, a, a^{2}, - - - - - -, a^{n}, - - - - -$ is

$$G(x) = 1 + ax + a^{2}x^{2} + - - - + a^{n}x^{n} + - - - = \frac{1}{1-ax}, \ |ax| < 1$$

Question of one mark:

***Example* 35.** Which one of the following is a closed form expression for the generating function for the sequence $\{a_n\}$ where

$a_n = 2n+3, \forall n = 1,2,3, - - - -$.

(A) $\dfrac{3}{(1-x)^{2}}$ (B) $\dfrac{3x}{(1-x)^{2}}$ (C) $\dfrac{2-x}{(1-x)^{2}}$ (D) $\dfrac{3-x}{(1-x)^{2}}$ [CS/IT 2018]

Answer: (D)

The generating function for 1 is $\dfrac{1}{1-x}$ and for n is $\dfrac{x}{(1-x)^{2}}$ so the

generating function for a_n is $G(x) = \dfrac{2x}{(1-x)^{2}} + \dfrac{3}{1-x} = \dfrac{3-x}{(1-x)^{2}}$

Question of two marks:

***Example* 36.** If the ordinary generating function of a sequence $\{a_n\}_{n=0}^{\infty}$ is

$\dfrac{1+z}{(1-z)^{3}}$, then $a_3 - a_0$ is

(A)   8 (B)   10   (C)   15   (D)   20   [CS/IT 2017]

Answer: (C)

$$\frac{1+z}{(1-z)^{3}} = (1+z)(1-z)^{-3} = (1+z)\left(1+{}^{3}C_{1}z+{}^{4}C_{2}z^{2}+{}^{5}C_{3}z^{3} + - -\right)$$

$a_0 = $ coefficient of $z^{0} = 1$

77

$a_3 = $ coefficient of $z^3 = {}^4C_2 + {}^5C_3 = 6 + 10 = 16$

So, $a_3 - a_0 = 16 - 1 = 15$

Example 37. Which one of the following is a closed form expression for the generating function for the sequence of squares

$0^2, 1^2, 2^2, 3^2, ----, n^2 ----$ is

(A) $\dfrac{x}{(1-x)^2}$ (B) $\dfrac{(1+x)}{(1-x)^3}$ (C) $\dfrac{x(1+x)}{(1-x)^3}$ (D) $\dfrac{x(1+x)}{(1-x)^2}$

Answer: (C)

The generating function for the sequence of squares

$0^2, 1^2, 2^2, 3^2, ----, n^2 ----$ is

$G(x) = 0^2.1 + 1^2.x + 2^2 x^2 + 3^2 x^3 + ---- + n^2 x^n + ----- (i)$

Now, $(1-x)^{-2} = 1 + 2x + 3x^2 + ---- + nx^{n-1} + ----$

$\Rightarrow x(1-x)^{-2} = x + 2x^2 + 3x^3 + ---- + nx^n + ----$

Differentiating, $\dfrac{(1+x)}{(1-x)^3} = 1 + 2^2 x + 3^2 x^2 + ----- + n^2 x^{n-1} + ----$

$\Rightarrow \dfrac{x(1+x)}{(1-x)^3} = x + 2^2 x^2 + 3^2 x^3 + ----- + n^2 x^n + ----(ii)$

Comparing (i) & (ii), $G(x) = \dfrac{x(1+x)}{(1-x)^3}$

EXERCISE

Question of one mark:

1. In a certain programming language, variable should be of length three and be made up of two letters followed by a digit or of length two made up of one letter followed by a digit. If repetition of letters is allowed then the possible number of variables is

(A)   4440   (B)   5550 (C)   6760 (D)   7020

2. The number of distinct four digit integers can be made from the digits 1, 3, 3, 7, 7 and 8 is

(A)   80   (B)   85   (C)   90   (D)   95

3. 5 balls are to be placed in 3 boxes. Each can hold all the 5 balls. If the balls and boxes are different, the number of ways the balls may be placed so that no box is left empty is
(A)   120   (B)   150   (C)   180   (D)   100

4. Let a_n represents the number of bit strings of length n containing two consecutive 1 s. What is the recurrence relation for a_n ? (A)

$a_{n-2} + a_{n-1} + 2^{n-2}$ (B) $a_{n-2} + 2a_{n-1} + 2^{n-2}$

(C) $2a_{n-2} + a_{n-1} + 2^{n-2}$ (D) $2a_{n-2} + 2a_{n-1} + 2^{n-2}$ [CS/IT 2015]

5. In the recurrence relation $a_n = 2a_{n-1} + 5a_{n-2} - 6a_{n-3}$, the value of a_6 is
(A)   234   (B)   244   (C)   254   (D)   264

Answers:

1. (D)   2. (C)   3. (B)   4. (A)   5. (C)

Chapter 5

LINEAR ALGEBRA

The concept of Linear Algebra is based on Matrices-their Eigen Values, Eigen Vectors and Vector Spaces.

1. *Matrix:* A rectangular array of $m \times n$ elements (elements $\in F$, the field of real numbers) arranged in m rows and n columns is called a matrix of order $m \times n$.

The element placed in the i-th row and the j-th column is called the (i, j) *-th* element of the matrix for i = 1, 2, 3, ------, m; j = 1, 2, 3, --------, n.

A matrix of order $m \times n$ is written as $(a_{ij})_{m \times n}$.

If $m = n$ i.e., the number of rows = number of columns, the matrix is a square matrix of order $n \times n$ or simply n.

$$A = \begin{pmatrix} a_1 & a_2 & a_3 \\ b_1 & b_2 & b_3 \\ c_1 & c_2 & c_3 \end{pmatrix}$$ is a square matrix of order 3 × 3 or, simply 3.

Null matrix: A square matrix all of whose elements are zero is called a *Null Matrix*.

Identity (or, Unit matrix): A square matrix whose leading diagonal elements are each 1 and all other elements are zero is called an *Identity (Or, Unit) Matrix*. It is denoted by I_n.

Idempotent Matrix: A matrix A is said to be *Idempotent* if $A^2 = A$

Nilpotent Matrix: A matrix A is said to be *Nilpotent of index p* if $A^p = O$ (the null matrix) for the least positive integer p.

2. *Determinant of a Square matrix:* The square array of a square matrix A determines a determinant which may be expanded to get a real number. The determinant is called the *Determinant of the square matrix A* and is denoted by $|A|$ or, *det A*.

3. *Singular and Non-singular matrices:* A square matrix A is said to be *Singular* if *det A* = 0 and *Non-singular* if *det A* ≠ 0.

4. *Triangular Matrix*: A square matrix is called a *Triangular Matrix* if the elements either below or above the leading diagonal elements are all zeroes. *Triangular Matrices* are of two kinds:

i) *Lower Triangular Matrix*: A square matrix is called a *Lower Triangular Matrix(L)* if all elements **above** the leading diagonal elements are zeroes.

Example: $\begin{pmatrix} 7 & 0 \\ -2 & 8 \end{pmatrix}$, $\begin{pmatrix} 5 & 0 & 0 \\ 9 & 1 & 0 \\ -1 & 6 & 3 \end{pmatrix}$

ii) *Upper Triangular Matrix*: A square matrix (U) is called an upper *Triangular Matrix* if all elements **below** the leading diagonal elements are zeroes.

Example: $\begin{pmatrix} 7 & 5 \\ 0 & 8 \end{pmatrix}$, $\begin{pmatrix} 5 & 3 & -1 \\ 0 & 1 & 10 \\ 0 & 0 & 3 \end{pmatrix}$

5. *LU Decomposition*: A non-singular matrix [A] can be written as $A = LU$ where L is unit lower triangular and U is upper triangular.

6. *Transpose of a matrix*: A matrix obtained by interchanging its rows into corresponding columns is called the Transpose of the original matrix. The transpose of a matrix A is denoted by A' or, A^t. Also, $(A^t)^t = A$. In general, $A \neq A^t$.

7. *Symmetric and Skew symmetric matrices*: A square matrix A is said to be *Symmetric* if $A^t = A$ and *Skew-symmetric* if $A^t = -A$.

Properties:

i) Every square matrix A can be expressed uniquely as a sum of a *Symmetric* and a *Skew-symmetric* matrix i.e. $A = \frac{1}{2}(A + A') + \frac{1}{2}(A - A')$

where $\frac{1}{2}(A + A')$ is *Symmetric* and $\frac{1}{2}(A - A')$ is *Skew-symmetric*.

ii) All positive integral powers of a symmetric matrix are symmetric.

iii) All positive *odd* integral powers of a skew-symmetric matrix are skew-symmetric and all positive *even* integral powers of a skew-symmetric matrix is symmetric.

81

iv) If A and B are symmetric matrices of same order, **AB - BA** is skew-symmetric and **AB + BA** is symmetric.

v) If A and B are symmetric matrices of same order, then AB is *symmetric* if and only if **AB = BA.**

vi) The matrix B^tAB is symmetric or *skew*-symmetric according as A is symmetric or skew-symmetric.

8. *Orthogonal matrix:* A square matrix A is said to be *Orthogonal* if $A^t \times A = A \times A^t = I$ (Unit matrix)

9. *Equality of two matrices:* Two matrices A and B are said to be *equal* if their (i) orders are equal and (ii) corresponding elements are equal.

If $A = (a_{ij})_{m \times n}$ and $B = (b_{ij})_{p \times q}$, then $A = B \Rightarrow m = p$, $n = q$ and $a_{ij} = b_{ij}$ for all i and j.

10. *Negative of a matrix:* If A be a matrix then $- A$ is a matrix of the same order of A but all of whose elements are negative of the corresponding

elements of A. If $A = \begin{pmatrix} a & c \\ b & d \end{pmatrix}$ then $- A = \begin{pmatrix} -a & -c \\ -b & -d \end{pmatrix}$

11. *Sum of two matrices:* Two matrices are conformable for addition if their orders are equal. The sum of two matrices $A = (a_{ij})_{m \times n}$ and $B = (b_{ij})_{m \times n}$ is a matrix of order $m \times n$ whose elements are the sum of the corresponding elements of A and B. So, $A + B = (c_{ij})_{m \times n} = B + A$, where $c_{ij} = a_{ij} + b_{ij}$ for $i = 1, 2, ---,m$ & $j = 1, 2, ---,n$.

12. *Multiplication of a matrix by a non-zero scalar:* The product of a matrix A with a scalar k ($\neq 0$) is a matrix kA of the same order as A all of whose elements are k times the corresponding elements of A.
If $A = (a_{ij})_{m \times n}$, then $kA = (ka_{ij})_{m \times n}$

13. *Multiplication of two matrices:* Two matrices $A = (a_{ij})_{m \times n}$ and $B = (b_{ij})_{n \times p}$ are conformable for the product AB if the number of columns of A = the number of rows of B. In that case, the order of the product matrix will be $(m \times p)$.

The (i, j)-th element of AB = Sum of products of the elements of the i-th row of A with the corresponding elements of the k-th column of B.
Also $(AB)' = B'. A'$.

Example: If $A = \begin{pmatrix} a_1 & a_2 & a_3 \\ b_1 & b_2 & b_3 \end{pmatrix}_{2\times3}$ and $B = \begin{pmatrix} c_1 & d_1 \\ c_2 & d_2 \\ c_3 & d_3 \end{pmatrix}_{3\times2}$, then the product

$$AB = \begin{pmatrix} a_1c_1 + a_2c_2 + a_3c_3 & a_1d_1 + a_2d_2 + a_3d_3 \\ b_1c_1 + b_2c_2 + b_3c_3 & b_1d_1 + b_2d_2 + b_3d_3 \end{pmatrix}_{2\times2}$$

In general, *AB ≠ BA*, matrix multiplication is non-commutative.

For matrices conformable for multiplication, *A(BC) = (AB)C*.

The matrix A^2 is defined as $A^2 = A \times A$

14. *Conjugate of a matrix:* The matrix $\overline{A}$ obtained by replacing every element of a complex matrix A with the corresponding conjugate complex number is called the conjugate of A.

15. *Hermitian and Skew Hermitian Matrices:* A complex square matrix A is said to be *Hermitian* if $A = \overline{A}^{\,t}$ and *Skew Hermitian* if $A = -\overline{A}^{\,t}$, where $\overline{A}$ denotes the conjugate of A.

16. *Unitary Matrix:* A complex $n \times n$ matrix A is said to be *Unitary* if $A.\overline{A}^{\,t} = I_n$ (the identity matrix of order n).

17. *Adjoint (Or, Adjugate) of a matrix:* If $A = \begin{pmatrix} a_1 & a_2 & a_3 \\ b_1 & b_2 & b_3 \\ c_1 & c_2 & c_3 \end{pmatrix}$, then

$$Adj(A) = \begin{pmatrix} A_1 & A_2 & A_3 \\ B_1 & B_2 & B_3 \\ C_1 & C_2 & C_3 \end{pmatrix}' = \begin{pmatrix} A_1 & B_1 & C_1 \\ A_2 & B_2 & C_2 \\ A_3 & B_3 & C_3 \end{pmatrix}$$ where $A_1, B_2,$ --- are the

cofactors of $a_1, b_2,$ ---- in *det* A.

18. *Inverse of a matrix:* If A be a non-singular matrix of order n (i. e., *detA* ≠ 0) and B be another matrix of the same order as A such that *AB = BA = I_n* then B is called the inverse of A and is denoted by A^{-1}.

$$A^{-1} = \frac{adj(A)}{\det A}, \det A \neq 0.$$

If $A = \begin{pmatrix} a_1 & a_2 & a_3 \\ b_1 & b_2 & b_3 \\ c_1 & c_2 & c_3 \end{pmatrix} \neq O,$ then $A^{-1} = \dfrac{1}{\det A} \begin{pmatrix} A_1 & B_1 & C_1 \\ A_2 & B_2 & C_2 \\ A_3 & B_3 & C_3 \end{pmatrix}, \det A \neq 0$

where $A_1, B_2,$ --- are the cofactors of $a_1, b_2,$ ---- in det A.

19. Properties of matrices:

i) The inverse of a matrix, if exists, is unique.

ii) For a non-singular matrix $A, \left(A^{-1}\right)^{-1} = A.$

iii) For two non-singular matrices A & $B, (AB)^{-1} = B^{-1}A^{-1}.$

iv) For an orthogonal matrix $A, \det A = \pm 1$ and hence it is non-singular.

v) For a non-singular matrix $A,$ $A.A'$ is a symmetric matrix.

vi) For a non-singular matrix $A,$ $AB = AC \Rightarrow B = C.$

vii) Divisors of zero exist in Matrix Algebra i.e., $AB = O$ does not always imply either $A = O$ or $B = O.$

Example: $A = \begin{pmatrix} \sqrt{2} & 1 \\ 2 & \sqrt{2} \end{pmatrix} \neq O, B = \begin{pmatrix} -\sqrt{2} & 1 \\ 2 & -\sqrt{2} \end{pmatrix} \neq O$ but

$AB = \begin{pmatrix} 0 & 0 \\ 0 & 0 \end{pmatrix} = O.$

20. Rank of a matrix:
The positive integer r is said to be the rank of a matrix A if there exists at least one minor of A of order r that does not vanish and each minor of orders $(r + 1), (r + 2),$--- are 0. Or, the matrix A has r independent rows/columns. The rank of a matrix A is denoted as $r(A)$.

If A be a

i) Non-singular matrix (including *Identity matrix*) of order n, then $r(A) = n$

ii) Singular matrix of order n, then $r(A) < n$

iii) Null- matrix of any order, then $r(A) = 0$

iv) Rank of a matrix does not change due to elementary transformations i.e., row/column transformations.

21. Solution of Linear Simultaneous Equations:
The system of m linear simultaneous equations with n unknowns can be written as

$$a_{11}x_1 + a_{12}x_2 + - - - + a_{1n}x_n = b_1$$
$$a_{21}x_1 + a_{22}x_2 + - - - + a_{2n}x_n = b_2$$

$$a_{m1}x_1 + a_{m2}x_2 + - - - + a_{mn}x_n = b_m$$

The above set of equations is said to be

i) *consistent* if there is at least one solution of the set and

ii) *inconsistent* if there is no solution of the set.

The above set of equations can be written in matrix form $AX = B$, where

A is called the *coefficient matrix* and (A/B) is called the *augmented matrix*. r(A) = *r* and no. of unknowns = *n*

(a) *Homogeneous System:* If $b_1 = b_2 = - - - = 0$, the system is called *Homogeneous system* and is written as $AX = O$. It is always *consistent* i.e., *solvable*.

i) If *det* A ≠ 0 or if *r* = *n* it has the trivial solution (unique solution)

$$x_1 = x_2 = - = 0$$

ii) If *det* A = 0 or *r* < *n* it has non- trivial solutions (infinitely many or multiple solutions). The number of independent solutions = *(n − r)*.

(b) *Non-homogeneous System:* If at least one of $b_1, b_2, - - - ≠ 0$, the system is called non-homogeneous system and is written as $AX = B$.

Let r(A) = r & r(A/B)= R   It is consistent only if $r(A) = r(A/B) \Rightarrow r = R$

The *non-homogeneous system* has

i) a *unique solution* if $r(A) = r(A/B) =$ the number of variables

or, r = R = n or, *det* A ≠ 0

ii) infinitely many solutions if $r(A) = r(A/B) <$ the number of variables

or, r = R < n

iii) no solution if $r(A) ≠ r(A/B)$, r ≠ R

22. *Characteristic Equation*: If A be a square matrix of order *n* and I be a unit matrix of same order as A then the *Characteristic Equation* of A is given by $\det(A - \lambda I) = 0$.

23. *Cayley-Hamilton Theorem:* Every square matrix satisfies its own *Characteristic Equation.*

24. *Eigen Values*: The roots of the *Characteristic Equation* of the matrix A i.e., the roots of the equation $\det(A - \lambda I) = 0.$ are called *Eigen Values* of the matrix A.

Properties of Eigen Values:

i) The sum of the *Eigen Values* of a matrix A. is the sum of its principal diagonal elements i.e., the *trace* of A.

ii) The product of the *Eigen Values* of a matrix A is $\det A$.

iii) The *Eigen Values* of the matrix A and its transpose $A.^{T}$ are equal.

iv) The *Eigen Values* of the matrix $kA = k$ times of the *Eigen Values* of the matrix A.

v) If λ be an Eigen value of a matrix A, then λ^{k} is an eigen value of A^{k} where k is a positive integer.

vi) 0(zero) is an *Eigen Value* of any singular matrix A and 0(zero) is not an *Eigen Value* of any non-singular matrix A.

vii) If λ be an *Eigen Value* of any *non-singular matrix A* then $\dfrac{1}{\lambda}$ is an *Eigen Value* of A^{-1}.

viii) If λ be an *Eigen Value* of any *non-singular matrix A* then $\dfrac{\det(A)}{\lambda}$ is an *Eigen Value* of $adj(A)$.

ix) If λ be an *Eigen Value* of a real orthogonal matrix A then $\dfrac{1}{\lambda}$ is also an *Eigen Value* of A.

x) The *Eigen Values* of an upper triangular matrix are its leading diagonal elements.

xi) Eigen values of a real symmetric matrix are always real.

xii) If the Eigen values of a complex matrix are all real, then the matrix is Hermitian.

xiii) Eigen values of a Hermitian matrix are all real.

xiv) Eigen values of a skew-Hermitian matrix are purely imaginary or

0(zero).

xv) The Eigen values of an idempotent matrix (A^2 =A) is either 0 or 1.

xvi) If *A* and *B* are two square matrices of same order and B is non-singular, then $A \& B^{-1}AB$ have the same Eigen values.

xvii) If *A* and *B* are two square matrices of same order and B is non-singular, then $A^{-1}B \& BA^{-1}$ have the same Eigen values.

xviii) The modulus of each Eigen value of an orthogonal matrix is unity.

xix)   Similar matrices have the same eigen values.

25. *Eigen Vectors*: A non-null vector $X(X \neq O)$ satisfying the equation $(A - \lambda I)X = O$ (null matrix) is said to be an *Eigen Vector* of the square matrix A corresponding to the *Eigen Value* λ of the matrix A.

Properties of Eigen vectors:

i) Eigen vectors are non-zero vectors.

ii) Eigen vectors corresponding to two distinct Eigen values of a real symmetric matrix are orthogonal.

iii) If A be a square matrix, then there exists a unique Eigen value corresponding to each Eigen vector of A.

iv) If the Eigen values of a matrix are distinct, then the associated Eigen vectors are linearly independent.

WORKED OUT EXAMPLES:

Question of 1 mark:

Example 1. If any two columns of a determinant $P = \begin{vmatrix} 4 & 7 & 8 \\ 3 & 1 & 5 \\ 9 & 6 & 2 \end{vmatrix}$ are

interchanged, which one of the following statements regarding the value of the determinant is CORRECT?

(A) Absolute value remains unchanged but sign will change.

(B) Both absolute value and sign will change.

(C) Absolute value will change but sign will not change.

(D) Both absolute value and sign will remain unchanged.   [ME 2015]

Answer: (A)

It follows from the property of determinant.

Example 2. The determinant of matrix A is 5 and the determinant of matrix B is 40. The determinant of matrix AB is _____. [EC 2014]

Answer: 200

$|AB| = |A| . |B| = 5 \times 40 = 200$

Example 3. The determinant of matrix $\begin{pmatrix} 3 & 0 & 0 \\ 2 & 5 & 0 \\ 6 & -8 & -4 \end{pmatrix}$ is ___. [BT 2015]

Answer: - 60

$\begin{vmatrix} 3 & 0 & 0 \\ 2 & 5 & 0 \\ 6 & -8 & -4 \end{vmatrix} = 3(-20 - 0) = -60$

Example 4. If the matrix A be such that $A = \begin{bmatrix} 2 \\ -4 \\ 7 \end{bmatrix} . \begin{bmatrix} 1 & 9 & 5 \end{bmatrix}$ then the determinant of A is equal to _____. [CS/IT 2014]

Answer: 0

$A = \begin{bmatrix} 2 \\ -4 \\ 7 \end{bmatrix}_{3 \times 1} . \begin{bmatrix} 1 & 9 & 5 \end{bmatrix}_{1 \times 3} = \begin{bmatrix} 2 & 18 & 10 \\ -4 & -36 & -20 \\ 7 & 63 & 35 \end{bmatrix}_{3 \times 3}$

As the elements of the first and the second rows are proportional so $|A| = 0$

Example 5. Which one of the following does NOT equal $\begin{vmatrix} 1 & x & x^2 \\ 1 & y & y^2 \\ 1 & z & z^2 \end{vmatrix}$?

(A) $\begin{vmatrix} 1 & x(x+1) & x+1 \\ 1 & y(y+1) & y(y+1) \\ 1 & z(z+1) & z(z+1) \end{vmatrix}$ (B) $\begin{vmatrix} 1 & x+1 & x^2+1 \\ 1 & y+1 & y^2+1 \\ 1 & z+1 & z^2+! \end{vmatrix}$

(C) $\begin{vmatrix} 0 & x-y & x^2-y^2 \\ 0 & y-z & y^2-z^2 \\ 1 & z & z^2 \end{vmatrix}$ (D) $\begin{vmatrix} 2 & x+y & x^2+y^2 \\ 2 & y+z & y^2+z^2 \\ 1 & z & z^2 \end{vmatrix}$

Answer : (A)

In (B), $C_2 - C_1$ & $C_3 - C_1$ gives the given determinant.

In (C), $R_2 + R_3$ gives $\begin{vmatrix} 0 & x-y & x^2-y^2 \\ 1 & y & y^2 \\ 1 & z & z^2 \end{vmatrix}$, then $R_1 + R_2$ gives

the given determinant.

In (D) $R_2 - R_3$ gives $\begin{vmatrix} 2 & x+y & x^2+y^2 \\ 1 & y & y^2 \\ 1 & z & z^2 \end{vmatrix}$, then $R_1 - R_2$ gives the given

determinant.

But the determinant in (A) cannot be reduced to the given determinant.

Example 6. Perform the following operations on the following matrix:

$\begin{bmatrix} 3 & 4 & 45 \\ 7 & 9 & 105 \\ 13 & 2 & 195 \end{bmatrix}$

i) Add the third row to the second row

ii) Subtract the third column from the first column.

The determinant of the resultant matrix is _____ .   [CS/IT 2015]

Answer: 0

$\begin{bmatrix} 3 & 4 & 45 \\ 7 & 9 & 105 \\ 13 & 2 & 195 \end{bmatrix} \rightarrow \begin{bmatrix} 3 & 4 & 45 \\ 20 & 11 & 300 \\ 13 & 2 & 195 \end{bmatrix} \begin{bmatrix} R_2' = R_2 + R_3 \end{bmatrix}$

$$\rightarrow \begin{bmatrix} 3 & 4 & 42 \\ 20 & 11 & 280 \\ 13 & 2 & 182 \end{bmatrix} \quad \left[R_3^{'} = R_3 - R_1 \right]$$

The determinant of the matrix = 0
as the elements of the third and the first columns are proportional.

Example 7. The following set of three vectors $\begin{pmatrix} 1 \\ 2 \\ 1 \end{pmatrix}$, $\begin{pmatrix} x \\ 6 \\ x \end{pmatrix}$ & $\begin{pmatrix} 3 \\ 4 \\ 2 \end{pmatrix}$ is linearly

dependent if X is equal to

(A)　　0　　(B)　　1　　(C)　　2　　(D)　　3　　[CH 2015]

Answer: (D)

$$\begin{vmatrix} 1 & x & 3 \\ 2 & 6 & 4 \\ 1 & x & 2 \end{vmatrix} = 0 \Rightarrow 1(12 - 4x) - 2(2x - 3x) + 1(4x - 18) = 0 \Rightarrow x = 3$$

Example 8. The value of c for which the matrix product

$$\begin{pmatrix} 2 & 0 & 7 \\ 0 & 1 & 0 \\ 1 & -2 & 1 \end{pmatrix} \begin{pmatrix} -c & 14c & 7c \\ 0 & 1 & 6 \\ c & -4c & -2c \end{pmatrix}$$ is an identity matrix is

(A)　　1/2　　(B)　　1/3　　(C)　　1/4　　(D)　　1/5

Answer: (D)

$$\begin{pmatrix} 2 & 0 & 7 \\ 0 & 1 & 0 \\ 1 & -2 & 1 \end{pmatrix} \begin{pmatrix} -c & 14c & 7c \\ 0 & 1 & 6 \\ c & -4c & -2c \end{pmatrix} = \begin{pmatrix} 1 & 0 & 0 \\ 0 & 1 & 0 \\ 0 & 0 & 1 \end{pmatrix} \Rightarrow -2c + 7c = 1 \Rightarrow c = \frac{1}{5}$$

Example 9. The matrix P is the inverse of the matrix Q. If I denotes the identity matrix, which one of the following options is correct?

(A)　　PQ = I but QP ≠ I　　(B)　　QP = I but PQ ≠ I
(C)　　PQ = I and QP = I　　(D)　　PQ − QP = I　　　　[CE 2017]

Answer: (C)

It follows from the property of the inverse of the matrix.

Example 10. If $A^2 - A + I = O,$ then the inverse of the matrix A is

(A) $A - I$ (B) $I - A$ (C) $A + I$ (D) A

Answer: (B)

$$A^2 - A + I = O \Rightarrow A(A - I) = -I \Rightarrow A(I - A) = I \Rightarrow A^{-1} = I - A$$

Example 11. A real square matrix A is called Skew-symmetric if

(A) $A^T = A$ (B) $A^T = A^{-1}$ (C) $A^T = -A$ (D) $A^T = A + A^{-1}$ [ME 2016]

Answer: (C)

It follows from the definition 7 of skew-symmetric matrix.

Example 12. Consider the matrices $X_{(4 \times 3)}, Y_{(4 \times 3)}$ and $P_{(2 \times 3)}.$ The order of

$\left[P(X^T Y)^{-1} P^T \right]^T$ will be

(A)   (2 × 2)   (B)   (3 × 3)   (C)   (4 × 3)   (D)   3 × 4   [CS/IT 2005]

Answer: (A)

If O denotes the order of a determinant, then

$$O(X^T) = 3 \times 4, \ O(X^T Y) = 3 \times 3, \ O(X^T Y)^{-1} = 3 \times 3, \ O[P(X^T Y)^{-1}] = 2 \times 3$$

$$O(P^T) = 3 \times 2 \Rightarrow O\left[P(X^T Y)^{-1} P^T \right] = 2 \times 2 \Rightarrow O\left[P(X^T Y)^{-1} P^T \right]^T = 2 \times 2$$

Example 13. $\left[A \right]$ is a square matrix which is neither symmetric nor skew-symmetric and $\left[A \right]^T$ is its transpose. The sum and difference of the matrices are defined as $S = \left[A \right] + \left[A \right]^T$ and $D = \left[A \right] - \left[A \right]^T$ respectively. Which of the statements is TRUE?

(A) Both S and D are symmetric   (B) Both S and D are skew-symmetric

(C) S is skew-symmetric and D is symmetric

(D) S is symmetric and D is skew-symmetric   [CS/IT 2011]

Answer (D)

By property 7 i), every square matrix can be represented uniquely as the sum of a symmetric and skew-symmetric matrices.

Example 14. For an orthogonal matrix $Q,$ the valid equality is

(A) $Q^T = Q^{-1}$ (B) $Q^T = Q^{-1}$ (C) $Q^T = Q$ (D) $\det(Q) = 0$ [PI 2017]

Answer: (A)

$$Q.Q^T = I \ \& \ Q.Q^{-1} = I \Rightarrow Q.Q^T = Q.Q^{-1} \Rightarrow Q^T = Q^{-1}$$

Example 15. Let $M^4 = I$ (where I denotes the identity matrix) and $M \neq I, M^2 \neq I, M^3 \neq I$. Then for any natural number k, M^{-1} equals:

(A) M^{4k+1} (B) M^{4k+2} (C) M^{4k+3} (D) M^{4k} [EC 2016]

Answer:(C)

$$M.M^{4k+3} = M^{4k+4} = M^{4k}.M^4 = \left(M^4\right)^k.I = I^k = I \Rightarrow M^{-1} = M^{4k+3}$$

Example 16. If $A = \begin{bmatrix} 1 & 5 \\ 6 & 2 \end{bmatrix}$ and $B = \begin{bmatrix} 3 & 7 \\ 8 & 4 \end{bmatrix}$, AB^T is equal to

(A) $\begin{bmatrix} 38 & 28 \\ 32 & 56 \end{bmatrix}$ (B) $\begin{bmatrix} 3 & 40 \\ 42 & 8 \end{bmatrix}$ (C) $\begin{bmatrix} 43 & 27 \\ 34 & 50 \end{bmatrix}$ (D) $\begin{bmatrix} 38 & 32 \\ 28 & 56 \end{bmatrix}$ [CE 2017]

Answer: (A)

$$AB^T = \begin{bmatrix} 1 & 5 \\ 6 & 2 \end{bmatrix}.\begin{bmatrix} 3 & 8 \\ 7 & 4 \end{bmatrix} = \begin{bmatrix} 1.3+5.7 & 1.8+5.4 \\ 6.3+2.7 & 6.8+2.4 \end{bmatrix} = \begin{bmatrix} 38 & 28 \\ 32 & 56 \end{bmatrix}$$

Example 17. If A is a non-singular matrix of order n, , then $A.adjA =$

(A) A^{-1} (B) $|A|.I_n$ (C) I_n (D) $\dfrac{A}{|A|}$

Answer: (B)

From the definition 18 of the inverse of a non-singular matrix,

$$A^{-1} = \frac{adj A}{|A|}$$

$$\Rightarrow adj A = |A|.A^{-1} \Rightarrow A.adj A = A.|A|A^{-1} = |A|\left(A.A^{-1}\right) = |A|.I_n$$

Example 18. Which one of the following equations is a correct identity for arbitrary 3 × 3 real matrices P, Q, R?

(A) $P(Q + R) = PQ + RP$ (B) $(P-Q)^2 = P^2 - 2PQ + Q^2$

(C) $\det(P + Q) = \det P + \det Q$ (D) $(P+Q)^2 = P^2 + PQ + QP + Q^2$ [ME 2014]

Answer: (D)

Example 19. If $P = \begin{pmatrix} 1 & 1 \\ 2 & 2 \end{pmatrix}$, $Q = \begin{pmatrix} 2 & 1 \\ 2 & 2 \end{pmatrix}$, $R = \begin{pmatrix} 3 & 0 \\ 1 & 3 \end{pmatrix}$ then which one

of the following statements is TRUE?
(A)   PQ = PR (B)   QR = RP (C)   QP = RP (D)   PQ = QR [BT 2013]
Answer: (A)

$$PQ = \begin{pmatrix} 4 & 3 \\ 8 & 6 \end{pmatrix}, \quad PR = \begin{pmatrix} 4 & 3 \\ 8 & 6 \end{pmatrix}, \text{ So, } PQ = PR$$

The other products will be

$$QR = \begin{pmatrix} 7 & 3 \\ 8 & 6 \end{pmatrix}, \quad RP = \begin{pmatrix} 3 & 3 \\ 7 & 7 \end{pmatrix}, \quad QP = \begin{pmatrix} 4 & 4 \\ 6 & 6 \end{pmatrix}.$$

[For this problem it is not necessary to check the other options as only one option will be correct.]

Example 20. For matrices of same dimension M, N and scalar c which one of the properties DOES NOT ALWAYS hold?

(A) $\left(M^T\right)^T = M$

(B) $\left(cM\right)^T = cM^T$

(B) $\left(M+N\right)^T = M^T + N^T$ (D) $MN = NM$ [EC 2014]

Answer: (D)
Matrix multiplication is, in general, non-commutative.

Example 21. If A & B are square matrices of same order and A is symmetrical, then which one of the following matrices is symmetric?
(A) $B'AB$ (B) BAB' (C) $A'BA$ (D) ABA'
Answer: (A)

As A is symmetric, $A' = A$ Now, $(B'AB)' = B'A'(B')' = B'AB$

Example 22. If $A = \begin{pmatrix} 4 & 2 \\ 1 & 3 \end{pmatrix}$ then $A^2 + 3A$ will be

(A) $\begin{pmatrix} 30 & 20 \\ 10 & 20 \end{pmatrix}$ (B) $\begin{pmatrix} 28 & 10 \\ 4 & 18 \end{pmatrix}$ (C) $\begin{pmatrix} 31 & 13 \\ 7 & 21 \end{pmatrix}$ (D) $\begin{pmatrix} 20 & 10 \\ 5 & 15 \end{pmatrix}$ [BT 2015]

Answer:(A)

$$A^2 = \begin{pmatrix} 4 & 2 \\ 1 & 3 \end{pmatrix}\begin{pmatrix} 4 & 2 \\ 1 & 3 \end{pmatrix} = \begin{pmatrix} 18 & 14 \\ 7 & 11 \end{pmatrix}, \ A^2 + 3A = \begin{pmatrix} 18 & 14 \\ 7 & 11 \end{pmatrix} + \begin{pmatrix} 12 & 6 \\ 3 & 9 \end{pmatrix} = \begin{pmatrix} 30 & 20 \\ 10 & 20 \end{pmatrix}$$

Example 23. If $A = \begin{pmatrix} \alpha & 0 \\ 1 & 1 \end{pmatrix}, B = \begin{pmatrix} 1 & 0 \\ 5 & 1 \end{pmatrix}$ & $A^2 = B$, then the value of α is

(A)   1   (B)   -1   (C)   4   (D) no real value of α

Answer:(D)

$$A.A = B \Rightarrow \begin{pmatrix} \alpha & 0 \\ 1 & 1 \end{pmatrix}\begin{pmatrix} \alpha & 0 \\ 1 & 1 \end{pmatrix} = \begin{pmatrix} 1 & 0 \\ 5 & 1 \end{pmatrix} \Rightarrow \begin{pmatrix} \alpha^2 & 0 \\ \alpha+1 & 1 \end{pmatrix} = \begin{pmatrix} 1 & 0 \\ 5 & 1 \end{pmatrix} \Rightarrow \alpha^2 = 1 \ \& \ \alpha+1 = 5$$

They do not give a unique real value for α.

Example 24. For a given matrix $P = \begin{bmatrix} 4+3i & -i \\ i & 4-3i \end{bmatrix}$, where

$i = \sqrt{-1}$, the inverse of the matrix P is

(A) $\dfrac{1}{24}\begin{bmatrix} 4-3i & i \\ -i & 4+3i \end{bmatrix}$

(B) $\dfrac{1}{25}\begin{bmatrix} i & 4-3i \\ 4+3i & -i \end{bmatrix}$

(C) $\dfrac{1}{24}\begin{bmatrix} 4+3i & i \\ i & 4-3i \end{bmatrix}$

(D) $\dfrac{1}{25}\begin{bmatrix} 4+3i & -i \\ i & 4-3i \end{bmatrix}$ [ME 2015]

Answer: (A)

$\det P = (4+3i)(4-3i)+i^2 = 16+9-1 = 24$

$$P^{-1} = \frac{1}{\det P}(adj\ P) = \frac{1}{24}\begin{bmatrix} 4-3i & -i \\ i & 4+3i \end{bmatrix}' = \frac{1}{24}\begin{bmatrix} 4-3i & i \\ -i & 4+3i \end{bmatrix}$$

Example 25. Consider the matrix $J_6 = \begin{pmatrix} 0 & 0 & 0 & 0 & 0 & 1 \\ 0 & 0 & 0 & 0 & 1 & 0 \\ 0 & 0 & 0 & 1 & 0 & 0 \\ 0 & 0 & 1 & 0 & 0 & 0 \\ 0 & 1 & 0 & 0 & 0 & 0 \\ 1 & 0 & 0 & 0 & 0 & 0 \end{pmatrix}$ which is

obtained by reversing the order of the columns of the identity matrix I_6.

94

Let $P=I_6+\alpha J_6$, where α is a non-negative real number. The value of α for which $\det(P)=0$ is _____.   [EC 2014]

Answer: $\alpha=1$

$$P = I_6 + \alpha\, J_6 = \begin{pmatrix} 1 & 0 & 0 & 0 & 0 & \alpha \\ 0 & 1 & 0 & 0 & \alpha & 0 \\ 0 & 0 & 1 & \alpha & 0 & 0 \\ 0 & 0 & \alpha & 1 & 0 & 0 \\ 0 & \alpha & 0 & 0 & 1 & 0 \\ \alpha & 0 & 0 & 0 & 0 & 1 \end{pmatrix}$$

If $\alpha=1$, R_1 & R_6 will be identical. $\det(P)=0$

Example 26. Multiplication of matrices E & F is G. Matrices E & G are $E = \begin{pmatrix} \cos\theta & -\sin\theta & 0 \\ \sin\theta & \cos\theta & 0 \\ 0 & 0 & 1 \end{pmatrix}$ & $G = \begin{pmatrix} 1 & 0 & 0 \\ 0 & 1 & 0 \\ 0 & 0 & 1 \end{pmatrix}$

What is the matrix F ?

(A) $\begin{pmatrix} \cos\theta & -\sin\theta & 0 \\ \sin\theta & \cos\theta & 0 \\ 0 & 0 & 1 \end{pmatrix}$ (B) $\begin{pmatrix} \sin\theta & \cos\theta & 0 \\ -\cos\theta & \sin\theta & 0 \\ 0 & 0 & 1 \end{pmatrix}$

(C) $\begin{pmatrix} \cos\theta & \sin\theta & 0 \\ -\sin\theta & \cos\theta & 0 \\ 0 & 0 & 1 \end{pmatrix}$ (D) $\begin{pmatrix} \sin\theta & -\cos\theta & 0 \\ \cos\theta & \sin\theta & 0 \\ 0 & 0 & 1 \end{pmatrix}$ [ME 2006]

Answer: (C)
By matrix multiplication rule, (C) is the answer.

Example 27. If $A^3=O$, then $I+A+A^2$ is equal to

(A) $I+A$ (B) $(I+A)^{-1}$ (C) $I-A$ (D) $(I-A)^{-1}$

Answer: (D)

Now, $(I-A)(I+A+A^2)=I^3-A^3=I-A^3=I-O=I$

As $AB=I \Rightarrow B=A^{-1}$ so, $I+A+A^2=(I-A)^{-1}$

Example 28. If a matrix A can be reduced to the normal form
$\begin{pmatrix} I_r & O \\ O & O \end{pmatrix}$, then the rank of the matrix A is r.

(A) The statement is necessary but not sufficient.
(B) The statement is sufficient but not necessary.
(C) The statement is both necessary as well as sufficient.
(D) The statement is neither necessary nor sufficient.
Answer: (C)

Example 29. The necessary condition to diagonalise a matrix is that
(A) its eigen value should be distinct
(B) its eigen vectors should be independent
(C) its eigen values should be real
(D) the matrix is non-singular   [CS 2001]
Answer (D)

Example 30. The rank of the matrix $\begin{pmatrix} 1 & 2 & 3 \\ 2 & 3 & 4 \\ 3 & 4 & 5 \end{pmatrix}$ is

(A)   3   (B)   2   (C)   1   (D)   0
Answer: (B)

Let $A = \begin{pmatrix} 1 & 2 & 3 \\ 2 & 3 & 4 \\ 3 & 4 & 5 \end{pmatrix}$, then det A = 0, so $r(A) < 3$ Also, $\begin{vmatrix} 1 & 2 \\ 2 & 3 \end{vmatrix} = -1 \neq 0.$

So, A has at least one non-zero second order minor. r(A) = 2

Example 31. For $A = \begin{pmatrix} 1 & \tan x \\ -\tan x & 1 \end{pmatrix}$, the determinant of $A^T A^{-1}$ is

(A) $\sec^2 x$ (B) $\cos 4x$ (C) 1   (D) 0   [EC 2015]

96

Answer (C)

$$\left|A^T A^{-1}\right| = \left|A^T\right|\left|A^{-1}\right| = |A|\left|A^{-1}\right| = \left|A.A^{-1}\right| = |I| = 1$$

Example 32. If $U = \begin{pmatrix} 1 & 2 & 2 \\ -1 & 1 & 0 \\ -1 & -2 & -1 \end{pmatrix}$ & $(3\ \ 2\ \ 0)U\begin{pmatrix} 3 \\ 2 \\ 0 \end{pmatrix} = (a)$, then a equals

(A)   17   (B)   19   (C)   21   (D)   24

Answer: (B)

$$(3\ \ 2\ \ 0)U = (3\ \ 2\ \ 0)_{1\times 3}\begin{pmatrix} 1 & 2 & 2 \\ -1 & 1 & 0 \\ -1 & -2 & -1 \end{pmatrix}_{3\times 3} = (1\ \ 8\ \ 6)_{3\times 1}$$

Now, $(3\ \ 2\ \ 0)U\begin{pmatrix} 3 \\ 2 \\ 0 \end{pmatrix} = (1\ \ 8\ \ 6)_{1\times 3}.\begin{pmatrix} 3 \\ 2 \\ 0 \end{pmatrix}_{3\times 1} = (19)_{1\times 1} = (a)$

$\Rightarrow (19) = (a) \Rightarrow a = 19$

Example 33. If $\begin{pmatrix} 4 \\ 1 \\ 3 \end{pmatrix}.A = \begin{pmatrix} -4 & 8 & 4 \\ -1 & 2 & 1 \\ -3 & 6 & 3 \end{pmatrix}$, then $A =$

(A) $(-1\ \ 2\ \ 1)$ (B) $(1\ \ 2\ \ -1)$ (C) $(-1\ \ -2\ \ 1)$ (D) $(1\ \ 2\ \ 1)$

Answer: (A)

From the condition of matrix multiplication, the order of A should be 1×3.

Let $A = (x\ \ y\ \ z)$, then

$$\begin{pmatrix} 4 \\ 1 \\ 3 \end{pmatrix}.(x\ \ y\ \ z) = \begin{pmatrix} -4 & 8 & 4 \\ -1 & 2 & 1 \\ -3 & 6 & 3 \end{pmatrix} \Rightarrow \begin{pmatrix} 4x & 4y & 4z \\ x & y & z \\ 3x & 3y & 3z \end{pmatrix} = \begin{pmatrix} -4 & 8 & 4 \\ -1 & 2 & 1 \\ -3 & 6 & 3 \end{pmatrix},$$

Or, x = - 1, y = 2, z = 1 So, $A = (x\ \ y\ \ z) = (-1\ \ 2\ \ 1)$

Example 34. Let $A = \lfloor a_{i\,j} \rfloor, 1 \le i, j \le n$ with $n \ge 3$ and $a_{ij} = i.j$.

The rank of A is

(A) 0   (B) 1   (C) $n-1$ (D) n [CE 2015]

Answer: (B)

Taking $i = 2 \,\& \, j = 3$,

$$A = \begin{bmatrix} a_{11} & a_{12} & a_{13} \\ a_{21} & a_{22} & a_{23} \end{bmatrix} = \begin{bmatrix} 1.1 & 1.2 & 1.3 \\ 2.1 & 2.2 & 2.3 \end{bmatrix} = \begin{bmatrix} 1 & 2 & 3 \\ 2 & 4 & 6 \end{bmatrix} \Rightarrow r(A) = 1$$

Example 35. The rank of the matrix $\begin{bmatrix} 5 & 10 & 10 \\ 1 & 2 & 2 \\ 3 & 6 & 6 \end{bmatrix}$ is

(A)   0   (B)   1 (C)   2   (D)   3   [EC 2017]

Answer: (C)

$R_1 = 5(1 \quad 2 \quad 2)$ and $R_3 = 3(1 \quad 2 \quad 2) \Rightarrow R_1 \,\& \, R_3$ are dependent.

So, the matrix has two independent rows only. $r(A) = 2$

Example 36. Let $P = \begin{bmatrix} 1 & 1 & -1 \\ 2 & -3 & 4 \\ 3 & -2 & 3 \end{bmatrix} \,\& \, Q = \begin{bmatrix} -1 & -2 & -1 \\ 6 & 12 & 6 \\ 5 & 10 & 15 \end{bmatrix}$,

then the rank of $P + Q$ is _____.   [CS/IT 2017]

Answer: 2

$$P + Q = \begin{bmatrix} 0 & -1 & -2 \\ 8 & 9 & 10 \\ 8 & 8 & 8 \end{bmatrix} \rightarrow \begin{bmatrix} 0 & 1 & 2 \\ 0 & 1 & 2 \\ 8 & 8 & 8 \end{bmatrix} \begin{bmatrix} R_1' = (-1)R_1, R_2' = R_2 - R_3 \end{bmatrix}$$

$r(P + Q) = 2$ as the first and second rows are identical.

Example 37. Let A be an $n \times n$ matrix with rank r $(0 < r < n)$. Then $AX = O$ has p independent solutions, where p is

(A)   r   (B)   n   (C) $n - r$ (D) $n + r$ [IN 2015]

Answer: (C)

It follows from property 21 a) ii)

Example 38. The rank of the matrix $\begin{bmatrix} 1 & -1 & 0 & 0 & 0 \\ 0 & 0 & 1 & -1 & 0 \\ 0 & 1 & -1 & 0 & 0 \\ -1 & 0 & 0 & 0 & 1 \\ 0 & 0 & 0 & 1 & -1 \end{bmatrix}$ is _[EC 2017]

Answer: 4

$\begin{bmatrix} 1 & -1 & 0 & 0 & 0 \\ 0 & 0 & 1 & -1 & 0 \\ 0 & 1 & -1 & 0 & 0 \\ -1 & 0 & 0 & 0 & 1 \\ 0 & 0 & 0 & 1 & -1 \end{bmatrix} \rightarrow \begin{bmatrix} 0 & 0 & 0 & 0 & 0 \\ 0 & 0 & 1 & -1 & 0 \\ 0 & 1 & -1 & 0 & 0 \\ -1 & 0 & 0 & 0 & 1 \\ 0 & 0 & 0 & 1 & -1 \end{bmatrix}$ which has 4 ind. rows.

$\left[R_1' = R_1 + \left(R_2 + R_3 + R_4 + R_5 \right) \right]$ So, rank = 4

Example 39. Let three linear simultaneous non- homogeneous equations are given by $AX = B.$ If $r(A) = r(A/B) = 2,$ then the equations

(A) are inconsistent   (B) consistent and have a unique solution
(C) consistent and have only two solutions
(D) consistent and have infinite number of solutions
Answer: (D)
It follows from property 21. b) ii)

Example 40. Consider the systems, each consisting of m linear equations in n variables.
I. If m < n, then all such systems have a solution
II. If m > n, then none of these systems has a solution
III. If m = n, then there exists a system which has a solution
Which one of the following is CORRECT?
(A) I, II and III are true   (B) Only II and III are true
(C) Only III is true   (D) None of them is true   [CS/IT 2016]
Answer (C)
Example 41. The solutions to the system of equations

$$\begin{pmatrix} 2 & 5 \\ -4 & 3 \end{pmatrix} \begin{pmatrix} x \\ y \end{pmatrix} = \begin{pmatrix} 2 \\ -30 \end{pmatrix} \text{ is}$$

(A)   6, 2   (B) - 6, 2   (C)   - 6, - 2   (D)   6, - 2   [ME 2016]

Answer: (D)

The equations are $2x + 5y = 2, \; -4x + 3y = -30$

$x = 6, y = -2$ satisfy both the equations.

Example 42. The equation $\begin{pmatrix} 2 & -2 \\ 1 & -1 \end{pmatrix} \begin{pmatrix} x_1 \\ x_2 \end{pmatrix} = \begin{pmatrix} 0 \\ 0 \end{pmatrix}$ has

(A)   no solution   (B)   only one solution $\begin{pmatrix} x_1 \\ x_2 \end{pmatrix} = \begin{pmatrix} 0 \\ 0 \end{pmatrix}$

(C)   non-zero unique solution (D)   multiple solutions   [EE 2013]

Answer (D)

It is of the form $AX = O$

So, it is a set of homogeneous equations with det (A) = 0.

By formula 21(a), the system of equations has multiple solutions.

Example 43. Consider the following system of equations:

$3x + 2y = 1$ $4x + 7z = 1$ $x + y + z = 3$ $x - 2y + 7z = 0$

The number of solutions for this system is

(A)   1   (B)   0   (C)   2   (D)   3   [CS/IT 2014]

Answer: (A)

The equations in the matrix form:

$$\begin{bmatrix} 3 & 2 & 0 & - & 1 \\ 4 & 0 & 7 & - & 1 \\ 1 & 1 & 1 & - & 3 \\ 1 & -2 & 7 & - & 0 \end{bmatrix} \rightarrow \begin{bmatrix} 4 & 0 & 7 & - & 1 \\ 4 & 0 & 7 & - & 1 \\ 1 & 1 & 1 & - & 3 \\ 1 & -2 & 7 & - & 0 \end{bmatrix} \quad \left[R_1' = R_1 + R_4 \right]$$

Rank of the coefficient matrix = Rank of augmented matrix = No. of unknowns = 3   So, the solution is unique.

Example 44. $2x_1 + x_2 = 3$ & $5x_1 + bx_2 = 7.5$

The system of linear equations in two variables shown above will have infinite solutions, if and only if, b is equal to _____.[BT 2015]

100

Answer: 2.5

$$\frac{2}{5} = \frac{1}{b} \Rightarrow b = \frac{5}{2} = 2.5$$

Example 45. Let c_1, c_n be scalars not all zero such that the following

expression holds: $\sum_{i=1}^{n} c_i a_i = 0$ where a_i is column vectors in R^n.

Consider the set of linear equations $AX = B$ where

$A = [a_1, a_2, ---, a_n]$ and $B = \sum_{i=1}^{n} b_i$.

Then, the set of equations has

(A) a unique solution has $x = j_n$ where j denotes n dimensional vector

for all 1   (B) no solution

(C) infinitely many solutions (D) finitely many solutions[CS/IT 2017]

Answer: (C)

From the relation $\sum_{i=1}^{n} c_i a_i = 0$, $a_1, a_2, ---, a_n$ are linearly dependent.

For the set of linear equations $AX = B$

rank of coefficient matrix A = rank of coefficient matrix A/B $< n$

So, the set of equations has infinitely many solutions.

Example 46. Consider the following system of equations:

3x + 2y = 1   4x + 7z = 1

x + y + z = 3   x - 2y + 7z = 0

The number of solutions for this system is _____. [CS/IT 2014]

Answer: 1

The difference of the second equation from the first lead to the fourth.

So, two equations are x + y + z = 3, x - 2y + 7z = 0

$$A/B = \begin{bmatrix} 1 & 1 & 1 & 3 \\ 1 & -2 & 7 & 0 \end{bmatrix} \Rightarrow r(A) = r(A/B) = 2.$$

So, the system has a unique solution.

***Example* 47.** The value of c for which the following system of linear equations

$$\begin{bmatrix} 1 & 2 \\ 1 & 2 \end{bmatrix} \begin{bmatrix} x \\ y \end{bmatrix} = \begin{bmatrix} c \\ 4 \end{bmatrix}$$ has an infinite number of solutions is ___. [BT 2017]

Answer: 4

The equations are $x + 2y = c$ & $x + 2y = 4$

The equations has an infinite number of solutions if $\dfrac{1}{1} = \dfrac{2}{2} = \dfrac{c}{4} \Rightarrow c = 4$

***Example* 48.** Consider the following linear simultaneous equations (with c_1 & c_2 being constants):

$3x_1 + 2x_2 = c_1$

$4x_1 + x_2 = c_2$

The characteristic equation for these simultaneous equations is

(A) $\lambda^2 - 4\lambda - 5 = 0$ (B) $\lambda^2 - 4\lambda + 5 = 0$

(C) $\lambda^2 + 4\lambda - 5 = 0$ (D) $\lambda^2 + 4\lambda + 5 = 0$ [CE 2017]

Answer: (A)

The characteristic equation is $\begin{vmatrix} 3 - \lambda & 2 \\ 4 & 1 - \lambda \end{vmatrix} = 0 \Rightarrow \lambda^2 - 4\lambda - 5 = 0$

***Example* 49.** The characteristic equation of a (3×3) matrix P is defined

$\alpha(\lambda) = |\lambda I - P| = \lambda^3 + \lambda^2 + 2\lambda + 1 = 0$. If I denotes identity matrix,

then the inverse of the matrix P will be

(A) $P^2 + P + 2I$ (B) $P^2 + P + I$

(C) $-(P^2 + P + I)$ (D) $-(P^2 + P + 2I)$ [EE 2008]

Answer (D)

As P satisfies its characteristic equation

$$P^3 + P^2 + 2P + I = O \Rightarrow P(P^2 + P + 2I) = -I$$

Or, $P\{-\left(P^2 + P + 2I\right)\} = I \Rightarrow P^{-1} = -\left(P^2 + P + 2I\right)$

Example 50. Which one of the following statements is TRUE about every n × n matrix with only real eigen values?

(A) If the trace of the matrix is positive and the determinant of the matrix is negative, at least one of its eigen values is negative.

(B) If the trace of the matrix is positive, all its eigen values are positive.

(C) If the determinant of the matrix is positive, all its eigen values are positive.

(D) If the product of the trace and determinant of the matrix is positive, all its eigen values are positive. [CS/IT 2014]

Answer: (A)

Product of the eigen values = Determinant of the matrix

= a negative quantity. So, one of the eigen values must be negative.

Example 51. At least one eigen value of a singular matrix is

(A) positive (B) zero (C) negative (D) imaginary [ME 2015]

Answer (B)

It follows from property 24 vi) of eigen values.

Example 52. The eigen values of a skew-Hermitian matrix are

(A)   Purely imaginary   (B)   all real numbers

(C)   0 or purely imaginary   (D)   0 only

Answer: (C)

It follows from property 24. xiv)

Example 53. The eigen values of a real symmetric matrix are

(A) complex with non-zero positive imaginary part

(B) complex with non-zero negative imaginary part

(C) real   (D) purely imaginary [ME 2013] [PI 2013]

Answer (C)

It follows from the property 24 (x) of Eigen Values.

Example 54. If λ is an eigen value of a non-singular matrix A, then an eigen value of $adj\ A$ is

(A) $\dfrac{\lambda}{|A|}$ (B) $\dfrac{|A|}{\lambda}$ (C) $\lambda.|A|$ (D) $\dfrac{|A^{-1}|}{\lambda}$

Answer :(B)

It follows from the property 24. viii)

Example 55. If λ be an Eigen value of a non-singular matrix A, then $\dfrac{\det A}{\lambda}$ is an Eigen value of

(A) $adj\,(A)$ (B)   A   (C) A^{-1} (D) A^{T}

Answer (A)

As A is non- singular, $A^{-1} = \dfrac{adj\,(A)}{\det A}$. Now, eigen value of $A^{-1} = \dfrac{1}{\lambda}$

or, $\dfrac{1}{\lambda}$ = eigen value of $\dfrac{adj\,(A)}{\det A}$. or, eigen value of $adj\,(A) = \dfrac{\det A}{\lambda}$.

Example 56. If 1, - 4, 7 are the Eigen values of a matrix A and $B = P^{-1}AP$ then the Eigen values of B are

(A)   - 1, 4, - 7   (B) 1, -1/4, 1/7   (C)- 1, 1/4, -1/7   (D) 1, - 4, 7

Answer (D)

It follows from the property 24.xiii) of Eigen values.

Example 57. If A and B are two non-singular square matrices of same order 1, - 4, 7 are the Eigen values of a matrix $A^{-1}B$, then the Eigen values of BA^{-1} are

(A)   - 1, 4, - 7   (B) 1, -1/4, 1/7   (C)- 1, 1/4, -1/7   (D) 1, - 4, 7

Answer (D)

It follows from the property 24.xiv) of Eigen values.

Example 58. Two eigen values of a 3 × 3 real matrix P are $2+\sqrt{-1}$ and 3. The determinant of P is _____.   [CS/IT 2016]

Answer: 15

The third eigen value of P is $2+\sqrt{-1}$

det(P)= product of the eigen values = $3\left(2+\sqrt{-1}\right)\left(2-\sqrt{-1}\right)=15$

***Example* 59.** The larger of the two eigen values of the matrix

$\begin{bmatrix} 4 & 5 \\ 2 & 1 \end{bmatrix}$ is_____ [CS/IT 2015]

Answer: 6

The eigen values are given by $\begin{vmatrix} 4-\lambda & 5 \\ 2 & 1-\lambda \end{vmatrix} = 0 \Rightarrow \lambda = 6,-1$

***Example* 60.** The eigen values of $A = \begin{pmatrix} 1 & -4 \\ 2 & -3 \end{pmatrix}$ are

(A) $2 \pm i$ (B) $-1,-2$ (C) $-1 \pm 2i$ (D) non-existent [BT 2014]

Answer: (C)

The eigen values are given by

$|A - \lambda I| = 0 \Rightarrow \begin{vmatrix} 1-\lambda & -4 \\ 2 & -(3+\lambda) \end{vmatrix} = 0$

$\Rightarrow -(3+\lambda)(1-\lambda)+8 = 0 \Rightarrow \lambda^2 + 2\lambda + 5 = 0 \Rightarrow \lambda = -1 \pm 2i$

***Example* 61.** Suppose that the eigen values of matrix A are 1, 2, 4.

The determinant of $\left(A^{-1}\right)^T$ is _____. [CS/IT 2016]

Answer: 0.125

By property 24. vii), the eigen values of A^{-1} will be $\frac{1}{1}, \frac{1}{2}, \frac{1}{4}$

$\det\left(A^{-1}\right) = $ product of the eigen values of $A^{-1} = \frac{1}{8} = 0.125$

$\det\left(A^{-1}\right)^T = \det\left(A^{-1}\right) = 0.125$

***Example* 62.** Consider a 3 × 3 matrix with every element being equal to 1.
Its only non-zero eigen value is _____. [EE 2016]

Answer: 3

Sum of the eigen values = sum of the diagonal elements = 1 + 1 + 1 = 3
Product of the eigen values = det of the matrix = 0
As the matrix has only one non-zero eigen value, it is 3.

Example 63. The condition for which the eigen values of the matrix

$A = \begin{pmatrix} 2 & 1 \\ 1 & k \end{pmatrix}$ are positive, is

(A) $k > 1/2$ (B) $k > -2$ (C) $k > 0$ (D) $k < -1/2$

Answer: (A)

The Eigen values are given by the roots of the equation

$\begin{vmatrix} 2-x & 1 \\ 1 & k-x \end{vmatrix} = 0 \Rightarrow x^2 - (k+2)x + (2k-1) = 0$

Sum of the Eigen values $= k + 2$ and

the product of the Eigen values $= 2k - 1$

As the Eigen values are both positive, $k + 2 > 0$ & $2k - 1 > 0$

Both the relations are satisfied if $k > \dfrac{1}{2}$

Example 64. Eigen values of a matrix $S = \begin{pmatrix} 3 & 2 \\ 2 & 3 \end{pmatrix}$ are 5 and 1.

What are the eigen values of the matrix $S^2 = S.S$?

(A) 1 and 25 (B) 6 and 4 (C) 5 and 1 (D) 2 and 10   [ME 2006]

Answer (A)

It follows from the property 24(v) of Eigen values.

Example 65. If for a square A, $A^2 = A$, then the Eigen value/values of A is/are

(A)   0   (B)   1   (C)   either 0 or 1   (D)   both 0 & 1

Answer (C)

A is an idempotent matrix, it follows from property 24. Xiii).

Example 66. A real (4 × 4) matrix A satisfies the equation $A^2 = I$, where

I is an (4 × 4) identity matrix.

The positive eigen value of A is _____.   [EC 2014]

Answer: 1

The eigen value of $A^2 = I$ is 1.

So, the positive eigen value of A is $\sqrt{1} = 1$.

Example **67.** What are the eigen values of the following matrix?

$$\begin{bmatrix} 1 & 1 \\ -2 & 4 \end{bmatrix}$$

(A) 2 and 3   (B) - 2 and 3   (C) 2 and – 3   (D) - 2 and - 3 [BT 2015]

Answer: (A)

The eigen values are given by

$$\begin{vmatrix} 1-\lambda & 1 \\ -2 & 4-\lambda \end{vmatrix} = 0 \Rightarrow (1-\lambda)(4-\lambda)+2 = 0 \Rightarrow \lambda^2 - 5\lambda + 6 = 0 \Rightarrow \lambda = 2,3$$

Example **68.** The determinant of a 2 × 2 matrix is 50. If one eigen value of the matrix is 10, the other eigen value is _____.   [ME 2017]

Answer: 5

Product of the eigen values = determinant of the matrix = 50

Other eigen value = 50/10 = 5

Example **69.** Consider a matrix $A = uv^T$ where $u = \begin{pmatrix} 1 \\ 2 \end{pmatrix}, v = \begin{pmatrix} 1 \\ 1 \end{pmatrix}$. Note that

v^T denotes the transpose of v. The largest Eigen value of A is [CS/IT 2018]

Answer: 3

$$A = \begin{pmatrix} 1 \\ 2 \end{pmatrix}.(1 \quad 1) = \begin{pmatrix} 1 & 1 \\ 2 & 2 \end{pmatrix} \Rightarrow A \text{ is singular} \Rightarrow \text{ one Eigen value = 0}$$

The sum of the two Eigen values = trace = 1 + 2 = 3

The other Eigen value = 3 (largest).

Example **70.** If the sum of the diagonal elements of a 2 × 2 matrix is 6, then the maximum possible value of determinant of the matrix is

_____.   [EE 2015]

Answer: 9

If the eigen values be x & $y, x + y = 6$ Value of determinant of

the matrix $= z = xy = x(6-x) = 6x - x^2 \quad \dfrac{dz}{dx} = 0 \Rightarrow x = 3$

$z_{max} = 3.(6-3) = 9$

107

Example 71. The two Eigen values of the matrix $\begin{pmatrix} 2 & 1 \\ 1 & p \end{pmatrix}$ have a ratio of 3:1 for $p = 2$. What is another value of p for which the Eigen values have the same ratio of 3:1?

(A) – 2   (B)   1   (C)   7/3   (D)   14/3   [CE 2015]

Answer (D)

If the eigen values be x & $3x$, sum of the eigen values $= 4x = 2 + p$

Product of the eigen values $= 3x^2 = \begin{vmatrix} 2 & 1 \\ 1 & p \end{vmatrix} = 2p - 1$

Both the relations are satisfied for $p = \dfrac{14}{3}$ (other than $p = 2$)

Example 72. The eigen values of the matrix $\begin{pmatrix} 1 & 2 & 3 \\ 0 & 4 & 7 \\ 0 & 0 & 3 \end{pmatrix}$ are

(A) 1, 4, 3   (B) 3, 7, 3   (C) 7, 3, 2   (D) 1, 2, 3

Answer (A)

The given matrix is an upper triangular matrix.

Its eigen values are its diagonal elements i.e., 1, 4, 3

Example 73. The number of independent eigen values of

$\begin{pmatrix} 2 & 1 & 0 \\ 0 & 2 & 0 \\ 0 & 0 & 3 \end{pmatrix}$ are _____.   [ME 2016]

Answer: 2

The eigen values are given by

$\begin{vmatrix} 2-\lambda & 1 & 0 \\ 0 & 2-\lambda & 0 \\ 0 & 0 & 3-\lambda \end{vmatrix} = 0 \Rightarrow (2-\lambda)(2-\lambda)(3-\lambda) = 0 \Rightarrow \lambda = 2,3$

108

Example 74. The eigen values of the matrix given below are

$$\begin{pmatrix} 0 & 1 & 0 \\ 0 & 0 & 1 \\ 0 & -3 & -4 \end{pmatrix}$$

(A) (0, - 1, - 3) (B)   (0, - 2, - 3) (C) (0, 2, 3) (D) (0, 1, 3) [EE 2017]

Answer: (A)

The eigen values are given by

$$\begin{vmatrix} -\lambda & 1 & 0 \\ 0 & -\lambda & 1 \\ 0 & -3 & -(4+\lambda) \end{vmatrix} = 0 \Rightarrow \lambda\left(\lambda^2 + 4\lambda + 3\right) = 0 \Rightarrow \lambda = 0, -1, -3$$

Example 75. The eigen values of the matrix $\begin{pmatrix} 3 & -2 & 1 & 2 \\ 0 & 5 & 6 & -7 \\ 0 & 0 & -3 & 5 \\ 0 & 0 & 0 & 8 \end{pmatrix}$ are_____.

Answer: 3, 5, - 3, 8

The eigen values of a triangular matrix are its leading diagonal elements.
It follows from property 24. x)

Example 76. A system matrix is given as follows: $\begin{pmatrix} 0 & 1 & -1 \\ -6 & -11 & 6 \\ -6 & -11 & 5 \end{pmatrix}$

The absolute value of the ratio of the maximum eigen value to the
minimum eigen value is _____   [EE 2014]

Answer: 3:1

Sum of the eigen values = trace = 0- 11 + 5 = - 6

Product of the eigen values = det of the matrix

= 0 -1(- 30 + 36) -1(66 – 66) = - 6

The absolute value of the ratio of the maximum eigen value to the
minimum eigen value = 3:1

Example 77. The sum of the Eigen values of the matrix

$$\begin{pmatrix} 215 & 650 & 795 \\ 655 & 150 & 835 \\ 485 & 355 & 550 \end{pmatrix} \text{ is}$$

(A)   915   (B)   1355   (C)   1640   (D)   2180 [CE 2014]

Answer (A)

BY Property of Eigen values, the sum of the Eigen values of a matrix =
Trace of the matrix = 215 + 150 + 550 = 915

Example 78. The product of the eigen values of the matrix $\begin{bmatrix} 2 & 0 & 1 \\ 4 & -3 & 3 \\ 0 & 2 & -1 \end{bmatrix}$

(A)   - 6   (B)   2   (C)   6   (D)   - 2   [ME 2017]

Answer: (B)

Product of the eigen values

$$= \begin{vmatrix} 2 & 0 & 1 \\ 4 & -3 & 3 \\ 0 & 2 & -1 \end{vmatrix} = 2(3-6) - 0 + 1(8-0) = -6 + 8 = 2$$

Example 79. The value of X for which all the eigen values of the matrix

given below is real is $\begin{bmatrix} 10 & 5+j & 4 \\ x & 20 & 2 \\ 4 & 2 & -10 \end{bmatrix}$

(A) $5 + j$ (B) $5 - j$ (C) $1 + 5j$ (D) $1 - 5j$ [EC 2015]

Answer: (B)

By property 24. Xii), the matrix should be Hermitian, $A = \overline{A}^T$

$$\begin{bmatrix} 10 & 5+j & 4 \\ x & 20 & 2 \\ 4 & 2 & -10 \end{bmatrix} = \overline{\begin{bmatrix} 10 & 5-j & 4 \\ x & 20 & 2 \\ 4 & 2 & -10 \end{bmatrix}}^T = \begin{bmatrix} 10 & \overline{x} & 4 \\ 5-j & 20 & 2 \\ 4 & 2 & -10 \end{bmatrix} \Rightarrow x = 5-j$$

Example 80. The eigen values of the matrix given below are

$$\begin{pmatrix} 0 & 1 & 0 \\ 0 & 0 & 1 \\ 0 & -3 & -4 \end{pmatrix}$$

(A) (0, - 1, - 3) (B) (0, - 2, - 3) (C) (0, 2, 3) (D) (0, 1, 3) [EE 2017]

Answer: (A)

The eigen values are given by

$$\begin{vmatrix} -\lambda & 1 & 0 \\ 0 & -\lambda & 1 \\ 0 & -3 & -(4+\lambda) \end{vmatrix} = 0 \Rightarrow \lambda(\lambda^2 + 4\lambda + 3) = 0 \Rightarrow \lambda = 0, -1, -3$$

Example 81. The eigen values of the matrix $A = \begin{bmatrix} 1 & -1 & 5 \\ 0 & 5 & 6 \\ 0 & -6 & 5 \end{bmatrix}$ are

(A) - 1, 5, 6 (B) 1, -5 ± j.6 (C) 1, 5 ± j.6 (D) 1, 5, 5 [PI 2017]

Answer: (C)

The eigen values of the matrix A are given by $\begin{vmatrix} 1-\lambda & -1 & 5 \\ 0 & 5-\lambda & 6 \\ 0 & -6 & 5-\lambda \end{vmatrix} = 0$

$\Rightarrow (1-\lambda)\{(5-\lambda)^2 + 36\} = 0 \Rightarrow \lambda = 1 \ \& \ (5-\lambda)^2 = -36 \Rightarrow 5 - \lambda = \pm j.6$
$\Rightarrow \lambda = 1, 5 \pm j.6$

Example 82. Consider the matrix $P = \begin{bmatrix} \dfrac{1}{\sqrt{2}} & 0 & \dfrac{1}{\sqrt{2}} \\ 0 & 1 & 0 \\ -\dfrac{1}{\sqrt{2}} & 0 & \dfrac{1}{\sqrt{2}} \end{bmatrix}$

Which one of the following statements about P is INCORRECT?

(A) Determinant of P is equal to 1

(B) P is orthogonal

(C) Inverse of P is equal to its transpose

(D) All eigen values of P are real numbers [ME 2017]

Answer (D)

By property 24.xi), the eigen values of a real symmetric matrix are all real. P is not a real symmetric matrix.

Example 83. The matrix $A = \begin{bmatrix} a & 0 & 3 & 7 \\ 2 & 5 & 1 & 3 \\ 0 & 0 & 2 & 4 \\ 0 & 0 & 0 & b \end{bmatrix}$ has $\det(A) = 100$

& $trace\,(A) = 14$. The value of $|a-b|$ is _____. [EC 2016]

Answer: 3

$trace(A) = 14 \Rightarrow a + 5 + 2 + b = 14 \Rightarrow a + b = 7$

$\det(A) = 100 \Rightarrow a\begin{vmatrix} 5 & 1 & 3 \\ 0 & 2 & 4 \\ 0 & 0 & b \end{vmatrix} - 2\begin{vmatrix} 0 & 3 & 7 \\ 0 & 2 & 4 \\ 0 & 0 & b \end{vmatrix} = 100 \Rightarrow a[5(2b)] = 100 \Rightarrow ab = 10$

$(a-b)^2 = (a+b)^2 - 4ab = 49 - 40 = 9 \Rightarrow |a-b| = 3$

Example 84. Which one of the following statements is TRUE about every $n \times n$ matrix with only real Eigen value?
(A) If the trace of the matrix is positive and the determinant of the matrix is negative, at least one of its Eigen values is negative.
(B) If the trace of the matrix is positive, all its Eigen values are positive.
(C) If the determinant of the matrix is positive, all its Eigen values are positive.
(D) If the product of the trace and determinant of the matrix is positive, all its Eigen values are positive. [CS/IT 2014]
Answer (A)
Product of the eigen values = det A < 0 (given). So, at least one of the eigen values is negative.

Example 85. Which one of the following statements is NOT true for a square matrix?
(A) If A is upper triangular, the eigen values of A are the diagonal elements of it
(B) If A is real symmetric, the eigen values of A are always real and positive

112

(C) If A is real, the eigen values of A and AT are always the same
(D) If all the principal minors of A are positive, all the eigen values of A are also positive   [EC 2014]
Answer: (B)

$A = \begin{pmatrix} 1 & 4 \\ 4 & 1 \end{pmatrix}$ is a real symmetric matrix.

Its Eigen values are given by $\begin{vmatrix} 1-\lambda & 4 \\ 4 & 1-\lambda \end{vmatrix} = 0 \Rightarrow (1-\lambda)^2 - 16 = 0$

Or, $\lambda^2 - 2\lambda - 15 = 0 \Rightarrow \lambda = 5, -3$

So, the Eigen values are real but not always positive.

Example 86. Consider the matrix $\begin{pmatrix} 2 & 1 & 1 \\ 2 & 3 & 4 \\ -1 & -1 & -2 \end{pmatrix}$ whose eigen

values are 1, - 1 & 3. Then Trace of $A^3 - 3A^2$ is _____[IN 2016]

Answer: - 6

Eigen values of $A^3 - 3A^2$ are

$1^3 - 3.1^2, (-1)^3 - 3(-1)^2, (3)^3 - 3(3)^2 = -2, -4, 0.$

Trace = sum of the eigen values = - 2 - 4 + 0 = - 6

Example 87. For the matrix $\begin{bmatrix} 4 & 3 \\ 3 & 4 \end{bmatrix}$, if $\begin{pmatrix} 1 \\ 1 \end{pmatrix}$ is an eigen vector, then

the corresponding eigen value is _____.   [CH 2015]

Answer: 7

$AX = \lambda X \Rightarrow \begin{bmatrix} 4 & 3 \\ 3 & 4 \end{bmatrix} . \begin{bmatrix} 1 \\ 1 \end{bmatrix} = \lambda \begin{bmatrix} 1 \\ 1 \end{bmatrix} \Rightarrow \lambda = 7$

Example 88. The value of the dot product of the eigen vectors corresponding to any pair of different eigen values of a 4-by-4 symmetric positive definite matrix is

(A)   0   (B)   1   (C)   - 1   (D)   2   [CS/IT 2014]

Answer: (A)

The eigen vectors corresponding to any pair of different eigen values of

Eigen vector $= \{-4\alpha, 2\alpha, \alpha\} = \{\alpha(-4,2,1) / \alpha \neq 0, \alpha \in R\}$

Example 93. The value of p such that the vector $\begin{bmatrix} 1 \\ 2 \\ 3 \end{bmatrix}$ is an eigen vector

of the matrix $\begin{bmatrix} 4 & 1 & 2 \\ p & 2 & 1 \\ 14 & -4 & 10 \end{bmatrix}$ is_____.   [EC 2015]

Answer: 17

$$\begin{bmatrix} 4 & 1 & 2 \\ p & 2 & 1 \\ 14 & -4 & 10 \end{bmatrix}\begin{bmatrix} 1 \\ 2 \\ 3 \end{bmatrix} = \lambda \begin{bmatrix} 1 \\ 2 \\ 3 \end{bmatrix} \Rightarrow 12 = \lambda \ \& \ p+4+3 = 2\lambda = 24 \Rightarrow p = 17$$

Example 94. Consider a 3 × 3 real symmetric matrix S such that two of its eigen values are $a \neq 0$ & $b \neq 0$ with respective eigen vectors

$\begin{pmatrix} x_1 \\ x_2 \\ x_3 \end{pmatrix}, \begin{pmatrix} y_1 \\ y_2 \\ y_3 \end{pmatrix}$. If $a \neq b$, then $x_1 y_1 + x_2 y_2 + x_3 y_3$ equals

(A) a (B) b (C) ab (D)   0   [ME 2014]

Answer (D)

The eigen vectors corresponding to two distinct eigen values of a real symmetric matrix are orthogonal.

$$\begin{pmatrix} x_1 \\ x_2 \\ x_3 \end{pmatrix} \cdot \begin{pmatrix} y_1 \\ y_2 \\ y_3 \end{pmatrix} = (0) \Rightarrow x_1 y_1 + x_2 y_2 + x_3 y_3 = 0$$

Question of 2 marks:

Example 95. Given that $A = \begin{pmatrix} -5 & -3 \\ 2 & 0 \end{pmatrix}$ & $I = \begin{pmatrix} 1 & 0 \\ 0 & 1 \end{pmatrix}$, the value of A^3 is

(A) $15A + 12I$ (B) $19A + 30I$

(C) $17A + 15I$ (D) $17A + 21I$ [EE 2015]

116

Answer: (B)

The characteristic equation of the matrix A is given by

$$|A - \lambda I| = 0 \Rightarrow \begin{vmatrix} -5-\lambda & -3 \\ 2 & 0-\lambda \end{vmatrix} = 0 \Rightarrow \lambda^2 + 5\lambda + 6 = 0 \Rightarrow \lambda^2 = -5\lambda - 6$$

By Cayley Hamilton theorem, every square matrix satisfies its own *characteristic equation. So,* $A^2 = -5A - 6I$

Now, $A^3 = -5A^2 - 6I = -5(-5A - 6I) - 6A = 19A + 30I$

Example 96. The matrix $A = \begin{bmatrix} a & 0 & 3 & 7 \\ 2 & 5 & 1 & 3 \\ 0 & 0 & 2 & 4 \\ 0 & 0 & 0 & b \end{bmatrix}$ has

$\det(A) = 100$ and $trace\ (A) = 14$. The value of $|a-b|$ is __[EC 2016]

Answer: 3

$trace\ (A) = 14 \Rightarrow a + 5 + 2 + b = 14 \Rightarrow a + b = 7$

$$\det(A) = 100 \Rightarrow a\begin{vmatrix} 5 & 1 & 3 \\ 0 & 2 & 4 \\ 0 & 0 & b \end{vmatrix} - 2\begin{vmatrix} 0 & 3 & 7 \\ 0 & 2 & 4 \\ 0 & 0 & b \end{vmatrix} = 100 \Rightarrow a[5(2b)] = 100 \Rightarrow ab = 10$$

$$(a-b)^2 = (a+b)^2 - 4ab = 49 - 40 = 9 \Rightarrow |a-b| = 3$$

Example 97. In the *LU* decomposition of the matrix $\begin{pmatrix} 2 & 2 \\ 4 & 9 \end{pmatrix}$,

if the diagonal elements of *U* are both 1, then the lower diagonal   entry t_{22} of *L* is ---------.   [CS/ IT 2015]

Answer: 5

$$A = LU \Rightarrow \begin{pmatrix} 2 & 2 \\ 4 & 9 \end{pmatrix} = \begin{pmatrix} t_{11} & 0 \\ t_{21} & t_{22} \end{pmatrix}\begin{pmatrix} 1 & u_{12} \\ 0 & 1 \end{pmatrix} = \begin{pmatrix} t_{11} & t_{11}u_{12} \\ t_{21} & t_{21}u_{12} + t_{22} \end{pmatrix}$$

Or, $t_{11} = 2,\ t_{11}u_{12} = 2 \Rightarrow u_{12} = 1\ \&\ t_{21} = 4,\ t_{21}u_{12} + t_{22} = 9 \Rightarrow t_{22} = 5$

Example 98. For the matrix $A = \begin{pmatrix} 1 & 0 & 0 \\ 2 & 1 & 0 \\ 3 & 2 & 1 \end{pmatrix}$ If U_1, U_2, U_3 are column

matrices satisfying $AU_1 = \begin{pmatrix} 1 \\ 0 \\ 0 \end{pmatrix}, AU_2 = \begin{pmatrix} 2 \\ 3 \\ 0 \end{pmatrix}, AU_3 = \begin{pmatrix} 2 \\ 3 \\ 1 \end{pmatrix}$ and U is 3 × 3

matrix whose columns are U_1, U_2, U_3 then the value of the determinant
$|U|$ is

(A)　　3　　　　(B)　　- 3　　　(C)　　3/2　　(D)　　2　　[JECA 2013]

Answer: (A)

$A^{-1} = Adj\,(A)\,/\det A$ and

$\det A = 1 \; A^{-1} = \begin{pmatrix} 1 & 0 & 0 \\ -2 & 1 & 0 \\ 1 & -2 & 1 \end{pmatrix}, U_1 = A^{-1}\begin{pmatrix} 1 \\ 0 \\ 0 \end{pmatrix} = \begin{pmatrix} 1 & 0 & 0 \\ -2 & 1 & 0 \\ 1 & -2 & 1 \end{pmatrix}\begin{pmatrix} 1 \\ 0 \\ 0 \end{pmatrix} = \begin{pmatrix} 1 \\ -2 \\ 1 \end{pmatrix}$

Similarly, $U_2 = \begin{pmatrix} 2 \\ -1 \\ -4 \end{pmatrix}$ & $U_3 = \begin{pmatrix} 2 \\ -1 \\ -3 \end{pmatrix}$ So, $U = \begin{pmatrix} 1 & 2 & 2 \\ -2 & -1 & -1 \\ 1 & -4 & -3 \end{pmatrix}$

$\det U = \begin{vmatrix} 1 & 2 & 2 \\ -2 & -1 & -1 \\ 1 & -4 & -3 \end{vmatrix} = 1(3-4) - 2(6+1) + 2(8+1) = 3$

Example 99. $\begin{pmatrix} 2 & 1 \\ 3 & 2 \end{pmatrix} A \begin{pmatrix} -3 & 2 \\ 5 & -3 \end{pmatrix} = I_2,$ then the matrix A is equal to

(A) $\begin{pmatrix} 1 & 1 \\ 1 & 0 \end{pmatrix}$　　(B) $\begin{pmatrix} 1 & 1 \\ 0 & 1 \end{pmatrix}$　　(C) $\begin{pmatrix} 1 & 0 \\ 1 & 1 \end{pmatrix}$　　(D) $\begin{pmatrix} 0 & 1 \\ 1 & 1 \end{pmatrix}$

Answer (A)

Let $C = \begin{pmatrix} 2 & 1 \\ 3 & 2 \end{pmatrix}$ & $B = \begin{pmatrix} -3 & 2 \\ 5 & -3 \end{pmatrix}$ then $CAB = I_2$

Multiplying by C^{-1} in the beginning and by B^{-1} at the end, we get

$A = C^{-1}I_2B^{-1} = C^{-1}B^{-1} = (BC)^{-1}$ ---------------------(i)

Now, $BC = \begin{pmatrix} -3 & 2 \\ 5 & -3 \end{pmatrix}\begin{pmatrix} 2 & 1 \\ 3 & 2 \end{pmatrix} = \begin{pmatrix} 0 & 1 \\ 1 & -1 \end{pmatrix}$, $|BC| = -1$

From (i) $A = (BC)^{-1} = -1\begin{pmatrix} -1 & -1 \\ -1 & 0 \end{pmatrix}' = -\begin{pmatrix} -1 & -1 \\ -1 & 0 \end{pmatrix} = \begin{pmatrix} 1 & 1 \\ 1 & 0 \end{pmatrix}$

Example 100. The matrix $A = \begin{pmatrix} 2 & 1 \\ 4 & -1 \end{pmatrix}$ is decomposed into a product of lower triangular matrix [L] and an upper triangular matrix [U]. The properly decomposed [L] and [U] matrices respectively are

(A) $\begin{pmatrix} 1 & 0 \\ 4 & -1 \end{pmatrix}$ & $\begin{pmatrix} 1 & 1 \\ 0 & -2 \end{pmatrix}$ (B) $\begin{pmatrix} 2 & 0 \\ 4 & -1 \end{pmatrix}$ & $\begin{pmatrix} 1 & 1 \\ 0 & 1 \end{pmatrix}$

(C) $\begin{pmatrix} 1 & 0 \\ 4 & 1 \end{pmatrix}$ & $\begin{pmatrix} 2 & 0 \\ 1 & -1 \end{pmatrix}$ (D) $\begin{pmatrix} 2 & 0 \\ 4 & -3 \end{pmatrix}$ & $\begin{pmatrix} 1 & 0.5 \\ 0 & 1 \end{pmatrix}$ [EE 2011]

Answer: (D)

If we take

$[L] = \begin{pmatrix} 2 & 0 \\ 4 & -3 \end{pmatrix}$ & $[U] = \begin{pmatrix} 1 & 0.5 \\ 0 & 1 \end{pmatrix}$, then $[L][U] = \begin{pmatrix} 2 & 0 \\ 4 & -3 \end{pmatrix}\begin{pmatrix} 1 & 1.5 \\ 0 & 1 \end{pmatrix}$

$= \begin{pmatrix} 2 & 1 \\ 4 & -1 \end{pmatrix} = [A]$

Example 101. If $A + B = 2B'$ and $3A + 2B = I_3$ then $B =$

(A) $\frac{1}{3}I_3$ (B) $\frac{1}{4}I_3$ (C) $\frac{1}{5}I_3$ (D) $\frac{1}{6}I_3$

Answer: (C)

$3A + 3B = 6B'$ --------------------(i) and $3A + 2B = I_3$ --------------------(ii)

(i) – (ii) gives $B = 6B' - I_3$ or, $B' = \frac{1}{6}(B + I_3)$ ------(iii)

Also, $(B')' = \dfrac{1}{6}(B + I_3)' \Rightarrow B = \dfrac{1}{6}B' + \dfrac{1}{6}I_3 \Rightarrow B' = 6B - I_3$ as

$I_3' = I_3$ Putting the value of B' in (iii)

$6B - I_3 = \dfrac{1}{6}(B + I_3) \Rightarrow 36B - 6I_3 = B + I_3 \Rightarrow 35B = 7I_3 \Rightarrow B = \dfrac{1}{5}I_3$

Example 102. The skew-symmetric part of the matrix $A = \begin{pmatrix} 1 & 2 & 4 \\ 6 & 8 & 1 \\ 3 & 5 & 7 \end{pmatrix}$ is

(A) $\begin{pmatrix} 0 & 2 & 1/2 \\ -2 & 0 & 2 \\ 1/2 & -2 & 0 \end{pmatrix}$
(B) $\begin{pmatrix} 0 & -2 & 1/2 \\ -2 & 0 & -2 \\ 1/2 & 2 & 0 \end{pmatrix}$

(C) $\begin{pmatrix} 0 & -2 & 1/2 \\ 2 & 0 & -2 \\ -1/2 & 2 & 0 \end{pmatrix}$
(D) $\begin{pmatrix} 0 & 2 & 1/2 \\ -2 & 0 & -2 \\ -1/2 & 2 & 0 \end{pmatrix}$

Answer: (C)

$A = \begin{pmatrix} 1 & 2 & 4 \\ 6 & 8 & 1 \\ 3 & 5 & 7 \end{pmatrix} \Rightarrow A' = \begin{pmatrix} 1 & 6 & 3 \\ 2 & 8 & 5 \\ 4 & 1 & 7 \end{pmatrix}$ The skew-symmetric part of A

$= \dfrac{1}{2}(A - A') = \dfrac{1}{2}\begin{pmatrix} 0 & -4 & 1 \\ 4 & 0 & -4 \\ -1 & 4 & 0 \end{pmatrix} = \begin{pmatrix} 0 & -2 & 1/2 \\ 2 & 0 & -2 \\ -1/2 & 2 & 0 \end{pmatrix}$

Example 103. If A and B are two square matrices such that $B = -A^{-1}BA$ then $(A + B)^2 =$

(A) $A^2 + 2AB + B^2$
(B) $A^2 + AB + B^2$
(C) $A^2 + B^2$
(D) $A^2 - AB + B^2$

Answer: (C)

$B = -A^{-1}BA \Rightarrow AB = -AA^{-1}BA = -IBA = -BA$ as $AA^{-1} = I$
So, $AB + BA = O$

120

so, $(A+B)^2 = (A+B)(A+B) = AA + (AB+BA) + BB$

$= A^2 + O + B^2 = A^2 + B^2$

Example 104. If A be a skew-symmetric matrix of order n and P be a n × 1 matrix, then the matrix $P'AP$ is a

(A) Null matrix   (B) Identity matrix   (C)   A   (D) A^t

Answer: (A)

As the order of A is $n \times n$ and the order of P is $n \times 1$, so the order of AP is n × 1. Also, the order of P^t is $1 \times n$. Finally, the order of $P'AP$ will be 1×1.

So, the matrix $P'AP$ will consist of 1 element only.

As A is a skew-symmetric matrix, $A^t = -A$

$(P'AP)' = (AP)'.(P')' = (AP)'.P = P'A'.P = P'(-A).P = -(P'AP)$ S

o, $P'AP$ is skew-symmetric. The diagonal elements of a skew-symmetric matrix are 0. As $P'AP$ contains one element only, it is the diagonal element. So, $P'AP$ is a null matrix.

Example 105. If $A = \begin{pmatrix} \cos\theta & -\sin\theta & 0 \\ \sin\theta & \cos\theta & 0 \\ 0 & 0 & x \end{pmatrix}$ is orthogonal, then $x =$

(A) ± 1 (B) ± 2 (C) ± 3 (D) 0

Answer: (A)

As A is orthogonal, $A.A^t = I_3$

$\Rightarrow \begin{pmatrix} \cos\theta & -\sin\theta & 0 \\ \sin\theta & \cos\theta & 0 \\ 0 & 0 & x \end{pmatrix} . \begin{pmatrix} \cos\theta & \sin\theta & 0 \\ -\sin\theta & \cos\theta & 0 \\ 0 & 0 & x \end{pmatrix} = \begin{pmatrix} 1 & 0 & 0 \\ 0 & 1 & 0 \\ 0 & 0 & 1 \end{pmatrix}$

Or, $\begin{pmatrix} 1 & 0 & 0 \\ 0 & 1 & 0 \\ 0 & 0 & x^2 \end{pmatrix} = \begin{pmatrix} 1 & 0 & 0 \\ 0 & 1 & 0 \\ 0 & 0 & 1 \end{pmatrix} \Rightarrow x^2 = 1 \Rightarrow x = \pm 1$

Example 106. Let $A = \begin{pmatrix} 1 & -1 & 1 \\ 2 & 1 & -3 \\ 1 & 1 & 1 \end{pmatrix}$ & $B = \begin{pmatrix} 4 & 2 & 2 \\ -5 & 0 & a \\ 1 & -2 & 3 \end{pmatrix}$

121

If $B = 10A^{-1}$, then $a =$

(A)   2   (B)   - 1   (C)   - 2   (D)   5

Answer: (D)

$$B = 10A^{-1} \Rightarrow AB = 10I \Rightarrow \begin{pmatrix} 1 & -1 & 1 \\ 2 & 1 & -3 \\ 1 & 1 & 1 \end{pmatrix} \begin{pmatrix} 4 & 2 & 2 \\ -5 & 0 & a \\ 1 & -2 & 3 \end{pmatrix} = \begin{pmatrix} 10 & 0 & 0 \\ 0 & 10 & 0 \\ 0 & 0 & 10 \end{pmatrix}$$

$\Rightarrow 1.2 + (-1).a + 1.3 = 0 \Rightarrow a = 5$ [(Ist row of A)×(3rd column of B)]

Example 107. For a given matrix $P = \begin{pmatrix} 4 + 3i & -i \\ i & 4 - 3i \end{pmatrix}$, where

$i = \sqrt{-1}$ the inverse of the matrix P is

(A) $\dfrac{1}{24}\begin{pmatrix} 4 - 3i & i \\ -i & 4 + 3i \end{pmatrix}$ (B) $\dfrac{1}{25}\begin{pmatrix} i & 4 - 3i \\ 4 + 3i & -i \end{pmatrix}$

(C) $\dfrac{1}{24}\begin{pmatrix} 4 + 3i & -i \\ i & 4 - 3i \end{pmatrix}$ (D) $\dfrac{1}{25}\begin{pmatrix} 4 + 3i & -i \\ i & 4 - 3i \end{pmatrix}$ [ME 2015]

Answer: (A)

$\det P = (4 + 3i)(4 - 3i) + i^2 = 25 - 1 = 24$

$Adj (P) = \begin{pmatrix} 4 - 3i & -i \\ i & 4 + 3i \end{pmatrix}^T = \begin{pmatrix} 4 - 3i & i \\ -i & 4 + 3i \end{pmatrix} \Rightarrow P^{-1} = \dfrac{adj (P)}{\det P}$

$= \dfrac{1}{24}\begin{pmatrix} 4 - 3i & i \\ -i & 4 + 3i \end{pmatrix}$ [By property 18]

Example 108. There are two column vectors $X = \begin{pmatrix} x \\ 1 \end{pmatrix} \& \begin{pmatrix} 1 & 4 \\ 5 & 2 \end{pmatrix} X$.

If $X = \begin{pmatrix} x \\ 1 \end{pmatrix} \& \begin{pmatrix} 1 & 4 \\ 5 & 2 \end{pmatrix} X$. are parallel and θ be the angle between the two

column vectors, then the value of tan θ is

(A)   9   (B)   7   (C)   5   (D)   3

Answer: (A)

$X = \begin{pmatrix} x \\ 1 \end{pmatrix} \& \begin{pmatrix} 1 & 4 \\ 5 & 2 \end{pmatrix} X$. are parallel, so, $c\begin{pmatrix} x \\ 1 \end{pmatrix} = \begin{pmatrix} 1 & 4 \\ 5 & 2 \end{pmatrix}\begin{pmatrix} x \\ 1 \end{pmatrix}$ for a non-zero scalar c. $cx = x + 4$ & $c = 5x + 2$ On dividing,

$$x = \frac{x+4}{5x+2} \Rightarrow 5x^2 + 2x = x + 4 \Rightarrow 5x^2 + x - 4 = 0 \Rightarrow x = -1, \frac{4}{5}$$

The two column vectors are $\begin{pmatrix} -1 \\ 1 \end{pmatrix}$ & $\begin{pmatrix} 4/5 \\ 1 \end{pmatrix}$ i.e,

$$\begin{pmatrix} a_1 \\ b_1 \end{pmatrix} \& \begin{pmatrix} a_2 \\ b_2 \end{pmatrix} \quad \tan\theta = \left| \frac{a_1 b_2 - a_2 b_1}{a_1 a_2 + b_1 b_2} \right| = \frac{4/5 + 1}{-4/5 + 1} = 9$$

$\det A = 0 \Rightarrow 7t + 35 = 0 \Rightarrow t = -5 \Rightarrow r(A) = r(A/B) = 2$

Example 109. Let A be an $m \times n$ matrix and B an $n \times m$ matrix. It is given that determinant $(I_m + AB)$ = determinant $(I_n + BA)$, where

I_k is the k × k identity matrix. Using the above property, the determinant

of the matrix given below is $\begin{pmatrix} 2 & 1 & 1 & 1 \\ 1 & 2 & 1 & 1 \\ 1 & 1 & 2 & 1 \\ 1 & 1 & 1 & 2 \end{pmatrix}$

(A)   2   (B)   5   (C)   8   (D)   16   [EC 2013]

Answer: (B)

Let $A = \begin{pmatrix} 1 & 1 & 1 & 1 \end{pmatrix}_{1 \times 4}$ & $B = \begin{pmatrix} 1 \\ 1 \\ 1 \\ 1 \end{pmatrix}_{4 \times 1}$ So, $m = 1, n = 4$ & $I_1 = (1)$

$$I_4 = \begin{pmatrix} 1 & 0 & 0 & 0 \\ 0 & 1 & 0 & 0 \\ 0 & 0 & 1 & 0 \\ 0 & 0 & 0 & 1 \end{pmatrix}, \quad AB = (4) \& BA = \begin{pmatrix} 1 & 1 & 1 & 1 \\ 1 & 1 & 1 & 1 \\ 1 & 1 & 1 & 1 \\ 1 & 1 & 1 & 1 \end{pmatrix}$$

$\det (I_1 + AB) = \det (I_4 + BA) \Rightarrow \det(5) = \det (I_1 + AB)$

$$= \det\left(I_4 + BA\right) \Rightarrow \det(5) = \det \begin{pmatrix} 2 & 1 & 1 & 1 \\ 1 & 2 & 1 & 1 \\ 1 & 1 & 2 & 1 \\ 1 & 1 & 1 & 2 \end{pmatrix}$$

$$\Rightarrow \det \begin{pmatrix} 2 & 1 & 1 & 1 \\ 1 & 2 & 1 & 1 \\ 1 & 1 & 2 & 1 \\ 1 & 1 & 1 & 2 \end{pmatrix} = 5$$

Example 110. If the straight lines

$a_1 x + b_1 y + c_1 = 0$ & $a_2 x + b_2 y + c_2 = 0$ are coincident, then

the rank of the matrix $\begin{pmatrix} a_1 & b_1 & c_1 \\ a_2 & b_2 & c_2 \end{pmatrix}$ is

(A)   0   (B)   1   (C)   2   (D)   3

Answer: (B)

As $a_1 x + b_1 y + c_1 = 0$ & $a_2 x + b_2 y + c_2 = 0$ are coincident,

$\dfrac{a_1}{a_2} = \dfrac{b_1}{b_2} = t(\neq 0)$ say. i.e., $a_1 = a_2 t, b_1 = b_2 t$.

Now, $A = \begin{pmatrix} a_1 & b_1 & c_1 \\ a_2 & b_2 & c_2 \end{pmatrix}$ Its second order minors are

$$M_1 = \begin{vmatrix} a_1 & b_1 \\ a_2 & b_2 \end{vmatrix} = \begin{vmatrix} a_2 t & b_2 t \\ a_2 & b_2 \end{vmatrix} = t \begin{vmatrix} a_2 & b_2 \\ a_2 & b_2 \end{vmatrix} = 0$$

Similarly, $M_2 = \begin{vmatrix} a_1 & c_1 \\ a_2 & c_2 \end{vmatrix} = 0$ & $M_3 = \begin{vmatrix} b_1 & c_1 \\ b_2 & c_2 \end{vmatrix} = 0$

So, all the second order minors of A are 0. But its first order minors

are $a_1, b_1, a_2, b_2, c_1, c_2$ and at least one of them should be non-zero.

So, $r(A) = 1$.

Example 111. If the three distinct lines $a_i x + b_i y + c_i = 0, i = 1,2,3$ are concurrent, then the rank of the matrix $\begin{pmatrix} a_1 & b_1 & c_1 \\ a_2 & b_2 & c_2 \\ a_3 & b_3 & c_3 \end{pmatrix}$ is

(A)   0   (B)   1   (C)   2   (D)   3

Answer: (C)

Let $A = \begin{pmatrix} a_1 & b_1 & c_1 \\ a_2 & b_2 & c_2 \\ a_3 & b_3 & c_3 \end{pmatrix}$ As the lines are concurrent, det A = 0.

So, the only third order minor of A is 0 i.e., $r(A) < 3$.

As the lines intersect, so they are non-parallel i.e. their slopes are

not equal. Or, $-\dfrac{b_1}{a_1} \neq -\dfrac{b_2}{a_2} \neq -\dfrac{b_3}{a_3}$ -----(i)

A second order minor of A is $\begin{vmatrix} a_1 & b_1 \\ a_2 & b_2 \end{vmatrix} = a_1 b_2 - a_2 b_1 \neq 0$ by (i).

So, $r(A) = 2$

Example 112. If the equation $ax^2 + 2hxy + by^2 + 2gx + 2fy + c = 0$ represents a pair of straight lines, then the rank of the matrix

$\begin{pmatrix} a & h & g \\ h & b & f \\ g & f & c \end{pmatrix}$ is

(A)   1   (B) 2   (C)   greater than 2   (D)   less than 3

Answer (D)

Let $A = \begin{pmatrix} a & h & g \\ h & b & f \\ g & f & c \end{pmatrix}$

As $ax^2 + 2hxy + by^2 + 2gx + 2fy + c = 0$ represents a pair of straight lines, $abc + 2fgh - af^2 - bg^2 - ch^2 = 0$ i.e., det A = 0.

So, the only third order minor of A is 0, $r(A) < 3$

125

Example 113. If the equation $ax^2 + 2hxy + by^2 + 2gx + 2fy + c = 0$ represents a pair of coincident straight lines, then the rank of the matrix

$$\begin{pmatrix} a & h & g \\ h & b & f \\ g & f & c \end{pmatrix} \text{ is}$$

(A)　　1　　　(B)　　2　(C)　less than 2　　(D)　greater than 2

Answer (A)

Let $A = \begin{pmatrix} a & h & g \\ h & b & f \\ g & f & c \end{pmatrix}$ As $ax^2 + 2hxy + by^2 + 2gx + 2fy + c = 0$ represents

a pair of coincident straight lines, $abc + 2fgh - af^2 - bg^2 - ch^2 = 0$ ------(i)

and $h^2 = ab$, $f^2 = bc$, $g^2 = ca$, $gh = af$, $hf = bg$, $gf = ch$ ----(ii)

Using (i), $det\ A = 0$. So, the only third order matrix of A is 0, $r(A) < 3$

The different second order minors of A are

$$M_1 = \begin{vmatrix} a & h \\ h & b \end{vmatrix} = ab - h^2 = 0 \ M_2 = \begin{vmatrix} h & g \\ b & f \end{vmatrix} = hf - bg = 0$$

$$M_3 = \begin{vmatrix} a & g \\ h & f \end{vmatrix} = af - hg = 0 \text{ etc. by (ii). So, } r(A) < 2.$$

At least one of the a, b, c, f, g, h is non–zero. $r(A) = 1$.

Example 114. If the roots α, β, γ of the equation $x^3 + qx + r = 0$ be in

A. P., then the rank of the matrix $\begin{pmatrix} \alpha & \beta & \gamma \\ \beta & \gamma & \alpha \\ \gamma & \alpha & \beta \end{pmatrix}$ is

(A)　　0　　　(B)　　1　　　(C)　　2　　　(D)　　3

Answer: (C)

Here, $\Sigma\alpha = 0, \Sigma\alpha\beta = q, \alpha\beta\gamma = -r$ As α, β, γ are in A.P., $\alpha + \gamma = 2\beta$

Let $A = \begin{pmatrix} \alpha & \beta & \gamma \\ \beta & \gamma & \alpha \\ \gamma & \alpha & \beta \end{pmatrix}$ then

$$\det A = \begin{vmatrix} \alpha & \beta & \gamma \\ \beta & \gamma & \alpha \\ \gamma & \alpha & \beta \end{vmatrix} = \begin{vmatrix} \Sigma\alpha & \beta & \gamma \\ \Sigma\alpha & \gamma & \alpha \\ \Sigma\alpha & \alpha & \beta \end{vmatrix} = \begin{vmatrix} 0 & \beta & \gamma \\ 0 & \gamma & \alpha \\ 0 & \alpha & \beta \end{vmatrix} = 0 \text{ or, } r(A) < 3$$

As α, β, γ are in A.P., $\alpha + \gamma = 2\beta, \Sigma\alpha = 0 \Rightarrow 3\beta = 0 \Rightarrow \beta = 0$ and

naturally $\alpha + \gamma = 0 \Rightarrow \gamma = -\alpha$ & $\alpha \neq 0$ So, a second order minor of

A is $\begin{vmatrix} \alpha & \beta \\ \beta & \gamma \end{vmatrix} = \alpha\gamma - \beta^2 = -\alpha^2 \neq 0 \Rightarrow r(A) = 2$

Example 115. The value of a for which the system of equations $3x - y + z = 0$, $15x - 6y + 5z = 0$, $ax - 2y + 2z = 0$ has a non-trivial solution is

(A)   10   (B)   8   (C)   6   (D)   4

Answer (C)

The coefficient matrix $A = \begin{pmatrix} 3 & -1 & 1 \\ 15 & -6 & 5 \\ a & -2 & 2 \end{pmatrix}$ The system of homogeneous

equations will have a non-trivial solution if $\det A = 0$ i.e., $a = 6$

Example 116. Consider a system of linear equations:
$3x - y + 4z = 3, x + 2y - 3z = -2$ & $6x + 5y + tz = -3$
The value of t for which the system has infinitely many solutions is _.

Answer: - 5

$$A = \begin{pmatrix} 3 & -1 & 4 \\ 1 & 2 & -3 \\ 6 & 5 & t \end{pmatrix} \Rightarrow \det(A) = 7t + 35 = 0 \Rightarrow t = -5$$

Example 117. If the following system has non-trivial solution
$px + qy + rz = 0, \quad qx + ry + pz = 0, \quad rx + py + qz = 0$
Then which one of the following options is TRUE?

(A) $p - q + r = 0$ or $p = q = -r$
(B) $p + q - r = 0$ or $p = -q = r$
(C) $p + q + r = 0$ or $p = q = r$
(D) $p - q + r = 0$ or $p = -q = -r$ [CS/IT 2015]

Answer (C)

The system of homogeneous equations has non-trivial solution if

$$\begin{vmatrix} p & q & r \\ q & r & p \\ r & p & q \end{vmatrix} = 0 \Rightarrow \begin{vmatrix} p+q+r & q & r \\ p+q+r & r & p \\ p+q+r & p & q \end{vmatrix} = 0$$

$$(p+q+r)\begin{vmatrix} 1 & q & r \\ 1 & r & p \\ 1 & p & q \end{vmatrix} = 0 \Rightarrow p+q+r = 0 \text{ or } \begin{vmatrix} 1 & q & r \\ 1 & r & p \\ 1 & p & q \end{vmatrix} = 0$$

$$\text{or} \begin{vmatrix} 1 & q & r \\ 0 & r-q & p-r \\ 0 & p-r & q-p \end{vmatrix} = 0 \Rightarrow (r-q)(q-p)-(p-r)^2 = 0$$

$$\Rightarrow p^2 + q^2 + r^2 - pq - qr - rp = 0$$

$$\frac{1}{2}\{(p-q)^2 + (q-r)^2 + (r-p)^2\} = 0 \Rightarrow p = q = r$$

Example 118. The following simultaneous
equations $x + y + z = 3, x + 2y + 3z = 4$ & $x + 4y + kz = 6$ will NOT
have a unique solution for $k =$
(A)   0   (B)   5   (C)   6   (D)   7 [CS/IT 2008]
Answer (D)
The equations will not have a unique solution if the coefficient

determinant $\begin{vmatrix} 1 & 1 & 1 \\ 1 & 3 & 3 \\ 1 & 4 & k \end{vmatrix} = 0 \Rightarrow k - 7 = 0 \Rightarrow k = 7$

Example 119. Consider a system of linear
equations: $x - 2y + 3z = -1, x - 3y + 4z = 1$ & $-2x + 4y - 6z = k$
The value of k for which the system has infinitely many solutions is _.
Answer: 2

Coefficient matrix $A = \begin{pmatrix} 1 & -2 & 3 \\ 1 & -3 & 4 \\ -2 & 4 & -6 \end{pmatrix}$ As the elements of the first and

third rows are proportional, $\det A = 0$. So, $r(A) = 2$

Augmented matrix $A/B = \begin{pmatrix} 1 & -2 & 3 & -1 \\ 1 & -3 & 4 & 1 \\ -2 & 4 & -6 & k \end{pmatrix}$

For infinitely many solutions, $r(A) = r(A/B) = 2$

SO, the elements of the first and third rows of A/B should be

proportional. $\dfrac{1}{-2} = \dfrac{-1}{k} \Rightarrow k = 2$

Example 120. The system of linear equations $\begin{pmatrix} 2 & 1 & 3 \\ 3 & 0 & 1 \\ 1 & 2 & 5 \end{pmatrix}\begin{pmatrix} a \\ b \\ c \end{pmatrix} = \begin{pmatrix} 5 \\ -4 \\ 14 \end{pmatrix}$ has

(A) a unique solution   (B) infinitely many solutions
(C) no solution   (D) exactly two solutions   [EC 2014]
Answer (B)

$A/B = \begin{pmatrix} 2 & 1 & 3 & 5 \\ 3 & 0 & 1 & -4 \\ 1 & 2 & 5 & 14 \end{pmatrix} \rightarrow \begin{pmatrix} -1 & 1 & 2 & 9 \\ 2 & -2 & -4 & -18 \\ 1 & 2 & 5 & 14 \end{pmatrix} [R_1^{'} = R_1 - R_2, R_2^{'} = R_2 - R_3]_T$

he first and the second rows are proportional.
So, $r(A) = r(A/B) = 2 <$ the no. of variables (= 3)
The system of linear equations has infinitely many solutions.

Example 121. The system of equations
$x + y + z = 6, x + 4y + 6z = 20, x + 4y + \lambda z = \mu$ has NO solution for
values of λ and μ given by
(A) $\lambda = \mu = 6$, 20 (B) $\lambda = \mu \neq 6$, 20 (C) $\lambda \neq \mu = 6$, 20 (D) $\lambda \neq \mu \neq 6$, 20[EC 14]
Answer: - (B)

The coefficient matrix $A = \begin{pmatrix} 1 & 1 & 1 \\ 1 & 4 & 6 \\ 1 & 4 & \lambda \end{pmatrix}$ and the corresponding augmented matrix $A / B = \begin{pmatrix} 1 & 1 & 1 & | & 6 \\ 1 & 4 & 6 & | & 20 \\ 1 & 4 & \lambda & | & \mu \end{pmatrix}$

$\rightarrow \begin{pmatrix} 1 & 0 & 1 & | & 4 \\ 0 & 1 & 1 & | & -1 \\ 0 & 0 & 0 & | & 2 \end{pmatrix}$ $\left[R_1' = R_1 - 2R_2, R_3' = R_3 - 2R_2 \right]$

Rank of A = 2, Rank of A/B =3, so, Rank of A ≠ Rank of A/B
Hence the system has no solution.

Example 122. Consider the following set of algebraic equations
$x_1 + 2x_2 + 3x_3 = 2$, $x_2 + x_3 = -1$, $2x_2 + 2x_3 = 0$. The system has
(A) A unique solution   (B) No solution
(C) An infinite number of solutions (D) Only null solution [CH 2012]
Answer (B)

The coefficient matrix $A = \begin{pmatrix} 1 & 2 & 3 \\ 0 & 1 & 1 \\ 0 & 2 & 2 \end{pmatrix}$ and the corresponding

augmented matrix A / B is

$= \begin{pmatrix} 1 & 2 & 3 & | & 2 \\ 0 & 1 & 1 & | & -1 \\ 0 & 2 & 2 & | & 0 \end{pmatrix}$

$\rightarrow \begin{pmatrix} 1 & 0 & 1 & | & 4 \\ 0 & 1 & 1 & | & -1 \\ 0 & 0 & 0 & | & 2 \end{pmatrix}$ $\left[R_1' = R_1 - 2R_2, R_3' = R_3 - 2R_2 \right]$

Rank of A = 2, Rank of A/B =3, so, Rank of A ≠ Rank of A/B
Hence the system has no solution.

Example 123. The condition on a for which the system of equations
$x + y + z = 1$, $x + 2y + 4z = a$, $x + 4y + 10z = a^2$ is consistent is
(A) $a^2 - 3a + 2 = 0$ (B) $3a^2 + a + 2 = 0$
(C) $a^2 + 3a + 2 = 0$ (D) $3a^2 - a + 2 = 0$ [JECA 2013]

130

Answer (A)

The augmented matrix of the system

$$\text{is}\,(A, B) = \begin{pmatrix} 1 & 1 & 1 & : & 1 \\ 1 & 2 & 4 & : & a \\ 1 & 4 & 10 & : & a^2 \end{pmatrix}$$

$$\rightarrow \begin{pmatrix} 1 & 1 & 1 & : & 1 \\ 0 & 1 & 3 & : & a-1 \\ 0 & 3 & 9 & : & a^2-1 \end{pmatrix}, R_2' = R_2 - R_1, R_3' = R_3 - R_1$$

$$\rightarrow \begin{pmatrix} 1 & 1 & 1 & : & 1 \\ 0 & 1 & 3 & : & a-1 \\ 0 & 0 & 0 & : & a^2 - 3a + 2 \end{pmatrix}, R_3' = R_3 - 3R_2$$

So,$r(A) = 2$. For consistency, $r(A, B) = 2$ i.e., $a^2 - 3a + 2 = 0$

Example 124. For what values of a & b, the following equations have an infinite number of

solutions? $x + y + z = 5;\ \ x + 3y + 3z = 9;\ \ x + 2y + az = b$

(A)   2, 7   (B)   3, 8   (C)   8, 3   (D)   7, 2   [CS/IT 2007]

Answer: (A)

$$A/B = \begin{bmatrix} 1 & 1 & 1 & : & 5 \\ 1 & 3 & 3 & : & 9 \\ 1 & 2 & a & : & b \end{bmatrix} \rightarrow \begin{bmatrix} 1 & 1 & 1 & : & 5 \\ 0 & 2 & 2 & : & 4 \\ 0 & 1 & a-1 & : & b-5 \end{bmatrix}$$

If $a = 2$, then 2nd and 3rd rows become dependent. So, $r(A) = 2$

For infinite number of solutions $r(A/B) = 2 \Rightarrow b = 7$

Example 125. The values of (a, b) for which the system of equations

$x + y + z = 1$, $x + 2y - z = b$, $5x + 7y + az = b^2$ has no solution are

(A) $a = 1, b \neq -1$ (B) $a = 1, b \neq 3$ (C) $a = 1, b \neq 3, -1$ (D) $a = 1, b \neq 3, b = -1$

Answer (C)

The augmented matrix of the system

$$\text{is}\,(A, B) = \begin{pmatrix} 1 & 1 & 1 & : & 1 \\ 1 & 2 & -1 & : & b \\ 5 & 7 & a & : & b^2 \end{pmatrix}$$

$$\rightarrow \begin{pmatrix} 1 & 1 & 1 & : & 1 \\ 0 & 1 & -2 & : & b-1 \\ 0 & 2 & a-5 & : & b^2-5 \end{pmatrix}, R_2{}' = R_2 - R_1, R_3{}' = R_3 - 5R_1$$

$$\rightarrow \begin{pmatrix} 1 & 1 & 1 & : & 1 \\ 0 & 1 & -2 & : & b-1 \\ 0 & 0 & a-1 & : & b^2-2b-3 \end{pmatrix}, R_3{}' = R_3 - 2R_2$$

If $a = 1$, $b^2 - 2b - 3 \neq 0$ i.e., $b \neq -1, 3$ then $r(A) = 2$ but $r(A, B) = 3$

So, $r(A) \neq r(A, B)$ i.e., system has no solution.

Example 126. The values of (a, b) for which the system of equations $x + y + z = 1$, $x + 2y - z = b$, $5x + 7y + az = b^2$ has infinitely many solutions are

(A) $a = 1, b \neq -1$ (B) $a = 1, b \neq 3$ (C) $a \neq 1, b = 3, -1$ (D) $a = 1, b = 3, -1$

Answer (D)

The augmented matrix of the system

$$\text{is } (A, B) = \begin{pmatrix} 1 & 1 & 1 & : & 1 \\ 1 & 2 & -1 & : & b \\ 5 & 7 & a & : & b^2 \end{pmatrix}$$

$$\rightarrow \begin{pmatrix} 1 & 1 & 1 & : & 1 \\ 0 & 1 & -2 & : & b-1 \\ 0 & 2 & a-5 & : & b^2-5 \end{pmatrix}, R_2{}' = R_2 - R_1, R_3{}' = R_3 - 5R_1$$

$$\rightarrow \begin{pmatrix} 1 & 1 & 1 & : & 1 \\ 0 & 1 & -2 & : & b-1 \\ 0 & 0 & a-1 & : & b^2-2b-3 \end{pmatrix}, R_3{}' = R_3 - 2R_2$$

If $a = 1$, $b^2 - 2b - 3 = 0$ i.e., $b = -1, 3$ then $r(A) = 2$ but $r(A, B) = 2$

So, $r(A) = r(A, B) <$ the number of variables (n) i.e., system has infinitely many solutions.

Example 127. A set of linear simultaneous algebraic equations is represented in the matrix form as shown below:

$$\begin{pmatrix} 0 & 0 & 0 & 4 & 13 \\ 2 & 5 & 5 & 2 & 10 \\ 0 & 0 & 2 & 5 & 3 \\ 0 & 0 & 0 & 4 & 5 \\ 2 & 3 & 2 & 1 & 5 \end{pmatrix} \begin{pmatrix} x_1 \\ x_2 \\ x_3 \\ x_4 \\ x_5 \end{pmatrix} = \begin{pmatrix} 46 \\ 161 \\ 61 \\ 30 \\ 81 \end{pmatrix}$$

The value (rounded off to the next integer) of x_3 is_____.   [CH 2016]

Answer: 15

From the first & fourth rows $4x_4 + 13x_5 = 46$ & $4x_4 + 5x_5 = 30$

Solving, $x_4 = 5$ & $x_5 = 2$.

From the third row, $2x_3 + 5x_4 + 3x_5 = 61$ $x_3 = 15$

Example 128. If $\lambda^4 - 6\lambda^2 + 9\lambda - 4$ is the characteristic polynomial of a square matrix A, then $A^{-1} =$

(A) $A^3 - 6A + 9I$

(B) $\dfrac{1}{4}A^3 - \dfrac{3}{2}A + \dfrac{9}{4}I$

(C) $\dfrac{1}{4}A^3 + \dfrac{3}{2}A - \dfrac{9}{4}I$

(D) $-\dfrac{1}{4}A^3 + \dfrac{3}{2}A + \dfrac{9}{4}I$

Answer: (B)

By Cayley Hamilton theorem,

$A^4 - 6A^2 + 9A - 4I = O \Rightarrow A(A^3 - 6A + 9I) = 4I$

or, $A\left(\dfrac{1}{4}A^3 - \dfrac{3}{2}A + \dfrac{9}{4}I\right) = I \Rightarrow A^{-1} = \dfrac{1}{4}A^3 - \dfrac{3}{2}A + \dfrac{9}{4}I$

Example 129. Consider the matrix $A = \begin{bmatrix} 50 & 70 \\ 70 & 80 \end{bmatrix}$ whose eigen vectors corresponding to the eigen values λ_1 & λ_2 are

$X_1 = \begin{bmatrix} 70 \\ \lambda_1 - 50 \end{bmatrix}$ & $X_2 = \begin{bmatrix} \lambda_2 - 80 \\ 70 \end{bmatrix}$, respectively.

The value of $X_1^T . X_2$ is _.   [ME 2017]

Answer: 0

133

$\lambda_1 + \lambda_2$ = trace of the matrix

$$= 50 + 80 = 130 \quad X_1^T . X_2 = \begin{bmatrix} 70 & \lambda_1 - 50 \end{bmatrix} \begin{bmatrix} \lambda_2 - 80 \\ 70 \end{bmatrix}$$

$$= 70(\lambda_1 + \lambda_2) - 5600 - 3500 = 70 \times 130 - 9100 = 0$$

Example 130. For the matrix A satisfying the equation given below,

the eigen values are $[A] \begin{bmatrix} 1 & 2 & 3 \\ 7 & 8 & 9 \\ 4 & 5 & 6 \end{bmatrix} = \begin{bmatrix} 1 & 2 & 3 \\ 4 & 5 & 6 \\ 7 & 8 & 9 \end{bmatrix}$

(A) $(1, -j, j)$ (B) $(1, 1, 0)$ (C) $(1, 1, -1)$ (D) $(1, 0, 0)$ [IN 2014]

Answer (C)

$$[A] \begin{bmatrix} 1 & 2 & 3 \\ 7 & 8 & 9 \\ 4 & 5 & 6 \end{bmatrix} = \begin{bmatrix} 1 & 2 & 3 \\ 4 & 5 & 6 \\ 7 & 8 & 9 \end{bmatrix} \Rightarrow |A| \begin{vmatrix} 1 & 2 & 3 \\ 7 & 8 & 9 \\ 4 & 5 & 6 \end{vmatrix} = \begin{vmatrix} 1 & 2 & 3 \\ 4 & 5 & 6 \\ 7 & 8 & 9 \end{vmatrix}$$

$$\Rightarrow |A| \begin{vmatrix} 1 & 2 & 3 \\ 7 & 8 & 9 \\ 4 & 5 & 6 \end{vmatrix} = - \begin{vmatrix} 1 & 2 & 3 \\ 7 & 8 & 9 \\ 4 & 5 & 6 \end{vmatrix} \Rightarrow |A| = -1$$

Product of the eigen values of $A = |A| = -1$ So, (C) is the correct option.

Example 131. Consider the following 2 × 2 matrix $A = \begin{pmatrix} 1 & 4 \\ b & a \end{pmatrix}$ where two elements are unknown and are marked by a & b. The eigen values of this matrix are - 1 & 7. What are the values of a & b?

(A) $a = 6, b = 4$ (B) $a = 4, b = 6$
(C) $a = 3, b = 5$ (D) $a = 5, b = 3$ [CS/IT 2015]

Answer (D)

Sum of eigen values = trace of the matrix   Or, - 1 + 7 = 1 + a or a = 5
Product of the eigen values = det A or, - 1 × 7 = a − 4b
or, - 7 = 5 − 4b or, b = 3   So, a = 5, b = 3

Example 132. The minimum and the maximum eigen values of the matrix
$$\begin{bmatrix} 1 & 1 & 3 \\ 1 & 5 & 1 \\ 3 & 1 & 1 \end{bmatrix}$$ are -2 and 6 respectively. What is the other eigen value?

(A) 5 (B) 3 (C) 1 (D) -1 [CS/IT 2008]

Answer (B)

If the other eigen value = a then $a - 2 + 6 =$ trace $= 1 + 5 + 1 \Rightarrow a = 3$

Example 133. For a given matrix $A = \begin{bmatrix} 2 & -2 & 3 \\ -2 & -1 & 6 \\ 1 & 2 & 0 \end{bmatrix}$, one of the eigen

values is 3. The other two eigen values are

(A) 2, -5 (B) 3, -5 (C) 2, 5 (D) 3, 5 [CS/IT 2006]

Answer (B)

If the other eigen values = a, b then $a + b + 3 =$ trace

$= 2 - 1 + 0 \Rightarrow a + b = -2$ and $|A| = -45 =$ product of the

$= a \times b \times 3 \Rightarrow ab = -15$ The other two eigen values = 3, -5

Example 134. If the sum of the diagonal elements of a 2 × 2 matrix   is
- 6, then the maximum possible value of the determinant of the
matrix is _____.   [EE 2015]

Answer : 18

If a, b be the eigen values of the matrix A,

sum of the eigen values = trace of the matrix $\Rightarrow a + b = -6$

Now, det A = product of the eigen values

$= ab = \frac{1}{2}\{(a + b)^2 - (a^2 + b^2)\} = \frac{1}{2}\{36 - (a^2 + b^2)\}$

For maximum possible value of det A, $a^2 + b^2 = 0 \Rightarrow (\det A)_{max} = 18$

Example 135. The value of x for which the matrix

$$A = \begin{pmatrix} 3 & 2 & 4 \\ 9 & 7 & 13 \\ -6 & -4 & -9+x \end{pmatrix}$$ has zero as an Eigen value is ___ [EC 2016]

Answer: $x = 1$

As 0 is an Eigen value, the matrix A is singular, $|A| = 0$

$\Rightarrow 3(-63 + 7x + 52) - 2(-81 + 9x + 78) + 4(-36 + 42) = 0$

$\Rightarrow 21x - 33 - 18x + 6 + 24 = 0 \Rightarrow x = 1$

Example 136. Let the eigen values of a 2 × 2 matrix A be 1, - 2 with eigen vectors x_1, x_2 respectively. Then the eigen values and eigen vectors of the matrix $A^2 - 3A + 4I$ would, respectively, be

(A)   2, 14; x_1, x_2 (B)   2, 14, $x_1 + x_2$, $x_1 - x_2$

(C)   2, 0, x_1, x_2 (D)   2, 0, $x_1 + x_2$, $x_1 - x_2$ [EE 2016]

Answer: (A)

The eigen values of A are 1, - 2 (distinct).

By property 25. ii) Eigen vectors corresponding to two distinct Eigen values of a real symmetric matrix are orthogonal, so, $x_1 . x_2 = 0$.

The eigen values of

$A^2 - 3A + 4I$ are $(1)^2 - 3.1 + 4.1$, $(-2)^2 - 3(-2) + 4.1 = 2,14$

Answer should be (A) or (B). By property 25. ii) Eigen vectors corresponding to two distinct Eigen values of a real symmetric matrix are orthogonal i.e., $x_1 . x_2$ should be 0.

Example 137. If $A = \begin{pmatrix} 2 & 3 & 4 \\ 0 & 4 & 2 \\ 0 & 0 & 3 \end{pmatrix}$ the eigen values of $adj\ A$ are ___.

Answer: 6, 8, 12

A being a triangular matrix, its eigen values are its leading diagonal elements i.e., 2, 4, 3 and $|A| = 24$

The eigen values of $adj\ A$ are $\dfrac{|A|}{\lambda}$ = 6, 8, 12.

Example 138. If the characteristic polynomial of a 3 × 3 matrix M over R (the set of real numbers) is $\lambda^3 - 4\lambda^2 + a\lambda + 30, a \in R$ and one eigen value of M is 2, then the largest among the absolute values of the eigen values of M is _____.   [CS/IT 2017]

Answer: 5

The characteristic equation of M is $\lambda^3 - 4\lambda^2 + a\lambda + 30 = 0$

$\lambda = 2$ is a root of the equation, $2^3 - 4.2^2 + a.2 + 30 = 0 \Rightarrow a = -11$

The characteristic equation becomes $\lambda^3 - 4\lambda^2 - 11\lambda + 30 = 0$

As $\lambda = 2$ is a root so, $\lambda - 2$ is a factor of the equation.

$\lambda^2(\lambda - 2) - 2\lambda(\lambda - 2) - 15(\lambda - 2) = 0 \Rightarrow (\lambda - 2)(\lambda^2 - 2\lambda - 15) = 0$

$\Rightarrow (\lambda - 2)(\lambda - 5)(\lambda + 3) = 0 \Rightarrow \lambda = 2, 5, -3$

Largest among the absolute values of the eigen values = 5

Example 139. Consider a 2 × 2 square matrix $\begin{pmatrix} \sigma & x \\ \omega & \sigma \end{pmatrix}$ where x is

unknown. If the Eigen values of the matrix A are

$(\sigma + j\omega)$ and $(\sigma - j\omega)$, then X is equal to

(A) $+ j\omega$ (B) $- j\omega$ (C) $+\omega$ (D) $-\omega$ [EC 2016]

Answer (D)

$\begin{vmatrix} \sigma & x \\ \omega & \sigma \end{vmatrix}$ = product of the Eigen values $= (\sigma + j\varpi)(\sigma - j\omega)$

or, $\sigma^2 - \omega x = \sigma^2 + \omega^2 \Rightarrow -\omega x = \omega^2 \Rightarrow x = -\omega$

Example 140. Let A be the 2 × 2 matrix with elements

$a_{11} = a_{12} = a_{21} = 1$ & $a_{22} = -1$.

Then the eigen values of the matrix A^{19} are

(A) $1024, -1024$ (B) $1024\sqrt{2}, -1024\sqrt{2}$

(C) $4\sqrt{2}, -4\sqrt{2}$ (D) $512\sqrt{2}, -512\sqrt{2}$

Answer (D)

$A = \begin{pmatrix} 1 & 1 \\ 1 & -1 \end{pmatrix}$ The eigen values of A are given by

137

$$\det(A - \lambda I) = 0 \Rightarrow \begin{vmatrix} 1-\lambda & 1 \\ 1 & -1-\lambda \end{vmatrix} = 0 \Rightarrow \lambda^2 - 2 = 0 \Rightarrow \lambda = \pm\sqrt{2}$$

The characteristic equation of A is given by $\lambda^2 - 2 = 0$

By Caley-Hamilton theorem – Every matrix satisfies its own characteristic equation. So,

$$A^2 - 2I = 0 \Rightarrow A^2 = 2I \Rightarrow A^{19} = (A^2)^9 \times A = (2I)^9 \times A = 512A$$

So, the eigen values of

$$A^{19} = 512 \times (eigen\ values\ of\ A) = 512\sqrt{2}, -512\sqrt{2}$$

Example 141. A matrix has Eigen values – 1 and – 2. The corresponding

Eigen vectors are $\begin{pmatrix} 1 \\ -1 \end{pmatrix} \& \begin{pmatrix} 1 \\ -2 \end{pmatrix}$ respectively. The matrix is

(A) $\begin{pmatrix} 1 & 1 \\ -1 & -2 \end{pmatrix}$ (B) $\begin{pmatrix} 1 & 2 \\ -2 & -4 \end{pmatrix}$ (C) $\begin{pmatrix} -1 & 0 \\ 0 & -2 \end{pmatrix}$ (D) $\begin{pmatrix} 0 & 1 \\ -2 & -3 \end{pmatrix}$

Answer: (D)

Let $A = \begin{pmatrix} a & b \\ c & d \end{pmatrix}$ For the eigen value – 1 & eigen vector $\begin{pmatrix} 1 \\ -1 \end{pmatrix}$

$$\begin{pmatrix} a+1 & b \\ c & d+1 \end{pmatrix}\begin{pmatrix} 1 \\ -1 \end{pmatrix} = \begin{pmatrix} 0 \\ 0 \end{pmatrix} \Rightarrow a+1-b = 0 ---(i) \& c-d-1 = 0 ---(ii)$$

For the eigen value – 2 & eigen vector $\begin{pmatrix} 1 \\ -2 \end{pmatrix}$

$$\begin{pmatrix} a+2 & b \\ c & d+2 \end{pmatrix}\begin{pmatrix} 1 \\ -2 \end{pmatrix} = \begin{pmatrix} 0 \\ 0 \end{pmatrix} \Rightarrow a+2-2b = 0 ---(iii) \& c-2d-4 = 0 ---(iv)$$

Solving, $(i) \& (iii), a = 0, b = 1$ and solving $(ii) \& (iv), c = -2, d = -3$

Example 142. For the matrix $\begin{pmatrix} 5 & 3 \\ 1 & 3 \end{pmatrix}$ one of its normalized eigen vector is

given as

(A) $\begin{pmatrix} 1/2 \\ \sqrt{3}/2 \end{pmatrix}$ (B) $\begin{pmatrix} 1/\sqrt{2} \\ -1/\sqrt{2} \end{pmatrix}$ (C) $\begin{pmatrix} 3/\sqrt{10} \\ -1/\sqrt{10} \end{pmatrix}$ (D) $\begin{pmatrix} 1/\sqrt{5} \\ 2/\sqrt{5} \end{pmatrix}$ [ME 2012]

Answer (B)

The eigen values of the matrix are given by $\begin{vmatrix} 5-\lambda & 3 \\ 1 & 3-\lambda \end{vmatrix} = 0$

Or, $(5-\lambda)(3-\lambda) - 3 = 0 \Rightarrow \lambda^2 - 8\lambda + 12 = 0 \Rightarrow \lambda = 2,6$

The eigen vector X corresponding to the eigen value $\lambda = 2$ is given by

$(A - 2I))X = O \Rightarrow \begin{pmatrix} 3 & 3 \\ 1 & 1 \end{pmatrix}\begin{pmatrix} x_1 \\ x_2 \end{pmatrix} = \begin{pmatrix} 0 \\ 0 \end{pmatrix} \Rightarrow 3x_1 + 3x_2 = 0 \ \& \ x_1 + x_2 = 0$

$\Rightarrow x_2 = -x_1$. The eigen vector $= \begin{pmatrix} x_1 \\ -x_1 \end{pmatrix}$ So, the normalized

eigen vector $= \begin{pmatrix} x_1/(\sqrt{x_1^2 + (-x_1)^2}) \\ -x_1/\sqrt{(x_1^2 + (-x_1)^2)} \end{pmatrix} = \begin{pmatrix} 1/\sqrt{2} \\ -1/\sqrt{2} \end{pmatrix}$

Example 143. The matrix $A = \begin{bmatrix} a & 0 & 3 & 7 \\ 2 & 5 & 1 & 3 \\ 0 & 0 & 2 & 4 \\ 0 & 0 & 0 & b \end{bmatrix}$ has

$\det(A) = 100 \ \& \ trace(A) = 14$. The value of $|a - b|$ is _.[EC 2016]

Answer: 3

$trace(A) = 14 \Rightarrow a + 5 + 2 + b = 14 \Rightarrow a + b = 7$

$\det(A) = 100 \Rightarrow a\begin{vmatrix} 5 & 1 & 3 \\ 0 & 2 & 4 \\ 0 & 0 & b \end{vmatrix} - 2\begin{vmatrix} 0 & 3 & 7 \\ 0 & 2 & 4 \\ 0 & 0 & b \end{vmatrix} = 100 \Rightarrow a[5(2b)] = 100 \Rightarrow ab = 10$

$(a - b)^2 = (a + b)^2 - 4ab = 49 - 40 = 9 \Rightarrow |a - b| = 3$

Example 144. Consider the following (2 × 2) matrix $\begin{pmatrix} 4 & 0 \\ 0 & 4 \end{pmatrix}$. Which one

of the following vectors is NOT a valid Eigen vector of the above matrix?

(A) $\begin{pmatrix} 1 \\ 0 \end{pmatrix}$ (B) $\begin{pmatrix} -2 \\ 1 \end{pmatrix}$ (C) $\begin{pmatrix} 4 \\ -3 \end{pmatrix}$ (D) $\begin{pmatrix} 0 \\ 0 \end{pmatrix}$ [CH 2012]

Answer (D)

The Eigen vector of a matrix should be a non-null vector.

Alt. process: If the non-null vector X is an Eigen vector of the matrix A, then $AX = \lambda X$.

Option (A): $\begin{pmatrix} 4 & 0 \\ 0 & 4 \end{pmatrix}\begin{pmatrix} 1 \\ 0 \end{pmatrix} = \begin{pmatrix} 4 \\ 0 \end{pmatrix} = 4\begin{pmatrix} 1 \\ 0 \end{pmatrix}$. True .Eigen value = 4

Option (B): $\begin{pmatrix} 4 & 0 \\ 0 & 4 \end{pmatrix}\begin{pmatrix} -2 \\ 1 \end{pmatrix} = \begin{pmatrix} -8 \\ 4 \end{pmatrix} = 4\begin{pmatrix} -2 \\ 1 \end{pmatrix}$. True. Eigen value = 4

Option (C): $\begin{pmatrix} 4 & 0 \\ 0 & 4 \end{pmatrix}\begin{pmatrix} 4 \\ -3 \end{pmatrix} = \begin{pmatrix} 16 \\ -12 \end{pmatrix} = 4\begin{pmatrix} 4 \\ -3 \end{pmatrix}$. True. Eigen value = 4

Option (D): False, as the given vector is a null vector.

Example 145. The smallest and largest Eigen values of the

matrix $\begin{pmatrix} 3 & -2 & 2 \\ 4 & -4 & 6 \\ 2 & -3 & 5 \end{pmatrix}$ are

(A) 1.5 & 2.5 (B) 0.5 & 2.5 (C) 1.0 & 3.0 (D) 1.0 & 2.0   [CE 2015]
Answer (D)

If the eigen values are a, b, c then $a + b + c = 3 - 4 + 5 = 4 - -(i)$
and $abc =$ the det. of the given matrix = 2 $- - - (ii)$
For option (A), sum of two given eigen values = 4, so from (i), the third is 0 which contradicts (ii).
For option (A), sum of two given eigen values = 3, so from (i), the third is 1 which contradicts (ii).
For option (C), sum of two given eigen values = 4, so from (i), the third is 0 which contradicts (ii).
For option (D), sum of two given eigen values = 3, so from (i), the third is 1 which follows (ii) as $1 \times 2 \times 1 = 2$

Example 146. For the matrix $\begin{pmatrix} 4 & 3 \\ 3 & 4 \end{pmatrix}$, if $\begin{pmatrix} 1 \\ 1 \end{pmatrix}$ is an Eigen vector then the

corresponding Eigen value is _____ .   [CH 2015]
Answer 7

Let $A = \begin{pmatrix} 4 & 3 \\ 3 & 4 \end{pmatrix}$, $X = \begin{pmatrix} 1 \\ 1 \end{pmatrix}$ If the corresponding Eigen value is λ, then

$$AX = \lambda X \Rightarrow \begin{pmatrix} 4 & 3 \\ 3 & 4 \end{pmatrix} \cdot \begin{pmatrix} 1 \\ 1 \end{pmatrix} = \lambda \begin{pmatrix} 1 \\ 1 \end{pmatrix} \Rightarrow \begin{pmatrix} 7 \\ 7 \end{pmatrix} = \begin{pmatrix} \lambda \\ \lambda \end{pmatrix} \Rightarrow \lambda = 7$$

Example 147. In the *LU* decomposition of the matrix $\begin{pmatrix} 2 & 2 \\ 4 & 9 \end{pmatrix}$, if the

diagonal elements of *U* are both 1, then the lower diagonal entry t_{22} of *L*

is _____.

Answer: 5

$$A = LU \Rightarrow \begin{pmatrix} 2 & 2 \\ 4 & 9 \end{pmatrix} = \begin{pmatrix} t_{11} & 0 \\ t_{21} & t_{22} \end{pmatrix} \begin{pmatrix} 1 & u_{12} \\ 0 & 1 \end{pmatrix} = \begin{pmatrix} t_{11} & t_{11}u_{12} \\ t_{21} & t_{21}u_{12} + t_{22} \end{pmatrix}$$

$t_{11} = 2$, $t_{11}u_{12} = 2 \Rightarrow u_{12} = 1$ & $t_{21} = 4$, $t_{21}u_{12} + t_{22} = 9 \Rightarrow t_{22} = 5$

Example 148. The product of the non-zero eigen values of the matrix

$$\begin{bmatrix} 1 & 0 & 0 & 0 & 1 \\ 0 & 1 & 1 & 1 & 0 \\ 0 & 1 & 1 & 1 & 0 \\ 0 & 1 & 1 & 1 & 0 \\ 1 & 0 & 0 & 0 & 1 \end{bmatrix} \text{ is } _____.$$

[CS/IT 2014]

Answer: 6

$$\begin{bmatrix} 1 & 0 & 0 & 0 & 1 \\ 0 & 1 & 1 & 1 & 0 \\ 0 & 1 & 1 & 1 & 0 \\ 0 & 1 & 1 & 1 & 0 \\ 1 & 0 & 0 & 0 & 1 \end{bmatrix} \rightarrow \begin{bmatrix} 2 & 0 & 0 & 0 & 1 \\ 3 & 1 & 1 & 1 & 0 \\ 3 & 1 & 1 & 1 & 0 \\ 3 & 1 & 1 & 1 & 0 \\ 2 & 0 & 0 & 0 & 1 \end{bmatrix}$$ [adding all the columns in first]

$$\rightarrow \begin{bmatrix} 2 & 0 & 0 & 0 & 1 \\ 3 & 1 & 1 & 1 & 0 \\ 0 & 0 & 0 & 0 & 0 \\ 0 & 0 & 0 & 0 & 0 \\ 0 & 0 & 0 & 0 & 0 \end{bmatrix} \quad \left[R_3' = R_3 - R_2, R_4' - R_2, R_5' = R_5 - R_1 \right]$$

The non-zero eigen values are 2 and 3, product = 6

Example 149. Let A be $m \times n$ real valued square symmetric matrix of rank 2 with expression given below: $\sum_{i=1}^{n} \sum_{j=1}^{n} A_{ij}^2 = 50.$

Consider the following statements:
(i) One Eigen value must be in [- 5, 5].
(ii) The Eigen value with the largest magnitude must be strictly greater than 5.
Which of the above statements about Eigen values of A is/are necessarily CORRECT?

(A) Both (i) and (ii) (B) (i) only
(C) (ii) only (D) N either (i) nor (ii)   [CS/IT 2017]
Answer: (B)

As the rank of A is 2, so $(n-2)$ Eigen values are zero.

Let the Eigen values of A are $\lambda_1, \lambda_2, ----, \lambda_n$.

Given, $\sum_{i=1}^{n} \sum_{j=1}^{n} A_{ij}^2 = 50 ------(i)$

But $\sum_{i=1}^{n} \sum_{j=1}^{n} A_{ij}^2 =$ Trace of $A.A^T =$ Trace of A^2

$= \lambda_1^2 + \lambda_2^2 + 0 + 0 + -- +0 = \lambda_1^2 + \lambda_2^2 ----(ii)$

[As A is symmetric, $A = A^T$]

From (i) & (ii), $\lambda_1^2 + \lambda_2^2 = 50$ which states that one of the Eigen values lies in [-5, 5] so, (i) is TRUE. But (ii) is not necessarily TRUE as the Eigen values may be 5, 5.

Example 150. The maximum value of "a" such that the matrix

$$\begin{bmatrix} -3 & 0 & -2 \\ 1 & -1 & 0 \\ 0 & a & -2 \end{bmatrix}$$ has three linearly independent real eigen vectors is

(A) $\dfrac{2}{3\sqrt{3}}$ (B) $\dfrac{1}{3\sqrt{3}}$ (C) $\dfrac{1+2\sqrt{3}}{3\sqrt{3}}$ (D) $\dfrac{1+\sqrt{3}}{3\sqrt{3}}$ [EE 2015]

Answer (B)

The eigen values of the matrix are given by

$$\begin{vmatrix} \lambda+3 & 0 & -2 \\ 1 & \lambda+1 & 0 \\ 0 & a & \lambda+2 \end{vmatrix} = 0$$

$$\Rightarrow \lambda^3 + 6\lambda^2 + 11\lambda + (6-2a) = 0 - - - - - (i)$$

The eigen vectors will be independent if the eigen values are distinct, i.e., The roots of equation (i) are all unequal.

Comparing equation (i) with the standard cubic

$$ax^3 + 3bx^2 + 3cx + d = 0, \ a=1, b=2, c=11/3, d=6-2a$$
$$H = ac - b^2 = -1/3, G = a^2 d - -abc + 2b^3 = -2a$$

The condition for unequal roots is

$$G^2 + 4H^3 < 0 \Rightarrow 4a^2 + 4\left(-\frac{1}{27}\right) < 0 \Rightarrow a^2 < \frac{1}{27} \Rightarrow a < \frac{1}{3\sqrt{3}}$$

EXERCISE

1. Consider the following matrix: $A = \begin{pmatrix} 2 & 3 \\ x & y \end{pmatrix}$. If the eigen values of A are 4 and 8 then

(A) $x = 4, y = 10$ (B) $x = 5, y = 8$

(C) $x = -3, y = 9$ (D) $x = -4, y = 10$ [CS/IT 2010]

2. Eigen values of a real symmetric matrix are always

(A) positive   (B) negative   (C) real   (D) complex

3. One of the eigen values of the matrix $A = \begin{pmatrix} 10 & -4 \\ 18 & -12 \end{pmatrix}$ is

(A)   2   (B)   4   (C)   6   (D)   8   [BT 2013]

4. One of the eigen vectors of the matrix $A = \begin{pmatrix} 2 & 2 \\ 1 & 3 \end{pmatrix}$ is

(A) $\begin{Bmatrix} 2 \\ -1 \end{Bmatrix}$ (B) $\begin{Bmatrix} 2 \\ 1 \end{Bmatrix}$ (C) $\begin{Bmatrix} 4 \\ 1 \end{Bmatrix}$ (D) $\begin{Bmatrix} 1 \\ -1 \end{Bmatrix}$

5. The Eigen values of the matrix $\begin{pmatrix} 1 & -4 \\ 2 & -3 \end{pmatrix}$ are

(A) $2 \pm I$ (B) $-1, -2$ (C) $-1 \pm 2i$ (D) non-existent   [BT 2014]

6. Given the matrices $J = \begin{pmatrix} 3 & 2 & 1 \\ 2 & 4 & 2 \\ 1 & 2 & 6 \end{pmatrix}$ & $K = \begin{pmatrix} 1 \\ 2 \\ -1 \end{pmatrix}$, the product

$K^T JK$ is -------------   [CE 2014]

7. Which of the following statements are TRUE?

P. The Eigen values of a symmetric matrix are real

Q. The value of the determinant of an orthogonal matrix can only be +1

R. The transpose of a square matrix A has the same eigen values as those of A

S. The inverse of an 'n × n' matrix exists if and only if the rank < n

(A)   P and Q only   (B)   P and R only

(C)   Q and R only   (D)   P and S only   [CH 2013]

8. Two matrices $A \& B$ are given below:

$$A = \begin{pmatrix} p & q \\ r & s \end{pmatrix}, B = \begin{pmatrix} p^2 + q^2 & pr + qs \\ pr + qs & r^2 + s^2 \end{pmatrix}$$ If the rank of matrix A is N,

then the rank of the matrix B is

(A)   N/2   (B)   N − 1   (C)   N   (D)   N + 1

9. If 2, 3, 5 are eigen values of A and $B = P^{-1}AP$ where $\det(P) \neq 0$, then the eigen values of B are

(A)   2, 3, 5 (B)   - 2, -3, - 5 (C)   0, 1, 2 (D)   none of these

10. If 2, 3, 5 are eigen values of a 3rd order matrix A, the value of det(A) is

(A)   10   (B)   30   (C)   - 10   (D)   - 30

11. If V is a non-zero vector of dimension 3 × 1, the matrix $A = vv^T$ has a rank _____.   [IN 2017]

12. The minimum and the maximum eigen values of the matrix

$$\begin{bmatrix} 1 & 1 & 3 \\ 1 & 5 & 1 \\ 3 & 1 & 1 \end{bmatrix}$$ are − 2 and 6 respectively. What is the other eigen value?

(A)   5   (B)   3   (C)   1   (D)   - 1   [CS/IT 2007]

13. Consider the 5 × 5 matrix $A = \begin{bmatrix} 1 & 2 & 3 & 4 & 5 \\ 5 & 1 & 2 & 3 & 4 \\ 4 & 5 & 1 & 2 & 3 \\ 3 & 4 & 5 & 1 & 2 \\ 2 & 3 & 4 & 5 & 1 \end{bmatrix}$. It is given that

A has only one real eigen value. Then the eigen value of A is
(A)   - 2.5   (B)   0   (C)   15   (D)   25   [EC 2017]

14. If a matrix has a non-zero minor of order r, then its rank is
(A) r (B) $\leq r$ (C) $\geq r$ (D) $> r$

15. If V is a non-zero vector of dimension 3 × 1, the matrix $A = vv^T$ has a
rank _____.   [IN 2017]

Let A be $n \times n$ real valued square symmetric matrix of rank 2 with

$$\sum_{i=1}^{n} \sum_{j=1}^{n} \left(A_{ij}\right)^2 = 50.$$ Consider the following statements

(I) One eigen value must be in [- 5, 5]
(II) The eigen value with the largest magnitude must be strictly greater
than 5
Which of the above statements about eigen values of A is/are
necessarily CORRECT?
(A) Both (I) and (II) (B) I only (C) II only (D) Neither I nor II [CS/IT 2017]
Answer: (B)
Answers:

1. (D)   2. (C)   3. (C)   4. (A) 5. (C)   6. (23)   7. (B) 8. (C)   9. (A)
10. (B)   11. (3)   12. (B)   13. (C)   14. (C) 15. (3)

1. Sequence & Series

Question of One mark:

Example 1. Which of the following is not true for the sequence $\left\{\dfrac{1}{n^2}+1\right\}$

(A)   convergent   (B)   bounded
(C)   monotonic decreasing   (D)   oscillatory
Answer: (D)

It is of the form $\left\{\dfrac{1}{1^2}+1,\dfrac{1}{2^2}+1,---\right\}$. It converges to 1. Convergent

sequence is bounded. Monotone decreasing also, but not oscillatory.

Example 2. Indicate the correct statement from the following:
(A)   A bounded sequence is convergent
(B)   A monotonic sequence is convergent
(C)   A bounded and monotonic sequence is convergent
(D)   A convergent sequence may not be bounded
Answer: (C)

Example 3. The sequence $\left\{x_n\right\}$, where $x_n=\left(1+\dfrac{1}{n}\right)^n$

(A)   diverges   (B)   converges to 0
(C)   converges to 1   (D)   converges to e
Answer: (D)

Example 4. The sequence $\left\{x_n\right\}$, where $x_n=\log\left(\dfrac{1}{n}\right)$

(A)   diverges   (B)   converges to 0

(C)   converges to 1   (D)   converges to e

Answer: (A)

Example 5. The sequence $\{x_n\}$, where $x_n = \coth n$

(A)   diverges   (B)   converges to 0

(C)   converges to 1   (D)   oscillates finitely

Answer: (C)

$$\lim_{n\to\infty} \coth n = \lim_{n\to\infty} \frac{\cosh n}{\sinh n} = \lim_{n\to\infty}\left(\frac{e^n + e^{-n}}{e^n - e^{-n}}\right) = \lim_{n\to\infty}\left(\frac{1 + e^{-2n}}{1 - e^{-2n}}\right) = 1$$

Example 6. The series $\displaystyle\sum_{n=1}^{\infty}\frac{1}{n^p}$ converges for

(A) $p < 1$ (B) $p \le 1$ (C) $p > 1$ (D) $p \ge 1$

Answer: (C)

It is $p-$series.

Example 7. Which of the following statements is true for the series given below?

$$S_n = 1 + \frac{1}{\sqrt{2}} + \frac{1}{\sqrt{3}} + \frac{1}{\sqrt{4}} + ---+ \frac{1}{\sqrt{n}}$$

(A) S_n converges to $\log(\sqrt{n})$ (B) S_n converges to $\sqrt{n}$

(C) S_n converges to $\exp(\sqrt{n})$ (D) S_n diverges   [BT 2014]

Answer (D)

The series is of the form $\displaystyle\sum\frac{1}{n^{1/2}} = \sum\frac{1}{n^p} \Rightarrow p = \frac{1}{2} < 1.$

So, the p-series diverges.

Example 8. The value of $\displaystyle\sum_{n=0}^{\infty} n.\left(\frac{1}{2}\right)^n$ is _____.   [EC 2015]

Answer: 2

$$\sum_{n=0}^{\infty} n.\left(\frac{1}{2}\right)^n = 0 + 1.\frac{1}{2} + 2.\left(\frac{1}{2}\right)^2 + 3.\left(\frac{1}{2}\right)^3 + ---$$

$$= \frac{1}{2}\left[1 + 2.\frac{1}{2} + 3.\left(\frac{1}{2}\right)^2 + ---\right] = \frac{1}{2}\left(1 - \frac{1}{2}\right)^{-2} = \frac{1}{2}\left(\frac{1}{2}\right)^{-2} = \frac{1}{2}\times 2^2 = 2$$

148

Example 9. The summation of series $S = 2 + \dfrac{5}{2} + \dfrac{8}{2^2} + \dfrac{11}{2^3} + - - - \infty$ is

(A) 4.50 (B) 6.0 (C) 6.75 (D) 10.0   [CS/IT 2004]

Answer (D)

Let $s = \dfrac{5}{2} + \dfrac{8}{2^2} + \dfrac{11}{2^3} + - - - \infty - - - (i)$

$\dfrac{1}{2}s = \qquad \dfrac{5}{2^2} + \dfrac{8}{2^3} + \dfrac{11}{2^4} + - - - \infty - - - (ii)$

$(i) - (ii), \dfrac{1}{2}s = \dfrac{5}{2} + \dfrac{3}{2^2} + \dfrac{3}{2^3} + - - - = \dfrac{5}{2} + 3\left(\dfrac{1}{2^2} + \dfrac{1}{2^3} + - - -\right)$

$= \dfrac{5}{2} + 3\left(\dfrac{1/4}{1 - 1/2}\right) = \dfrac{5}{2} + \dfrac{3}{2} = 4 \Rightarrow s = 8$ So, $S = 2 + s = 10$

Example 10. Consider the following infinite series:

$1 + r + r^2 + r^3 + - - - \infty$

If $r = 0.3$, then the sum of the infinite series is ____.   [BT 2015]

Answer: 1.43

It is an infinite G. P. with $r = 0.3 < 1$ and $a = 1$.

Sum $= \dfrac{a}{1 - r} = \dfrac{1}{1 - 0.3} = 1.43$ *Question of 2 marks:*

Example 11. Let $S = \displaystyle\sum_{n=0}^{\infty} n\alpha^n, |\alpha| < 1.$ The value of α in the range

$0 < \alpha < 1$ such that $S = 2\alpha$ is -------.   [EE 2016]

Answer: 0.29

$2\alpha = \alpha + 2\alpha^2 + 3\alpha^3 + - - - \Rightarrow 2 = 1 + 2\alpha + 3\alpha^2 + - -$

$2 = (1 - \alpha)^{-2} \Rightarrow (1 - \alpha) = \dfrac{1}{\sqrt{2}} \Rightarrow \alpha = 1 - \dfrac{1}{\sqrt{2}} = 1 - \dfrac{\sqrt{2}}{2} = 0.29$

Example 12. The value of $\displaystyle\lim_{n\to\infty}\left[\dfrac{\sqrt{1} + \sqrt{2} + \sqrt{3} + - - + \sqrt{n}}{n\sqrt{n}}\right]$ is

(A)   0   (B)   1   (C) $\dfrac{2}{3}$ (D) $\dfrac{3}{2}$

Answer: (C)

$$\lim_{n\to\infty}\left[\frac{\sqrt{1}+\sqrt{2}+\sqrt{3}+---+\sqrt{n}}{n\sqrt{n}}\right]=\lim_{n\to\infty}\left\{\frac{1}{n}\cdot\sum\sqrt{\frac{r}{n}}\right\}=\int_0^1\sqrt{x}\,dx$$

$$=\frac{2}{3}\left[x^{3/2}\right]_0^1=\frac{2}{3}$$

Example 13. $\displaystyle\sum_{x=1}^{\infty}\frac{1}{x(x+1)}=$ [CS/IT 2013]

Answer: 1

$$\sum_{x=1}^{\infty}\frac{1}{x(x+1)}=\sum_{x=1}^{\infty}\left(\frac{1}{x}-\frac{1}{x+1}\right)=\left(1-\frac{1}{2}\right)+\left(\frac{1}{2}-\frac{1}{3}\right)+----=1$$

Example 14. A sequence $x[n]$ is specified as

$$\begin{bmatrix}x[n]\\x[n-1]\end{bmatrix}=\begin{bmatrix}1&1\\1&0\end{bmatrix}^n\begin{bmatrix}1\\0\end{bmatrix},n\ge2.$$

The initial conditions are $x[0]=1,x[1]=1$ & $x[n]=0,n<0$.
The value of $x[12]$ is _____. [EC 2016]

Answer: 233

$$n=2,\begin{bmatrix}x[2]\\x[1]\end{bmatrix}=\begin{bmatrix}1&1\\1&0\end{bmatrix}^2\begin{bmatrix}1\\0\end{bmatrix}=\begin{bmatrix}2&1\\1&1\end{bmatrix}\begin{bmatrix}1\\0\end{bmatrix}=\begin{bmatrix}2\\1\end{bmatrix}\Rightarrow x[2]=2,x[1]=1$$

$$n=3,\begin{bmatrix}x[3]\\x[2]\end{bmatrix}=\begin{bmatrix}1&1\\1&0\end{bmatrix}^3\begin{bmatrix}1\\0\end{bmatrix}=\begin{bmatrix}2&1\\1&1\end{bmatrix}\begin{bmatrix}1&1\\1&0\end{bmatrix}\begin{bmatrix}1\\0\end{bmatrix}=\begin{bmatrix}2\\1\end{bmatrix}$$

$$=\begin{bmatrix}3&2\\2&1\end{bmatrix}\begin{bmatrix}1\\0\end{bmatrix}=\begin{bmatrix}3\\2\end{bmatrix}\Rightarrow x[3]=3,x[2]=2$$

So, $x[n]=x[n-1]+x[n-2]---(i)$ Putting
$n=4,5,---,12$ $x[4]=x[3]+x[2]=3+2=5$
$x[5]=x[4]+x[3]=5+3=8$ $x[6]=x[5]+x[4]=8+5=13$
$x[7]=x[6]+x[5]=13+8=21$ $x[8]=x[7]+x[6]=21+13=34$
$x[9]=x[8]+x[7]=34+21=55$ $x[10]=x[9]+x[8]=55+34=89$
$x[11]=x[10]+x[9]=89+55=144$ $x[12]=x[11]+x[10]=144+89=233$

Example 15. Two sequences $[a,b,c]$ and $[A,B,C]$ are related as,

$$\begin{bmatrix} A \\ B \\ C \end{bmatrix} = \begin{bmatrix} 1 & 1 & 1 \\ 1 & W_3^{-1} & W_3^{-2} \\ 1 & W_3^{-2} & W_3^{-4} \end{bmatrix} \begin{bmatrix} a \\ b \\ c \end{bmatrix} \quad \text{where } W_3 = e^{j\frac{2\pi}{3}} . \text{If another sequence}$$

$[p,q,r]$ is derived as $\begin{bmatrix} p \\ q \\ r \end{bmatrix} = \begin{bmatrix} 1 & 1 & 1 \\ 1 & W_3^1 & W_3^2 \\ 1 & W_3^2 & W_3^4 \end{bmatrix} \begin{bmatrix} 1 & 0 & 0 \\ 0 & W_3^2 & 0 \\ 0 & 0 & W_3^4 \end{bmatrix} \begin{bmatrix} A/3 \\ B/3 \\ C/3 \end{bmatrix}$

then the relationship between the sequences $[p,q,r]$ and $[a,b,c]$ is

(A) $[p,q,r] = [b,a,c]$ (B) $[p,q,r] = [b,c,a]$

(C) $[p,q,r] = [c,a,b]$ (D) $[p,q,r] = [c,b,a]$ [EC 2015]

Answer: (C)

$W_3^3 = \cos 2\pi + i \sin 2\pi = 1 \Rightarrow W_3^{-1} = W_3^2 \ etc. \textbf{and}$

$1 + W + W^2 = 0 \ A = a + b + c, B = a + bW^2 + cW, C = a + bW + cW^2$

$$\begin{bmatrix} p \\ q \\ r \end{bmatrix} = \frac{1}{3} \begin{bmatrix} 1 & W^2 & W \\ 1 & 1 & 1 \\ 1 & W & W^2 \end{bmatrix} \begin{bmatrix} A \\ B \\ C \end{bmatrix} = \begin{bmatrix} A + BW^2 + CW \\ A + B + C \\ A + BW + CW^2 \end{bmatrix}$$

$A + BW^2 + CW = (a+b+c) + (aW^2 + bW + c) + (aW + bW^2 + c) = 3c$

$A + B + C = 3a, \ A + BW + CW^2 = 3b \Rightarrow [p \quad q \quad r] = [c \quad a \quad b]$

$$\bigcirc \text{ 2. Function}$$

Question of 1 mark:

Example 16. The function $x + x + x + - - - x \ times$, is defined

(A) at all real values of x

(B) only at positive integer values of x

(C) only at negative integer values of x

(D) only at rational values of x [PI 2015]

Answer: (A)

For x times x should be a positive integer.

Example **17.** Choose the correct set of functions which are linearly dependent?

(A) $\sin x, \sin^2 x$ and $\cos^2 x$ (B) $\cos x, \sin x$ and $\tan x$

(C) $\cos 2x, \sin^2 x$ and $\cos^2 x$ (D) $\cos 2x, \sin x$ and $\cos x$ [PI 2013]

Answer: (C)

$\cos 2x = \cos^2 x - \sin^2 x$

Example **18.** A non-zero polynomial $f(x)$ has roots at $x = 1, x = 2$ and $x = 3$. Which one of the following must be TRUE?

(A) $f(0).f(4) < 0$ (B) $f(0).f(4) > 0$

(C) $f(0) + f(4) < 0$ (D) $f(0) + f(4) > 0$ [CS/IT 2014]

Answer (A)

$f(x) = a(x-1)(x-2)(x-3), a > 0 \Rightarrow f(0) < 0 \ \& \ f(4) > 0$
$\Rightarrow f(0).f(4) < 0$

Example **19.** Consider the function $f(x) = 1 - |x|, -1 \le x \le 1$. The value of x at which the function attains its minimum and the maximum value are

(A)   0, - 1   (B)   - 1, 0   (C)   0, 1   (D)   - 1, 2   [CS/IT 2015]

Answer (C)

In $-1 \le x \le 1.$, max. $|x| = 1$ and min. $|x| = 0$

In $-1 \le x \le 1.$, max. $f(x) = 1 - 0 = 1$ and min. $f(x) = 1 - 1 = 0$

Example **20.** Consider the following statements about the linear dependence of the real valued functions $y_1 = 1, y_2 = x, y_3 = x^2$ in the field of real numbers

I. $y_1, y_2 \ \& \ y_3$ are linearly independent on $-1 \le x \le 0$

II. $y_1, y_2 \ \& \ y_3$ are linearly dependent on $0 \le x \le 1$

III. $y_1, y_2 \ \& \ y_3$ are linearly independent on $0 \le x \le 1$

IV. $y_1, y_2 \ \& \ y_3$ are linearly dependent on $-1 \le x \le 0$

Which one among the following are correct?

(A) Both I and II are true   (B)   Both I and III are true

(C) Both II and IV are true   (D)   Both III and IV are true   [EC 2017]

Answer: (B)

Example 21. If $g(x) = 1 - x$ & $h(x) = \dfrac{x}{x-1}$ then $\dfrac{g(h(x))}{h(g(x))} =$

(A) $\dfrac{h(x)}{g(x)}$ (B) $-\dfrac{1}{x}$ (C) $\dfrac{g(x)}{h(x)}$ (D) $\dfrac{x}{(1-x)^2}$ [CS/IT 2015]

Answer (A)

$$g(h(x)) = g\left(\frac{x}{x-1}\right) = 1 - \frac{x}{x-1} = -\frac{1}{x-1} = \frac{1}{g(x)}$$

$$h(g(x)) = h(1-x) = \frac{1-x}{1-x-1} = \frac{x-1}{x} = \frac{1}{h(x)} \Rightarrow \frac{g(h(x))}{h(g(x))} = \frac{h(x)}{g(x)}$$

Example 22. If x varies from -1 to $+3$, which one of the following

describes the behaviour of the function $f(x) = x^3 - 3x^2 + 1$?

(A) $f(x)$ increases monotonically

(B) $f(x)$ increases, then decreases and increases again

(C) $f(x)$ decreases, then increases and decreases again

(D) $f(x)$ increases and then decreases   [EC 2016]

Answer: (B)

$f'(x) = 3x^2 - 6x = 3x(x-2)$

$f'(x) > 0, \quad -1 < x < 0 \Rightarrow f(x)$ increases.

$f'(x) < 0, \quad\;\; 0 < x < 2 \Rightarrow f(x)$ decreases.

$f'(x) > 0, \quad\;\; 2 < x < 3 \Rightarrow f(x)$ increases.

Example 23. $2x^2 + y^2 = 34,\; x + 2y = 11.$

The value of $(x + y)$ is _____. [EE 2017]

Answer: 7

$2x^2 + y^2 = 34$ is satisfied by $x^2 = 9, y^2 = 16 \Rightarrow x = \pm 3, y = \pm 4$

But $x + 2y = 11$ is satisfied by $x = 3, y = 4 \Rightarrow x + y = 7$

Example 24. Let $y^2 - 2y + 1 = x$, $\sqrt{x} + y = 5$. The value of

$x + \sqrt{y}$ equals ____.   [EE 2017]

Answer: 5.73

$y^2 - 2y + 1 = (5 - y)^2 \Rightarrow y = 3 \Rightarrow x = 4$, $\quad x + \sqrt{y} = 4 + \sqrt{3} = 5.73$

Question of 2 marks:

Example 25. The coefficient of x^{12} in $(x^3 + x^4 + x^5 + x^6 + \text{---})^3$

is____.   [CS/IT 2016]

Answer: 10

$(x^3 + x^4 + x^5 + x^6 + \text{---})^3 = x^9(1 + x + x^2 + \text{---})^3 = x^9(1 - x)^{-3}$

$= x^9(1 + 3x + 6x^2 + 10x^3 + \text{---}) \Rightarrow$ Coefficient of x^{12} is 10.

Remember:

$$(1 - x)^{-n} = 1 + nx + \frac{n(n+1)}{2!}x^2 + \frac{n(n+1)(n+2)}{3!}x^3 + \text{---}$$

$$(1 + x)^{-n} = 1 - nx + \frac{n(n+1)}{2!}x^2 - \frac{n(n+1)(n+2)}{3!}x^3 + \text{---}$$

Example 26. For non-zero x, $af(x) + bf\left(\dfrac{1}{x}\right) = \dfrac{1}{x} - 25$ where

$a \neq b$. Then $\displaystyle\int_1^2 f(x)dx = $

(A) $\dfrac{1}{a^2 - b^2}\left[a(\ln 2 - 25) + \dfrac{47b}{2}\right]$

(B) $\dfrac{1}{a^2 - b^2}\left[a(2\ln 2 - 25) - \dfrac{47b}{2}\right]$

(C) $\dfrac{1}{a^2 - b^2}\left[a(2\ln 2 - 25) + \dfrac{47b}{2}\right]$

(D) $\quad \dfrac{1}{a^2-b^2}\left[a(\ln 2-25)-\dfrac{47b}{2}\right]$ $\qquad$ [CS/IT 2015]

Answer: (A)

$$af(x)+bf\left(\dfrac{1}{x}\right)=\dfrac{1}{x}-25----(i)\quad af\left(\dfrac{1}{x}\right)+bf(x)=x-25----(ii)$$

Solving $(i)\,\&\,(ii),\ (a^2-b^2)f(x)=a\left(\dfrac{1}{x}-25\right)-b(x-25)$

$$(a^2-b^2)\int_1^2 f(x)dx=\left[a(\ln x-25x)-b\left(\dfrac{x^2}{2}-25x\right)\right]_1^2$$

$$\Rightarrow\int_1^2 f(x)dx=\dfrac{1}{a^2-b^2}\left[a(\ln 2-25)+\dfrac{47b}{2}\right]$$

3. Limit

1. *Some Standard Limits:*

(i) $\displaystyle\lim_{x\to a}\dfrac{x^n-a^n}{x-a}=na^{n-1}$

(ii) $\displaystyle\lim_{x\to 0}\dfrac{(1+x)^n-1}{x}=n$

(iii) $\displaystyle\lim_{x\to 0}\dfrac{\log(1+x)}{x}=1$

(iv) $\displaystyle\lim_{x\to 0}\dfrac{e^x-1}{x}=1$

(v) $\displaystyle\lim_{x\to 0}\dfrac{a^x-1}{x}=\log a, a>0$

(vi) $\displaystyle\lim_{x\to 0}(1+x)^{1/x}=e$

(vii) $\displaystyle\lim_{x\to 0}\dfrac{x}{\sin x}=1$

(viii) $\displaystyle\lim_{x\to 0}\dfrac{x}{\tan x}=1$

(ix) $\displaystyle\lim_{n\to\pm\infty}\left(1+\dfrac{1}{n}\right)^n=e$ (x) $\displaystyle\lim_{n\to\infty}\dfrac{x^n}{n!}=0$

(xi) $\displaystyle\lim_{n\to\infty}x^n=0$ if $-1<x<1$or $|x|<1$.

2. *L' Hospitals Rules:*

Rule I. If $f(x)\,\&\,\phi(x)$ be two functions such that

155

(i) $\lim\limits_{x \to a} f(x) = \lim\limits_{x \to a} \phi(x) = 0$ (ii) $f'(x)$ & $\phi'(x)$ both exist and

$\phi'(x) \neq 0, \; \forall x \in \left]a - \delta, a + \delta\right[, \; \delta > 0$

(iii) $\lim\limits_{x \to a} \dfrac{f'(x)}{\phi'(x)}$ exists, then $\lim\limits_{x \to a} \dfrac{f(x)}{\phi(x)} = \lim\limits_{x \to a} \dfrac{f'(x)}{\phi'(x)}$

Rule II. If $f(x)$ & $\phi(x)$ be two functions such that

(i) $\lim\limits_{x \to a} f(x) = \lim\limits_{x \to a} \phi(x) = \infty$ (ii) $f'(x)$ & $\phi'(x)$ both exist and finite

number), then $\lim\limits_{x \to a} \dfrac{f(x)}{\phi(x)} = \lim\limits_{x \to a} \dfrac{f'(x)}{\phi'(x)}$.

Question of one mark:

Example 27. $\lim\limits_{x \to 0} \dfrac{\sin(x)}{x}$ is _____.　　　　　　　[BT 2017]

Answer: 1

Example 28. The value of $\lim\limits_{x \to 0} \dfrac{\tan x}{x}$ is _____.　　　　[CH 2017]

Answer: 1

Example 29. $\lim\limits_{x \to 4} \dfrac{\sin(x - 4)}{x - 4} = $ _____.　　　　　[CS/IT 2016]

Answer: 1

$x - 4 = z$　　As $x \to 4, z \to 0$ $\lim\limits_{x \to 4} \dfrac{\sin(x - 4)}{x - 4} = \lim\limits_{z \to 0} \dfrac{\sin z}{z} = 1$

Example 30. $\lim\limits_{x \to 0} \dfrac{\sin x^0}{x}$ equals

(A)　0　　　(B)　1　　　(C)　$\dfrac{\pi}{180}$　　　(D)　$\dfrac{180}{\pi}$

Answer: (C)

$1^0 = \left(\dfrac{\pi}{180}\right)^c \Rightarrow x^0 = \left(\dfrac{\pi x}{180}\right)^c$

The given limit $= \lim\limits_{x \to 0} \dfrac{\sin x^0}{x} = \lim\limits_{x \to 0} \dfrac{\sin\left(\dfrac{\pi x}{180}\right)}{x} = \lim\limits_{x \to 0} \dfrac{\sin\left(\dfrac{\pi x}{180}\right)}{(\pi x)/180} \cdot \dfrac{\pi}{180}$

Let $y = \dfrac{\pi x}{180}$, $x \to 0 \Rightarrow y \to 0$

The given limit $= \dfrac{\pi}{180}.\lim\limits_{y \to 0} \dfrac{\sin y}{y} = \dfrac{\pi}{180}.1 = \dfrac{\pi}{180}$

Example 31. $\lim\limits_{\theta \to 0} \dfrac{\sin(\theta/2)}{\theta}$ is

(A) 0.5   (B) 1   (C) 2   (D) not defined   [EC 2007]

Answer: (A)

$$\lim\limits_{\theta \to 0} \dfrac{\sin(\theta/2)}{\theta} = \lim\limits_{\theta/2 \to 0} \dfrac{\sin(\theta/2)}{(\theta/2) \times 2} = \dfrac{1}{2} \times \lim\limits_{\theta/2 \to 0} \dfrac{\sin(\theta/2)}{(\theta/2)} = \dfrac{1}{2}$$

Example 32. $\lim\limits_{x \to 0} \dfrac{\tan x}{x^2 - x}$ is equal to _____.   [CE 2017]

Answer: -1

$$\lim\limits_{x \to 0} \dfrac{\tan x}{x^2 - x} = \lim\limits_{x \to 0} \dfrac{\tan x}{x(x-1)} = \lim\limits_{x \to 0} \dfrac{\tan x}{x}.\dfrac{1}{x-1} = 1.(-1) = -1$$

Example 33. $\lim\limits_{x \to \infty} \dfrac{\sin x}{x}$

(A) exists and the value is 1   (B) exists and the value is 0

(C) exists and the value is -1 (D) does not exist

Answer: (B)

Let $x = \dfrac{1}{y}$, $x \to \infty \Rightarrow y \to 0$

$$\lim\limits_{x \to \infty} \dfrac{\sin x}{x} = \lim\limits_{y \to 0} \left(y.\sin\dfrac{1}{y} \right) = 0 \text{ as } \left| \sin\dfrac{1}{y} \right| \le 1$$

Example 34. If $y = e^{-x^2}$, then the value of $\lim\limits_{x \to \infty} \dfrac{1}{x}\dfrac{dy}{dx}$ is __. [BT 2018]

Answer: 0

$$\dfrac{1}{x}\dfrac{dy}{dx} = \dfrac{1}{x}(-2x)e^{-x^2} = -2e^{-x^2} \Rightarrow \lim\limits_{x \to \infty} \dfrac{1}{x}\dfrac{dy}{dx} = -2\lim\limits_{x \to \infty} e^{-x^2} = 0$$

Example 35. $\lim\limits_{x \to \infty} x^{1/x}$ is

(A) ∞ (B) 0   (C) 1   (D) Not defined   [CS/IT 2015]

Answer (C)

$$\lim_{x \to \infty} \log(x^{1/x}) = \lim_{x \to \infty} \frac{\log x}{x} = \lim_{x \to \infty} \frac{1/x}{1} = 0 = \log 1 \Rightarrow \lim_{x \to \infty} x^{1/x} = 1$$

Example 36. The limit of the function $\lim\limits_{x \to \infty}\left(1 + \dfrac{1}{x}\right)^{x}$ is

(A) $\ln 2$ (B) 1.0   (C) e (D) ∞ [EC 2014]

Answer (C)

It follows from definition.

Example 37. $\lim\limits_{x \to \infty}\left(1 + \dfrac{1}{x}\right)^{x}$ is equal to

(A) e^{-2} (B) e (C) 1   (D) e^{2} [CE 2015]

Answer (B)

It follows from definition.

Example 38. $\lim\limits_{x \to \infty}\left(1 + \dfrac{1}{x}\right)^{2x}$ is equal to

(A) e^{-2} (B) e (C) 1   (D) e^{2} [CE 2015]

Answer (D)

$$\lim_{x \to \infty}\left(1 + \frac{1}{x}\right)^{2x} = \lim_{x \to \infty}\left\{\left(1 + \frac{1}{x}\right)^{x}\right\}^{2} = e^{2}$$

Example 39. The limit of the function

$\left(1 + \dfrac{x}{n}\right)^{n}$ as $n \to \infty$ is

(A) $\ln x$ (B) $\ln\dfrac{1}{x}$ (C) e^{-x} (D) e^{x} [BT 2015]

Answer (D)

$$\lim_{n \to \infty}\left(1 + \frac{x}{n}\right)^{n} = \lim_{z \to 0}(1 + z)^{x/z}$$

$$\left[\frac{x}{n} = z \ as \ n \to \infty, z \to 0\right] = \lim_{z \to 0}\left[(1+z)^{1/z}\right]^x = e^x$$

Example 40. The value of $\lim_{x \to 1} \dfrac{x^7 - 2x^5 + 1}{x^3 - 3x^2 + 2}$ is

(A)   0   (B)   - 1   (C)   1   (D)   does not exist   [CS/IT 2017]

Answer: (C)

$$\lim_{x \to 1} \frac{x^7 - 2x^5 + 1}{x^3 - 3x^2 + 2} \quad \left[form \ \frac{0}{0}\right]$$

$$= \lim_{x \to 1} \frac{7x^6 - 10x^4}{3x^2 - 6x} = \lim_{x \to 1} \frac{7x^5 - 10x^3}{3x - 6} = \frac{7 - 10}{3 - 6} = 1$$

Example 41. $\lim_{x \to \infty}\left(1 + \dfrac{a}{x}\right)^x$ equals

(A) a^e (B) e^a (C)   1   (D) a

Answer (B)

$$\lim_{x \to \infty}\left(1 + \frac{a}{x}\right)^x = \lim_{z \to 0}(1+z)^{a/z} \quad \left[\frac{a}{x} = z \ as \ x \to \infty, z \to 0\right]$$

$$= \lim_{z \to 0}\left[(1+z)^{1/z}\right]^a = e^a$$

Example 42. $\lim_{x \to \infty} \dfrac{x + \sin x}{x}$ equals to

(A) $-\infty$ (B)   0   (C)   1   (D) ∞ [CE 2014]

Answer (C)

Let $x = \dfrac{1}{y}$ as $x \to \infty, y \to 0$

$$\lim_{x \to \infty} \frac{x + \sin x}{x} = \lim_{y \to 0} \frac{1/y + \sin(1/y)}{1/y} = \lim_{y \to 0}\left(1 + y\sin\frac{1}{y}\right) = 1 = 0 = 1$$

Example 43. $\lim_{x \to 0} \dfrac{x - \sin x}{1 - \cos x}$ is

(A)   0   (B)   1   (C)   3   (D)   not defined   [ME 2014]

Answer: (A)

$$\frac{x - \sin x}{1 - \cos x} = \frac{x - \left(x - \frac{x^3}{3!} + - -\right)}{1 - \left(1 - \frac{x^2}{2!} + - -\right)} = \frac{x^3()}{x^2()} = x(). \text{ So, limit} = 0$$

Example 44. The value of $\displaystyle\lim_{x\to 0} \frac{x^3 - \sin x}{x}$ is

(A)   0   (B)   3   (C)   1   (D)   - 1   [ME 2017]

Answer: (D)

$$\lim_{x\to 0} \frac{x^3 - \sin x}{x} = \lim_{x\to 0}\left(x^2 - \frac{\sin x}{x}\right) = 0 - 1 = -1$$

Example 45. $\displaystyle\lim_{\alpha\to 0} \frac{x^\alpha - 1}{\alpha}$ is equal to

(A)   log x   (B)   0   (C)   x log x   (D) ∞   [CE 2014]

Answer: (A)

Example 46. Evaluate $\displaystyle\lim_{x\to\infty} x \tan\frac{1}{x}$ is

(A)   ∞   (B)   1   (C)   0   (D)   - 1   [BT 2013]

Answer: (B)

$$\lim_{x\to\infty} x \tan\frac{1}{x} = \lim_{x\to\infty} \frac{\tan(1/x)}{(1/x)} \left[x = \frac{1}{y}, x\to\infty \Rightarrow y\to 0\right]$$

$$= \lim_{y\to 0} \frac{\tan y}{y} = 1$$

Example 47. The value of $\displaystyle\lim_{x\to\infty}(1 + x^2)e^{-x}$ is

(A)   0   (B)   1/2   (C)   1   (D)   ∞   [CS/IT 2015]

Answer: (A)

$$\lim_{x\to\infty}(1 + x^2)e^{-x} = \lim_{x\to\infty}\frac{1 + x^2}{e^x} = \lim_{x\to\infty}\frac{2x}{e^x} = \lim_{x\to\infty}\frac{2}{e^x} = 0$$

Example 48. The limit of the function $e^{-2t} \sin(t)$ as $t \to \infty$ is [BT 2014]

Answer: 0

As $\sin t \leq 1$, $e^{-2t} . \sin t \leq e^{-2t} \to 0$ as $t \to \infty$

So, $e^{-2t} \sin(t) \to 0$ as $t \to \infty$

Example 49. $\lim\limits_{x \to 0}\left(\dfrac{1 - \cos x}{x^2} \right)$ is

(A) $\dfrac{1}{4}$ (B) $\dfrac{1}{2}$ (C) 1 (D)   2   [ME 2012]

Answer (B)

$$\lim_{x \to 0}\left(\frac{1 - \cos x}{x^2} \right) = \lim_{x \to 0}\left(\frac{\sin x}{2x} \right) = \frac{1}{2}\lim_{x \to 0} \frac{\sin x}{2} = \frac{1}{2} \times 1 = \frac{1}{2}$$

Example 50. The value of $\lim\limits_{x \to 0} \dfrac{1 - \cos(x^2)}{2x^4}$ is

(A) 0   (B) 1/2   (C) 1/4   (D) undefined   [ME 2015]

Answer (C)

Let $[x^2 = z]$ then

$$\lim_{x \to 0} \frac{1 - \cos(x^2)}{2x^4} = \lim_{z \to 0} \frac{1 - \cos z}{2z^2}$$

$$= \lim_{z \to 0} \frac{2\sin^2(z/2)}{2z^2} = \lim_{z \to 0} \frac{\sin^2(z/2)}{4.(z/2)^2} = \frac{1}{4} \times 1 = \frac{1}{4}$$

Example 51. $\lim\limits_{x \to 0} \dfrac{e^{1/x}}{e^{1/x} + 1}$ equals

(A)   0   (B)   1   (C)   e   (D) does not exist

Answer: (D)

When $x \to 0^-$, $\dfrac{1}{x} \to -\infty$, $e^{1/x} \to 0$

Left Hand Limit $= \lim\limits_{x \to 0^-} \dfrac{e^{1/x}}{e^{1/x}+1} = \dfrac{0}{0+1} = 0$

As $x \to 0^+$, $\dfrac{1}{x} \to \infty$, $-\dfrac{1}{x} \to -\infty$, $e^{-1/x} \to 0$

Right Hand Limit $= \lim\limits_{x \to 0^+} \dfrac{e^{1/x}}{e^{1/x}+1} = \lim\limits_{x \to 0^+} \dfrac{1}{1+e^{-1/x}} = \dfrac{1}{1+0} = 1$

Left Hand Limit $\neq$ Right Hand Limit i.e., the limit does not exist.

Example 52. $\lim\limits_{n \to \infty} \dfrac{1^2 + 2^2 + ---- + n^2}{n^3}$ equals

(A) $\dfrac{1}{3}$ (B) $\dfrac{1}{2}$ (C) $\dfrac{1}{6}$ (D) $\dfrac{1}{4}$

Answer: (A)

Limit $= \lim\limits_{n \to \infty} \dfrac{1^2 + 2^2 + ---- + n^2}{n^3} = \lim\limits_{n \to \infty} \dfrac{n(n+1)(2n+1)/6}{n^3}$

$= \dfrac{1}{6} \lim\limits_{n \to \infty} \dfrac{(n+1)(2n+1)}{n^2} = \dfrac{1}{6} \lim\limits_{n \to \infty} \left(1 + \dfrac{1}{n}\right)\left(2 + \dfrac{1}{n}\right) = \dfrac{1}{6} \cdot 1.2 = \dfrac{1}{3}$

$\left[As\ n \to \infty,\ \dfrac{1}{n} \to 0 \right]$

Example 53. $\lim\limits_{n \to \infty} \dfrac{2^{n+1} + 3^{n+1}}{2^n + 3^n}$ equals

(A)   2   (B) 3   (C)   5   (D)   0

Answer: (B)

As $\dfrac{2}{3} < 1$, so $\lim\limits_{n \to \infty} \left(\dfrac{2}{3}\right)^n = 0$ The given limit

$= \lim\limits_{n \to \infty} \dfrac{2.2^n + 3.3^n}{2^n + 3^n} = \lim\limits_{n \to \infty} \dfrac{2.\left(\frac{2}{3}\right)^n + 3.}{\left(\frac{2}{3}\right)^n + 1} = \lim\limits_{n \to \infty} \dfrac{2.0 + 3.}{0 + 1} = 3$

Example 54. $\lim_{x\to\infty}\left(\sqrt{x^2+x-1}-x\right)$ is

(A)　　0　(B)　　∞　　(C)　　1/2　　(D)　- ∞　[ME 2016]

Answer: (C)

Let $x=\dfrac{1}{y}$, $x\to\infty\Rightarrow y\to 0$

$$\lim_{x\to\infty}\left(\sqrt{x^2+x-1}-x\right)=\lim_{y\to 0}\left(\sqrt{\frac{1}{y^2}+\frac{1}{y}-1}-1\right)=\lim_{y\to 0}\left(\sqrt{\frac{1+y-y^2}{y^2}}-1\right)$$

$$=\lim_{y\to 0}\left(\frac{\sqrt{1+y-y^2}-1}{y}\right)=\lim_{y\to 0}\left(\frac{1+y-y^2-1}{y\left[\sqrt{1+y-y^2}+1\right]}\right)$$

$$=\lim_{y\to 0}\left(\frac{y(1-y)}{y\left[\sqrt{1+y-y^2}+1\right]}\right)=\lim_{y\to 0}\left(\frac{(1-y)}{\left[\sqrt{1+y-y^2}+1\right]}\right)=\frac{1}{1+1}=\frac{1}{2}$$

Example 55. $\lim_{n\to\infty}\left(\sqrt{n^2+n}-\sqrt{n^2+1}\right)$ is ___.　　　　　　[IN 2016]

Answer: 0.5

$$\lim_{n\to\infty}\left(\sqrt{n^2+n}-\sqrt{n^2+1}\right)=\lim_{n\to\infty}\frac{\left(n^2+n\right)-\left(n^2+1\right)}{\sqrt{n^2+n}+\sqrt{n^2+1}}$$

$$=\lim_{n\to\infty}\frac{n-1}{\sqrt{n^2+n}+\sqrt{n^2+1}}=\lim_{n\to\infty}\frac{1-1/n}{\sqrt{1+1/n}+\sqrt{1+1/n^2}}$$

$$=\frac{1-0}{\sqrt{1+0}+\sqrt{1+0}}=\frac{1}{2}\qquad n\to\infty,\frac{1}{n}\to 0$$

Example 56. If x is not an integer, then $\lim_{n\to\infty}\dfrac{1}{1+n\sin^2(\pi x)}=$

(A)　　0　(B)　1　(C)　- 1　(D) does not exist

Answer (A)

If x is not an integer, then $\sin(\pi x)\neq 0$

$$\lim_{n\to\infty}\frac{1}{1+n\sin^2(\pi x)}=\lim_{n\to\infty}\frac{1/n}{1/n+\sin^2(\pi x)}=\frac{0}{0+\sin^2(\pi x)}=0$$

163

Example 57. If [x] means the greatest integer ≤ x and m is any negative integer, then $\lim_{x \to m}(x - [x])$

(A)　　0　　(B)　　1　　(C)　　　-1　　(D)　　does not exist

Answer (D)

If m is a negative integer then [m + 0] = m and [m – 0] = m - 1

$$\lim_{x \to m+0}(x - [x]) = m - m = 0 \quad \& \quad \lim_{x \to m-0}(x - [x]) = m - (m-1) = 1$$

$$\lim_{x \to m+0}(x - [x]) \neq \lim_{x \to m-0}(x - [x]) \text{ So, } \lim_{x \to m}(x - [x]) \text{ does not exist.}$$

Example 58. If [x] means the greatest integer ≤x and n is an integer, then $\lim_{x \to n}(-1)^{[x]}$ equals

(A)　　　1　　　(B)　　　- 1　　　(C)　　0　(D) does not exist

Answer (D)

[n] is either an even integer or an odd integer. So, the value of the limit is either + 1 or – 1. The limit does not exist.

Example 59. If n is an integer, then $\lim_{x \to n}\dfrac{x - n}{\sin(\pi x)}$ equals

(A) $\dfrac{(-1)^n}{\pi}$　　　(B) $(-1)^n \pi$　　(C) 0　(D)　　does not exist

Answer (A)

Let $x - n = t,\ x \to n \Rightarrow t \to 0$ The given

$$\text{limit} = \lim_{t \to 0}\frac{t}{\sin(n\pi + \pi t)}$$

$$= \lim_{t \to 0}\frac{t}{(-1)^n \sin(\pi t)} = (-1)^n.\frac{1}{\pi}\lim_{\pi t \to 0}\frac{\pi t}{\sin(\pi t)} = (-1)^n.\frac{1}{\pi}.1 = \frac{(-1)^n}{\pi}$$

Example 60. $\lim_{x \to 0}(1 + 2x)^{\frac{x+3}{x}}$ equals

(A)　　　e^2　　　(B)　　　e^3　　　(C)　　　e^6　　(D) does not exist

Answer (C)

The given limit =

$$\lim_{x \to 0}(1 + 2x)^{\frac{x+3}{x}} = \lim_{x \to 0}(1 + 2x)^{1+\frac{3}{x}} = \lim_{x \to 0}(1 + 2x) \times \lim_{x \to 0}(1 + 2x)^{\frac{3}{x}}$$

$[2x = y \Rightarrow x = y/2 \ \& \ x \to 0 \Rightarrow y \to 0]$ The given limit

$$= 1 \times \lim_{y \to 0} (1 + y)^{6/y} = \lim_{y \to 0} \left[(1 + y)^{1/y}\right]^6 = \left[\lim_{y \to 0} (1 + y)^{1/y}\right]^6 = e^6$$

Example 61. $\lim\limits_{x \to 0} \dfrac{\log_e (1 + 4x)}{e^{3x} - 1}$ is equal to

(A) 0 (B) 1/12 (C) 4/3 (D) 1 [ME 2016]

Answer: (C)

$$\lim_{x \to 0} \frac{\log_e (1 + 4x)}{e^{3x} - 1} = \lim_{x \to 0} \frac{\log_e (1 + 4x)}{4x} \cdot \frac{3x}{e^{3x} - 1} \cdot \frac{4}{3} = 1.1.\frac{4}{3} = \frac{4}{3}$$

Example 62. $\lim\limits_{x \to 0} \dfrac{(1 + x)^n - 1}{x}$ is equal to

(A) 0 (B) 1 (C) n (D) e

Answer: (C)

$$\lim_{x \to 0} \frac{(1 + x)^n - 1}{x} = \lim_{x \to 0} \frac{(1 + nx + ---) - 1}{x} = \lim_{x \to 0} \frac{nx + ----}{x} = n$$

Example 63. $\lim\limits_{x \to 0} \left(\dfrac{\tan x}{x^2 - x}\right)$ is equal to _____. [CE 2017]

Answer: - 1

Applying L'Hospitals' rule $\lim\limits_{x \to 0} \left(\dfrac{\tan x}{x^2 - x}\right) = \lim\limits_{x \to 0} \left(\dfrac{\sec^2 x}{2x - 1}\right) = \dfrac{1}{0 - 1} = -1$

Example 64. $\lim\limits_{x \to a} \dfrac{x^n - a^n}{x - a} = na^{n-1}$ holds when

(A) $a > 0 \ \& \ n > 0$ only   (B) $a > 0 \ \& \ n$ is any real number
(C) $a \ \& \ n$ are any real numbers (D) $a \ \& \ n$ are positive integers only

Answer: (B)

Example 65. $\lim\limits_{x \to 1} (1 - x) \tan \dfrac{\pi x}{2}$ is equal to

(A)   0 (B)   1 (C) $\dfrac{\pi}{2}$ (D) $\dfrac{2}{\pi}$

Answer : (D)

$$1 - x = z, x \to 1 \Rightarrow z \to 0 \ \& \ \tan\frac{\pi x}{2} = \tan\left\{\frac{\pi}{2}(1-z)\right\} = \cot\frac{\pi z}{2}$$

$$\lim_{x \to 1}(1-x)\tan\frac{\pi x}{2} = \lim_{z \to 0} z.\cot\frac{\pi z}{2} = \frac{\pi z/2}{\tan \pi z/2} \times \frac{2}{\pi} = 1 \times \frac{2}{\pi} = \frac{2}{\pi}$$

Example 66. If $a > 0$, then $\lim_{x \to \infty}\left[x\left(a^{1/x} - 1\right)\right]$ equals

(A)   0   (B) a (C) $\log a$ (D) e^a

Answer (C)

Let $\frac{1}{x} = t, \ x \to \infty \Rightarrow t \to 0$ The given limit $= \lim_{t \to 0}\frac{a^t - 1}{t} = \log a$

Example 67. $\lim_{x \to 0}\dfrac{\sqrt{\cos x} - \sqrt[3]{\cos x}}{\sin^2 x}$ is equal to

(A) $-\dfrac{1}{12}$ (B) $-\dfrac{1}{6}$ (C) $\dfrac{1}{6}$ (D) $\dfrac{1}{12}$

Answer: (A)

$\cos x = t^6, \sin^2 x = 1 - \cos^2 x = 1 - t^{12} \ \& \ x \to 0 \Rightarrow t \to 1$

The given limit

$$= \lim_{t \to 1}\frac{t^3 - t^2}{1 - t^{12}} = -\lim_{t \to 1}\frac{t^2(t-1)}{t^{12} - 1} = -\lim_{t \to 1}\left(t^2 \times \frac{1}{(t^{12} - 1)/(t - 1)}\right)$$

$$= -1 \times \frac{1}{12} = -\frac{1}{12} \qquad \text{As} \ \lim_{x \to a}\frac{x^n - a^n}{x - a} = na^{n-1}$$

Example 68. If $f(x) = |x|$, then $\lim_{h \to 0}\dfrac{f(h) - f(0)}{h}$

(A)   equals 1   (B) equals 0 (C) equals -1 (D) does not exist

Answer: (D)

The given limit $= \lim_{h \to 0}\dfrac{|h| - 0}{h} = \lim_{h \to 0}\dfrac{|h|}{h} = +1 \ or -1$ according as $h > 0$ or

$h < 0$. So, the limit does not exist.

Example 69. The value of $\lim\limits_{x\to 0}\left(\dfrac{-\sin x}{2\sin x + x\cos x}\right)$ is ___. [ME 2015]

Answer: $-1/3$

$$\lim\limits_{x\to 0}\left(\dfrac{-\sin x}{2\sin x + x\cos x}\right) = \left(\dfrac{0}{0}\right)$$

$$= \lim\limits_{x\to 0}\left(\dfrac{-\cos x}{2\cos x + \cos x - x\sin x}\right) = -\dfrac{1}{3}$$

Example 70. Consider the space realization

$$\begin{bmatrix} \dot{x}_1(t) \\ \dot{x}_2(t) \end{bmatrix} = \begin{bmatrix} 0 & 0 \\ 0 & -9 \end{bmatrix}\begin{bmatrix} x_1(t) \\ x_2(t) \end{bmatrix} + \begin{bmatrix} 0 \\ 45 \end{bmatrix}u(t),\ \text{with the initial condition}$$

$$\begin{bmatrix} x_1(0) \\ x_2(0) \end{bmatrix} = \begin{bmatrix} 0 \\ 0 \end{bmatrix},\ \text{where } u(t) \text{ denotes the unit step function. The value}$$

of $\lim\limits_{t\to\infty}\sqrt{x_1{}^2(t) + x_2{}^2(t)}$ is _____. [EC 2017]

Answer: 5

$$u(t) = \begin{cases} 1, t > 0 \\ 0, t < 0 \end{cases}\quad \text{Now, } \dot{x}_1(t) = 0 \Rightarrow x_1(t) = c\ \&\ x_1(0) = 0 \Rightarrow c = 0$$

So, $x_1(t) = 0$

$$\dot{x}_2(t) = 0 - 9x_2(t) + 45u(t) \Rightarrow \dfrac{dx_2(t)}{dt} + 9x_2(t) = 45u(t)$$

It is in linear form.

I. F. $= e^{9t} \Rightarrow x_2(t)e^{9t} = 45\int u(t).e^{9t} + c$ as $u(t) = const.$

$$x_2(t)e^{9t} = 45.\dfrac{e^{9t}}{9}u(t) + c, x_2(0) = 0 \Rightarrow c = 0 \text{ as } u(0) = 0$$

So, $x_2(t) = 5u(t)$ As $t \to \infty, u(t) = 1$

$$\lim\limits_{t\to\infty}\sqrt{x_1{}^2(t) + x_2{}^2(t)} = \lim\limits_{t\to\infty}\sqrt{0 + 25.1} = 5$$

$$\boxed{\text{4. Continuity}}$$

Question of one mark

Example 71. The values of x, for which the function

$f(x) = \dfrac{x^2 - 3x - 4}{x^2 + 3x - 4}$ is NOT continuous are

(A) $4 \& -1$ (B) $4 \& 1$ (C) $-4 \& 1$ (D) $-4 \& -1$ [ME 2016]

Answer: (C)

$x^2 + 3x - 4 = 0 \Rightarrow (x+4)(x-1) = 0 \Rightarrow x = -4,1$

Example 72. Which one of the following functions is continuous at $x = 3$?

(A) $f(x) = \begin{cases} 2 & \text{if } x=3 \\ x-1 & \text{if } x>3 \\ \dfrac{x+3}{3} & \text{if } x<3 \end{cases}$ (B) $f(x) = \begin{cases} 4 & \text{if } x = 3 \\ 8 - x & \text{if } x \neq 3 \end{cases}$

(C) $f(x) = \begin{cases} x+3 & \text{if } x \leq 3 \\ x - 4 & \text{if } x > 3 \end{cases}$ (D) $f(x) = \dfrac{1}{x - 27}$ if $x \neq 3$ [CS/IT 2013]

Answer (A)

$f(x)$ will be continuous at $x = 3$ if $\lim\limits_{x\to 3-0} f(x) = \lim\limits_{x\to 3+0} f(x) = f(3)$ It is

possible only in option (A) as $\lim\limits_{x\to 3-0} f(x) = \lim\limits_{x\to 3+0} f(x) = f(3) = 2$

Example 73. Let $g(x) = \begin{cases} -x, & x \leq 1 \\ x+1, & x \geq 1 \end{cases}$ and $f(x) = \begin{cases} 1-x, & x \leq 0 \\ x^2, & x > 0 \end{cases}$

Consider the composition of f & g i.e., $(f \circ g)x = f(g(x))$. The

number of discontinuities in $(f \circ g)x$ present in the interval $(-\infty, 0)$ is

(A) 0 (B) 1 (C) 2 (D) 4 [EE 2017]

Answer: (A)

$(f \circ g)x = f(g(x)) = f(-x) = 1 + x$ - no discontinuity in $(-\infty, 0)$.

168

Example 74. Given the following statements about a function $f : R \rightarrow R$, select the right option:

P. If $f(x)$ is continuous at $x = x_0$, it is also differentiable at $x = x_0$

Q. If $f(x)$ is continuous at $x = x_0$, it may not be differentiable at $x = x_0$

R. If $f(x)$ is differentiable at $x = x_0$, it is also continuous at $x = x_0$

(A) P is true, Q is false, R is false   (B) P is false, Q is true, R is true
(C) P is false, Q is true, R is false   (D) P is true, Q is false, R is true

[EC 2016]

Answer: (B)

Question of two marks:

Example 75. Let $f(x) = x^{-1/3}$ and A denote the area of the region bounded by $f(x)$ and the X – axis, when x varies from $- 1$ to 1. Which of the following statements is/are TRUE?

I. f is continuous in [- 1, 1]   II. f is not bounded in [- 1, 1]

III. A is non-zero and finite

(A) II only (B) III only (C) II and III only (D) I, II and III [CS/IT 2013]

Answer (C)

f is not continuous at $x = 0 \in [-1,1] \Rightarrow$ I is false.

f is not bounded at [- 1, 1] as $f \rightarrow \infty$ at $x = 0 \in [-1,1] \Rightarrow$ II is TRUE.

$$A = \int_{-1}^{1} x^{-1/3} dx = \frac{3}{2} \left[x^{2/3} \right]_{-1}^{1}$$ which is non-zero and finite $\Rightarrow$ III is TRUE.

Example 76. A function $f(x)$ is defined as $f(x) = \begin{cases} \dfrac{1- \cos nx}{x \sin x}, & x \neq 0 \\ 2, & x = 0 \end{cases}$

If $f(x)$ is continuous at $x = 0$ then the value of n is

(A)   0   (B) ± 1 (C) ± 2 (D)   any integral value

Answer (C)

As $f(x)$ is continuous at

$x = 0$, $f(0) = \lim_{x \to 0} f(x)$

169

$$2 = \lim_{x\to 0}\frac{1-\cos nx}{x\sin x} = \lim_{x\to 0}\frac{2\sin^2\frac{nx}{2}}{x^2\frac{\sin x}{x}} = 2\lim_{x\to 0}\frac{\frac{n^2}{4}\cdot\left(\frac{\sin nx/2}{nx/2}\right)^2}{\frac{\sin x}{x}} = 2.\frac{\frac{n^2}{4}\times 1^2}{1} = \frac{n^2}{2}$$

So, $2 = \dfrac{n^2}{2} \Rightarrow n^2 = 4 \Rightarrow n = \pm 2$

Example 77. If $f(x) = \begin{cases} ax^2 + b, & x < 1 \\ x+2, & x \ge 1 \end{cases}$ then the values of a, b for

which $f(x)$ is continuous at $x = 1$ are

(A) $a = -1, b = 1$ (B) $a = 1, b = 1$

(C) $a = 0, b = 2$ (D) $a = 1, b = 2$

Answer (D)

$\lim_{x\to 1-0} f(x) = \lim_{x\to 1-0}\left(ax^2 + b\right) = a + b$ $\lim_{x\to 1+0} f(x) = \lim_{x\to 1+0}\left(x+2\right) = 3$ and

$f(1) = 3$ As $f(x)$ is continuous at

$x = 1$, $\lim_{x\to 1-0} f(x) = \lim_{x\to 1+0} f(x) = f(0) \Rightarrow a + b = 3$ It satisfies only the

third option.

Example 78. What should be the value of λ such that the function

defined below is continuous at $x = \dfrac{\pi}{2}$?

$$f(x) = \begin{cases} \dfrac{\lambda \cos x}{\pi/2 - x}, & x \ne \dfrac{\pi}{2} \\ 1, & x = \dfrac{\pi}{2} \end{cases}$$

(A)   0   (B) $2/\pi$ (C)   1   (D) $\pi/2$ [CS/IT 2011]

Answer (C)

$f(x)$ will be continuous at $x = \dfrac{\pi}{2}$ if $\lim_{x\to \pi/2} f(x) = f(\pi/2)$

Now, $\lim_{x\to \pi/2} f(x) = 1 \Rightarrow \lim_{z\to 0}\dfrac{\lambda\sin z}{z} = 1 \Rightarrow \lambda = 1$

$[\,\pi/2 - x = z, x \to \pi/2 \Rightarrow z \to 0 \ \& \ \cos x = \cos(\pi/2 - z) = \sin z\,]$

Example 79. A function $f(x)$ is defined as $f(x) = \dfrac{\sin 2x}{x}, x \neq 0$ is

continuous at $x = 0$ then the value of $f(0)$ is

(A)   1   (B)   1 / 2   (C)   2   (D)   0

Answer (C)

$$\lim_{x \to 0} f(x) = \lim_{x \to 0} \frac{\sin 2x}{x} = 2.\lim_{2x \to 0} \frac{\sin 2x}{2x} = 2 \times 1 = 2$$

As $f(x)$ is continuous at $x = 0$ then $f(0) = \lim_{x \to 0} f(x) = 2$

Example 80. A function $f(x)$ is defined as $f(x) = \dfrac{3x^3 + 5x^2 + 7x}{\sin x}, x \neq 0$

is continuous at $x = 0$ then the value of $f(0)$ is

(A)   1   (B)   3   (C)   5   (D)   7

Answer (D)

$$\lim_{x \to 0} f(x) = \lim_{x \to 0} \frac{3x^3 + 5x^2 + 7x}{\sin x} = \lim_{x \to 0} \frac{x(3x^2 + 5x + 7)}{\sin x}$$

$$= \lim_{x \to 0} \frac{x}{\sin x} \times \lim_{x \to 0} (3x^2 + 5x + 7) = 1 \times (7) = 7 \text{ As } f(x) \text{ is continuous at}$$

$x = 0$ then $f(0) = \lim_{x \to 0} f(x) = 7$

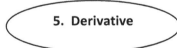

5. Derivative

Question of one mark:

Example 81. A function $f(x)$ is defined as

$$f(x) = \begin{cases} e^x, & x < 1 \\ \ln x + ax^2 + bx, & x \geq 1 \end{cases}, \text{ where } x \in R.$$

Which one of the following statements is TRUE?

(A) $f(x)$ is not differentiable at $x = 1$ for any values of a and b

(B) $f(x)$ is differentiable at $x = 1$ for the unique values of a and b

(C) $f(x)$ is differentiable at $x = 1$ for all values of a and b such that

171

$a + b = 1$

(D) $f(x)$ is differentiable at $x = 1$ for all values of a and b [EE 2017]

Answer: (B)

$\lim\limits_{x \to 1-0} f(x) = e, \ \lim\limits_{x \to 1+0} f(x) = a + b, \ f(1) = a + b$

$f(x)$ will be continuous at $x = 1$ if $a + b = e$ which will be true for the unique values of a and b.

Example 82. Consider the function $f(x) = |x|, \ -1 \le x \le 1$.

At the point $x = 0, \ f(x)$ is

(A)   Continuous and differentiable

(B)   Non-continuous and differentiable

(C)   Continuous and non-differentiable

(D)   Neither continuous nor differentiable   [ME 2012]

Answer (C)

$f(x) = |x|$ is continuous at $x = 0$ but

$f'(0) = \lim\limits_{h \to 0} \dfrac{f(0+h) - f(0)}{h} = \lim\limits_{h \to 0} \dfrac{f(h) - f(0)}{h} = \lim\limits_{h \to 0} \dfrac{|h| - 0}{h}$ which

does not exist.

Example 83. The function $f(x) = \begin{cases} \dfrac{x^2}{|x|}, & x \ne 0 \\ 0, & x = 0 \end{cases}$ is

(A)   Continuous and differentiable at $x = 0$

(B)   Non-continuous and differentiable at $x = 0$

(C)   Continuous and non-differentiable at $x = 0$

(D)   Neither continuous nor differentiable at $x = 0$

Answer (C)

$\lim\limits_{x \to 0+} f(x) = \lim\limits_{x \to 0-} f(x) = f(0) = 0 \Rightarrow f(x) =$ is continuous at $x = 0$ but

$f'(0) = \lim\limits_{h \to 0} \dfrac{f(0+h) - f(0)}{h} = \lim\limits_{h \to 0} \dfrac{f(h) - f(0)}{h} = \lim\limits_{h \to 0} \dfrac{h}{|h|}$ which does not exist.

Question of two marks:

Example 84. If $y = x^x$, then $\dfrac{dy}{dx}$ is

(A) $x^x(x-1)$ (B) x^{x-1}

(C) $x^x(1+\log x)$ (D) $e^x(1+\log x)$ [BT 2014]

Answer: (C)

$y = x^x$, $\log y = x\log x \Rightarrow \dfrac{1}{y}\dfrac{dy}{dx} = x.\dfrac{1}{x}+1.\log x = 1 + \log x$ or,

$\dfrac{dy}{dx} = y(1+\log x) = x^x(1+\log x)$

Example 85. A function $y = 5x^2 + 10x$ is defined over an open interval x = (1, 2). At least at one point in this interval, $\dfrac{dy}{dx}$ is exactly

(A) 20 (B) 25 (C) 30 (D) 35 [EE 2013]

Answer: (B)

$\dfrac{dy}{dx} = 10x + 10 = 25$ if $x = 1.5 \in (1,2)$

If $\dfrac{dy}{dx} = 10x + 10 = 20$ or 30 or $35, x \notin (1,2)$

Example 86. If $f(x) = R\sin\left(\dfrac{\pi x}{2}\right) + S, f'\left(\dfrac{1}{2}\right) = \sqrt{2}$ and

$\displaystyle\int_0^1 f(x)dx = \dfrac{2R}{\pi}$, then the constants R and S are respectively

(A) $\dfrac{2}{\pi}$ and $\dfrac{16}{\pi}$ (B) $\dfrac{2}{\pi}$ and 0 (C) $\dfrac{4}{\pi}$ and 0 (D) $\dfrac{4}{\pi}$ and $\dfrac{16}{\pi}$ [CS/IT 2017]

Answer: (C)

$f'(x) = R.\dfrac{\pi}{2}\cos\left(\dfrac{\pi x}{2}\right) \Rightarrow f'\left(\dfrac{1}{2}\right) = R.\dfrac{\pi}{2}.\dfrac{1}{\sqrt{2}} = \sqrt{2} \Rightarrow R = \dfrac{4}{\pi}$

$\displaystyle\int_0^1 f(x)dx = \left[-R.\dfrac{2}{\pi}.\cos\left(\dfrac{\pi x}{2}\right) + Sx\right]_0^1 = -0 + S + \dfrac{2R}{\pi} = \dfrac{2R}{\pi} \Rightarrow S = 0$

173

$$(R, S) = (4/\pi, 0)$$

5. Mean Value Theorems

Question of one mark:

Example 87. $f(x) = x^3 + 5$ in the interval $(1,4)$ at a value (rounded off to the second decimal place) of x equal to ____ .   [CH 2016]

Answer; 2.64

$$f'(c) = \frac{f(4) - f(1)}{4 - 1}, 1 < c < 4 \Rightarrow 3c^2 = \frac{69 - 6}{3} = 21 \Rightarrow c = 2.64$$

Example 88. Which of the following functions would have only odd powers of x in its Taylor series expansion about the point x=0?

(A) $\sin(x^3)$ (B) $\sin(x^2)$ (C) $\cos(x^3)$ (D) $\cos(x^2)$ [EC 2008]

Answer: (A)

$$\sin(x^3) = x^3 - \frac{x^9}{3!} + - - - -$$

Example 89. Let the function

$$f(\theta) = \begin{vmatrix} \sin\theta & \cos\theta & \tan\theta \\ \sin(\pi/6) & \cos(\pi/6) & \tan(\pi/6) \\ \sin(\pi/3) & \cos(\pi/3) & \tan(\pi/3) \end{vmatrix} \text{ where } \theta \in [\pi/6, \pi/3] \text{ and}$$

$f'(\theta)$ is the derivative of $f(\theta)$ w.r.t. θ. Which of the following statements is/are **TRUE**?

(I) There exists $\theta \in [\pi/6, \pi/3]$ such that $f'(\theta) = 0$.

(II) There exists $\theta \in [\pi/6, \pi/3]$ such that $f'(\theta) \neq 0$.

(A) (I) only   (B) (II) only

(C) Both (I) and (II)   (D) Neither(I) nor (II)   [CS/IT 2014]

Answer: (C)

$f(\theta)$ is continuous and derivable in $[\pi/6, \pi/3]$ and

$$f(\pi/6) = f(\pi/3) = 0.$$

By Rolle's theo. There exists some θ, $f'(\theta) = 0 \Rightarrow$ (I) is TRUE.

As $f(\theta)$ is not a constant function there may exist some

θ, $f'(\theta) \neq 0 \Rightarrow$ (II) is TRUE.

Example 90. Let the function

$$f(\theta) = \begin{bmatrix} \sin\theta & \cos\theta & \tan\theta \\ \sin(\pi/6) & \cos(\pi/6) & \tan(\pi/6) \\ \sin(\pi/3) & \cos(\pi/3) & \tan(\pi/3) \end{bmatrix} \text{ where } \theta \in \left[\pi/6, \pi/3\right] \text{ and}$$

$f'(\theta)$ is the derivative of $f(\theta)$ w.r.t. θ. Which of the following statements is/are **TRUE?**

(I) There exists $\theta \in \left[\pi/6, \pi/3\right]$ such that $f'(\theta) = 0$.

(II) There exists $\theta \in \left[\pi/6, \pi/3\right]$ such that $f'(\theta) \neq 0$.

(A) (I) only   (B) (II) only

(C) Both (I) and (II)   (D) Neither(I) nor (II) [CS/IT 2014]

Answer: (C)

$f(\theta)$ is continuous and derivable in $\left[\pi/6, \pi/3\right]$ and

$$f(\pi/6) = f(\pi/3) = 0.$$

By Rolle's theo. There exists some θ, $f'(\theta) = 0 \Rightarrow$ (I) is TRUE.

As $f(\theta)$ is not a constant function there may exist some

θ, $f'(\theta) \neq 0 \Rightarrow$ (II) is TRUE.

Example 91. If $f(x)$ is a real and continuous function of x, the Taylor series expansion of $f(x)$ about its minima will never have a term containing

(A)   first derivative   (B) second derivative

(C)   third derivative   (D) any higher derivative   [CH 2014]

Answer: (A)

$$f(x) = f(a) + (x-a)f'(a) + - - -$$

If $f(x)$ has a minima at $x = a$, then $f'(a) = 0$

***Example* 92.** A function $f(x) = 1 - x^2 + x^3$ is defined in the closed interval [-1, 1]. The value of x in the open interval $(-1,1)$ for which the mean value theorem is satisfied, is

(A) $-\dfrac{1}{2}$ (B) $-\dfrac{1}{3}$ (C) $\dfrac{1}{3}$ (D) $\dfrac{1}{2}$ [EC 2015]

Answer

(B) $\dfrac{f(b) - f(a)}{b - a} = f'(c) \Rightarrow \dfrac{f(1) - f(-1)}{1 - (-1)} = 3x^2 - 2x, -1 < x < 1$

$\dfrac{1 - (-1)}{2} = 3x^2 - 2x \Rightarrow 3x^2 - 2x - 1 = 0 \Rightarrow (x - 1)(3x + 1) = 0$ or,

$x = 1, -\dfrac{1}{3}$. As $-1 < x < 1$ so, $x = -\dfrac{1}{3}$

***Example* 93.** The Taylor series expansion of $3\sin x + 2\cos x$ is

(A) $2 + 3x - x^2 - \dfrac{x^3}{2} + - -$ (B) $2 - 3x + x^2 - \dfrac{x^3}{2} + - -$

(C) $2 + 3x + x^2 + \dfrac{x^3}{2} + - -$ (D) $2 - 3x - x^2 + \dfrac{x^3}{2} + - -$ [EC 2014]

Answer (A)

$3\sin x + 2\cos x = 3\left(x - \dfrac{x^3}{3!} + \dfrac{x^5}{5!} - - - \right) + 2\left(1 - \dfrac{x^2}{2!} + \dfrac{x^4}{4!} - - - \right)$

$= 2 + 3x - x^2 - \dfrac{x^3}{2} + - - -$

***Example* 94.** Let $f(x) = e^{x + x^2}$ for real x. From among the following, choose the Taylor series approximation of $f(x)$ around $x = 0$ which includes all powers of x less than or equal to 3.

(A) $1 + x + x^2 + x^3$ (B) $1 + x + \dfrac{3}{2}x^2 + x^3$

(C) $1 + x + \dfrac{3}{2}x^2 + \dfrac{7}{6}x^3$ (D) $1 + x + 3x^2 + 7x^3$ [EC 2017]

Answer: (C)

$$f(x) = e^{x+x^2} = e^x . e^{x^2} = \left(1 + x + \frac{x^2}{2} + \frac{x^3}{6} + - - \right)\left(1 + x^2 + \frac{x^4}{2} + - - \right)$$

$$= 1 + x + \frac{3}{2}x^2 + \frac{7}{6}x^3 \text{ with all powers of } x \text{ less than or equal to 3.}$$

6. Maxima & Minima

Question of one mark:

Example 95. At $x = 0$, the function $f(x) = |x|$ has

(A) a minimum   (B) a maximum   (C) a point of inflexion

(D) neither a maximum nor minimum   [ME 2015]

Answer: (A)

For all real values of x, the minimum value of $f(x) = |x|$ is 0 at $x = 0$.

Example 96. Let $f : (-1,1) \rightarrow R.$, where $f(x) = 2x^3 - x^4 - 10$. the minimum value of $f(x)$ is _____.   [IN 2016]

Answer: -13

$$f(x) = 2x^3 - x^4 - 10 = -\{x^4 - 2x^3\} - 10 = -\{(x^2 - x)^2 - x^2\} - 10$$
$$= x^2 - \{(x^2 - x)^2 + 10\}$$

It will be minimum if $\{(x^2 - x)^2 + 10\}$ is max. It is possible at $x = -1$

Min $f(x) = 1 - \{(1+1)^2 + 10\} = 1 - 14 = -13$

Example 97. While minimizing the function $f(x)$, the necessary and sufficient conditions for a point x_0 to be a minima are

(A) $f'(x_0) > 0$ & $f''(x_0) = 0$

(B) $f'(x_0) < 0$ & $f''(x_0) = 0$

(C) $f'(x_0) = 0$ & $f''(x_0) < 0$

(D) $f'(x_0) = 0$ & $f''(x_0) > 0$ [CE 2015]

Answer: (D)

Example 98. Consider the function $f(x) = 1 - |x|, -1 \le x \le 1$. The value of x at which the function attains a maximum and maximum value of the function, are

(A) 0, -1   (B) - 1,0   (C) 0, 1   (D) - 1, 2   [EE 2015]

Answer: (C)

$f(x) = 1 - |x|$ is maximum when $|x|$ is minimum $\Rightarrow |x| = 0 \Rightarrow x = 0$ and then $f(x) = 1$

Example 99. Consider the function $f(x) = \sin(x)$ in the interval $x \in (\pi/4, 7\pi/4)$. The number and location(s) of the local minima of this function are

(A) One, at $\pi/2$ (B) One, at $3\pi/2$
(C) Two, at $\pi/2$ and $3\pi/2$ (D)   Two, at $\pi/4$ and $3\pi/2$ [CS/IT 2014]

Answer (B)

$f'(x) = \cos(x), f'(x) = 0 \Rightarrow \cos(x) = 0 \Rightarrow x = \dfrac{\pi}{2}, \dfrac{3\pi}{2}$ in

$\left(\dfrac{\pi}{4}, \dfrac{7\pi}{4}\right)$. $f''(x) = -\sin(x) > 0$ at $x = \dfrac{3\pi}{2}$. $\Rightarrow f(x)$ has a local minima

at $x = 3\pi/2$.   As $\left[f''(\pi/2) = -1 < 0\right]$

Example 100. Consider the function $f(x) = 2x^3 - 3x^2$ in the domain [- 1, 2]. The global minimum of $f(x)$ is ___.   [ME 2016]

Answer: - 5

$f(x) = 2x^3 - 3x^2 \Rightarrow f'(x) = 6x^2 - 6x, \ f'(x) = 0 \Rightarrow x = 0,1$

Both $0,1 \in \left[-1, 2\right]$. The global minimum of $f(x)$ is

$\min\{f(-1), f(0), f(1), f(2)\} = \min\{-5, 0, -1, 4\} = -5$

Example 101. The global minimum value of the function $f(x) = $

$\dfrac{1}{3}x\left(x^2 - 3\right)$ in the interval $-100 \le x \le 100$ occurs at $x =$ [EC 2017]

Answer: -100.1 - - 99.1

$f'(x) = (x^2 - 1)$, $f'(x) = 0 \Rightarrow x = \pm 1$. Both the points lie in the interval $-100 \le x \le 100$. The global minimum of $f(x)$ is min $\{f(-100), f(-1), f(1), f(100)\} = f(-100)$ which occurs at $x = -100$.

Example 102. The minimum value of the function
$f(x) = x^3 - 3x^2 - 24x + 100$ in the interval [- 3, 3] is
(A)   20   (B)   28   (C)   16   (D)   32   [EE 2014]
Answer (B)
$f'(x) = 3x^2 - 6x - 24 = 3(x^2 - 2x - 8)$
Now, $f'(x) = 0 \Rightarrow x = -2,4$ only $-2 \in [-3,3]$
$f(-2) = 128, f(3) = 28, f(-3) = 118$
The minimum value $f(x)$ is 28

Example 103. At $x = 0$ the function $f(x) = x^3 + 1$ has
(A)   a maximum value   (B) a minimum value
(C)   a singularity   (D) a point of inflexion   [ME 2012]
Answer (D)
$f'(x) = 3x^2$, $f'(x) = 0 \Rightarrow x = 0$ $f''(x) = 6x$, $f''(0) = 0$
$f'''(x) = 6, f'''(0) = 6 \ne 0$ So, $x = 0$ is a point of inflexion.

Example 104. Consider the function $f(x) = x^2 - x - 2$. The maximum value of $f(x)$ in the closed interval $[-4, 4]$ is:

(A)   18 (B) 10 (C) -2.25 (D) indeterminate   [EC 2007]
Answer (B)
$f'(x) = 2x - 1$, $f'(x) = 0 \Rightarrow x = \dfrac{1}{2} \in [-4,4]$
$f(-4) = 18, f(1/2) = -2.25, f(4) = 10$ Maximum value = 18

Example 105. The maximum value of the function
$f(x) = \ln(1+x) - x$, (where $x > 0$) occurs at ____ .   [ECE 2014]
Answer: $x = 0$

$$f(x) = \ln(1+x) - x \Rightarrow f'(x) = \frac{1}{1+x} - 1, \ f'(x) = 0 \Rightarrow x = 0$$

$$f''(x) = -\frac{1}{(1+x)^2} - 0 \Rightarrow f''(0) = -1 < 0 \text{ So, } f(x) \text{ is max. at } x = 0$$

Example 106. The maximum value of the determinant among all 2 × 2 real symmetric matrices with trace 14 is _____.   [EC 2014]

Answer: 49

Let a 2 × 2 matrix be $\begin{pmatrix} x & z \\ z & t \end{pmatrix}$ The trace of the corresponding

determinant $= x + t = 14 \Rightarrow t = 14 - x$ and the value of the determinant

$$= D = xt - z^2 = x(14 - x) - z^2 = 14x - x^2 - z^2$$

$$\frac{\partial D}{\partial x} = 14 - 2x \ \& \ \frac{\partial D}{\partial z} = -2z$$

$$A = \frac{\partial^2 D}{\partial x^2} = -2, \ C = \frac{\partial^2 D}{\partial x \partial z} = 0, \ B = \frac{\partial^2 D}{\partial z^2} = -2$$

So,

$$\frac{\partial D}{\partial x} = 0 \Rightarrow x = 7 \ \& \ \frac{\partial D}{\partial z} = 0 \Rightarrow z = 0 \ \& \ t = 14 - x = 7$$

$$A < 0 \ \& \ AB - C^2 = 4 > 0 \Rightarrow D \text{ is max. at}$$

$$x = 7, z = 0, t = 7 \ D_{\max} = 7 \times 7 - 0 = 49$$

Example 107. Consider the equation $V = \dfrac{aS}{b + S + S^2 / c}$

Given $a = 4, b = 1, c = 9$, the positive value of S at which

V is maximum is _____.   [BT 2016]

Answer: 3

$$V = \frac{4S}{1 + S + S^2 / 9} = 36. \frac{S}{S^2 + 9S + 9} \Rightarrow \frac{dV}{dS} = 36. \frac{9 - S^2}{(S^2 + 9S + 9)^2}$$

For max. value of $V, \dfrac{dV}{dS} = 0 \Rightarrow S = 3 > 0$ At

$$S = 3, \frac{d^2 V}{dS^2} = 36. \frac{2(S^3 - 27S - 81)}{(S^2 + 9S + 9)^3} < 0 \Rightarrow V \text{ is max. at } S = 3 > 0.$$

Example 108. If $e^y = x^{1/x}$, then y has a

180

(A) maximum at $x = e$ (B) minimum at $x = e$

(C) maximum at $x = e^{-1}$ (D) minimum at $x = e^{-1}$ [EC 2010]

Answer: (A)

$$y = \frac{1}{x}\log x \Rightarrow \frac{dy}{dx} = \frac{1}{x^2}(1 - \log x), \frac{dy}{dx} = 0 \Rightarrow 1 - \log x = 0 \Rightarrow x = e$$

$$\frac{d^2 y}{dx^2} = -\frac{2}{x^3}(1 - \log x) + \frac{1}{x^2}\left(-\frac{1}{x}\right), \left[\frac{d^2 y}{dx^2}\right]_{x=e} = 0 - \frac{1}{e^3} < 0$$

So, the function is maximum at $x = e$.

Example 109. For real values of x, the minimum value of the function
$f(x) = \exp(x) + \exp(-x)$ is

(A)   2   (B)   1   (C)   0.5   (D)   0   [EC 2008]

Answer: (A)

$f'(x) = e^x - e^{-x} = 0 \Rightarrow e^{2x} = 1 = e^0 \Rightarrow x = 0$

$f''(x) = e^x + e^{-x} \Rightarrow f''(0) = 2 > 0$ **so,** $f(x)$ is minimum at $x = 0$.

The minimum value $= = f(0) = 1 + 1 = 2$

Example 110. The maximum value of the function of the solution $y(t)$
of the differential equation $y(t) + \ddot{y}(t) = 0$ with initial conditions
$\dot{y}(0) = 1$ & $y(0) = 1$, for $t \geq 0$ is

(A)   1   (B)   2   (C) π (D) $\sqrt{2}$ [IN 2015]

Answer: (D)

$y(t) + \ddot{y}(t) = 0 \Rightarrow y(t) = A\cos t + B\sin t, y(0) = 1 \Rightarrow A = 1$

$\dot{y}(t) = -A\sin t + B\cos t, \dot{y}(0) = 1 \Rightarrow B = 1$

So, $y(t) = \cos t + \sin t = \sqrt{2}\cos\left(t - \frac{\pi}{4}\right)$

As the max value of $\cos\left(t - \frac{\pi}{4}\right)$ is 1 so, Max $y(t) = \sqrt{2}$.

Example 111. The maximum area of (in sq. units) whose vertices lie
on the ellipse $x^2 + 4y^2 = 1$ is _____.   [EC 2015]

Answer: 1

The vertices are

$$\left(\cos\theta, \frac{1}{2}\sin\theta\right), \left(\cos\theta, -\frac{1}{2}\sin\theta\right), \left(-\cos\theta, \frac{1}{2}\sin\theta\right),$$

$$\left(-\cos\theta, -\frac{1}{2}\sin\theta\right).$$ Length of the sides $= 2\cos\theta, \sin\theta$

Area $= 2\cos\theta.\sin\theta = \sin 2\theta$ Max. area = 1

Example 112. The curve $y = x^4$ is
(A) concave up for all values of x (B) concave down for all values of x
(C) concave up only for positive values of x
(D) concave up only for negative values of x [PI 2015]
Answer: (A)

Example 113. Consider the function
$$f(x,y) = 4x^2 + 6y^2 - 8x - 4y + 8.$$ The optimal value of $f(x,y)$
(A) is a minimum equal to 10/3 (B) is a maximum equal to 10/3
(C) is a minimum equal to 8/3 (D) is a maximum equal to 8/3 [CS 2010]
Answer (A)
$f_x = 8x - 8, f_{xx} = 8, f_y = 12y - 4, f_{yy} = 12, f_{xy} = 0$
$f_x = 0, f_y = 0 \Rightarrow x = 1 \ \& \ y = 1/3$
$f_{xx} \cdot f_{yy} - \left(f_{xy}\right)^2 = 96 > 0 \ \& \ f_{xx} = 8 > 0$
So, $f(x,y)$ is minimum at (1, 1/3) The minimum value of $f(x,y)$ is
$$4 + \frac{2}{3} - 8 - \frac{4}{3} + 8 = \frac{10}{3}$$

Example 114. The contour on the x-y plane, where the partial derivative of $x^2 + y^2$ with respect to y is equal to the partial derivative of $6y + 4x$ with respect to x, is

(A) $y = 2$ (B) $x = 2$ (C) $x + y = 4$ (D) $x - y = 0$ [EC 2015]
Answer: (A)
$$\frac{\partial}{\partial y}\left(x^2 + y^2\right) = \frac{\partial}{\partial x}\left(6y + 4x\right) \Rightarrow 2y = 4 \Rightarrow y = 2$$

Example 115. Let $w = f(x, y)$ where x, y are functions of t. Then, according to chain rule $\dfrac{dw}{dt}$ is equal to

(A) $\dfrac{dw}{dx}.\dfrac{dx}{dt} + \dfrac{dw}{dy}.\dfrac{dy}{dt}$

(B) $\dfrac{\partial w}{\partial x}.\dfrac{\partial x}{\partial t} + \dfrac{\partial w}{\partial y}.\dfrac{\partial y}{\partial t}$

(C) $\dfrac{\partial w}{\partial x}.\dfrac{dx}{dt} + \dfrac{\partial w}{\partial y}.\dfrac{dy}{dt}$

(D) $\dfrac{\partial w}{dx}.\dfrac{\partial x}{\partial t} + \dfrac{\partial w}{dy}.\dfrac{\partial y}{\partial t}$ [CE 2017]

Answer: (C)

Example 116. If $u = \log(e^x + e^y)$, then $\dfrac{\partial u}{\partial x} + \dfrac{\partial u}{\partial y} =$

(A) $e^x + e^y$ (B) $e^x - e^y$ (C) $\dfrac{1}{e^x + e^y}$ (D) 1 [BT 2013]

Answer (D)

$$\dfrac{\partial u}{\partial x} = \dfrac{1}{e^x + e^y}.e^x \qquad \dfrac{\partial u}{\partial y} = \dfrac{1}{e^x + e^y}.e^y \qquad \dfrac{\partial u}{\partial x} + \dfrac{\partial u}{\partial y} = \dfrac{e^x + e^y}{e^x + e^y} = 1$$

Example 117. Consider a function $f(x, y, z)$ given by $f(x, y, z) = (x^2 + y^2 - 2z^2)(y^2 + z^2)$. The partial derivative of this function with respect to x at the point $x = 2, y = 1$ & $z = 3$ is ___.   [EE 2017]

Answer: 40

$$\dfrac{\partial f}{\partial x} = 2x(y^2 + z^2) \quad \dfrac{\partial f}{\partial x}(2,1,3) = 2.2(1^2 + 3^2) = 40$$

Example 118. The value of $\lim\limits_{(x,y) \to (0,0)} \dfrac{x^2 - xy}{\sqrt{x} - \sqrt{y}}$ is

(A)   0   (B)   1/2   (C)   1   (D) ∞ [PI 2015]

Answer: (A)

Let $y = mx$, as $x \to 0, y \to 0$ $\lim\limits_{(x,y) \to (0,0)} \dfrac{x^2 - xy}{\sqrt{x} - \sqrt{y}} = \lim\limits_{x \to 0} \dfrac{m^2 x^2 - mx^2}{\sqrt{x} - \sqrt{mx}}$

$= \lim\limits_{x \to 0}\left(x^{3/2}.\dfrac{m^2 - m}{1 - \sqrt{m}} \right) = 0$ for all real m. So, $\lim\limits_{(x,y) \to (0,0)} \dfrac{x^2 - xy}{\sqrt{x} - \sqrt{y}} = 0$

$$\boxed{\textbf{7. Applications}}$$

Question of one mark:

Example 119. The tangent to the curve represented by $y = x \ln x$ is required to have 45^0 inclination with the x − axis. The co-ordinates of the tangent point would be

(A) $(1,0)$ (B) $(0,1)$ (C) $(1,1)$ (D) $(\sqrt{2},\sqrt{2})$ [CE 2017]

Answer: (A)

$\dfrac{dy}{dx} = \ln x + 1 = \tan 45^0 = 1 \Rightarrow \ln x = 0 = \ln 1 \Rightarrow x = 1 \ \& \ y = 1.\ln 1 = 0$

So, the point is (1 0)

Question of two marks:

Example 120. A particle of mass 2 kg is travelling at a velocity of 1.5 m/s. A force $f(t) = 3t^2$ (in N) is applied to it in the direction of motion for a duration of 2 seconds where t denotes time in seconds. The velocity (in m/s up to one decimal place) of the particle immediately after the removal of the force is _____.   [CE 2017]

Answer: 5.5

Force = mass × acceleration $\Rightarrow 3t^2 = 2.\dfrac{dv}{dt} \Rightarrow \dfrac{dv}{dt} = \dfrac{3t^2}{2} \Rightarrow dv = \dfrac{3}{2}t^2 dt$

Integrating, $\displaystyle\int_{1.5}^{v} dv = \dfrac{3}{2}\int_0^2 t^2 dt \Rightarrow v - 1.5 = \dfrac{3}{2} \times \dfrac{1}{3} \times 8 = 4 \Rightarrow v = 5.5\, m/s$

Example 121. A particle, starting from origin at $t = 0\,s$, is travelling with velocity $v = \dfrac{\pi}{2}\cos\left(\dfrac{\pi}{2}t\right) m/s$. At $t = 3\,s$, the difference between the distance covered by the particle and the magnitude of the displacement of the particle from the origin is _____.   [EE 2014]

Answer: 2

If x be the displacement, $\dfrac{dx}{dt} = \dfrac{\pi}{2}\cos\left(\dfrac{\pi}{2}t\right)$ Integrating,

184

$$x = \sin\left(\frac{\pi}{2}t\right) + c \quad x = 0, t = 0 \Rightarrow c = 0. \text{ So, } x = \sin\left(\frac{\pi}{2}t\right)$$

At $t = 3, x = -1$ Required difference $= 1 - (-1) = 2$

Example 122. The initial velocity of an object is 40 m/s. The acceleration a of the object is given by the following expression: $a = -0.1v$, where v is the instantaneous velocity of the object. The velocity of the object after 3 seconds will be ____. [ME 2015]

Answer: 29.5 – 29.7

$$\frac{dv}{dt} = -0.1v \Rightarrow \frac{dv}{v} = -0.1dt \text{ Integrating, } \log v = -0.1t + c$$

Initially,

$$t = 0, v = 40 \Rightarrow c = \log 40 \Rightarrow \log v = -0.1t + \log 40 \Rightarrow v = 40e^{-0.1t}$$

For $t = 3, v = 40e^{-0.3} = 40 \times 0.7408 = 29.632$

Example 123. Consider the following state-space relation of a linear time-invariant system: $\dot{x}(t) = \begin{bmatrix} 1 & 0 \\ 0 & 2 \end{bmatrix} x(t), y(t) = c^T x(t), c = \begin{bmatrix} 1 \\ 1 \end{bmatrix}$ & $x(0) = \begin{bmatrix} 1 \\ 1 \end{bmatrix}$.

The value of $y(t)$ for $t = \log_e 2$ is _____. [EE 2016]

Answer: 6

Let $x(t) = \begin{bmatrix} x_1 \\ x_2 \end{bmatrix}$ then

$$\dot{x}(t) = \begin{bmatrix} 1 & 0 \\ 0 & 2 \end{bmatrix} x(t) \Rightarrow \begin{bmatrix} \dot{x}_1 \\ \dot{x}_2 \end{bmatrix} = \begin{bmatrix} 1 & 0 \\ 0 & 2 \end{bmatrix} \begin{bmatrix} x_1 \\ x_2 \end{bmatrix} \Rightarrow \frac{dx_1}{dt} = x_1, \frac{dx_2}{dt} = 2x_2$$

Integrating, $x_1 = c_1e^t$ & $x_2 = c_2e^{2t}$ $x(t) = \begin{bmatrix} c_1e^t \\ c_2e^{2t} \end{bmatrix}$ - - - (i) Now,

$$x(0) = \begin{bmatrix} 1 \\ 1 \end{bmatrix} \Rightarrow \begin{bmatrix} c_1 \\ c_2 \end{bmatrix} = \begin{bmatrix} 1 \\ 1 \end{bmatrix}$$

From (i), $x(t) = \begin{bmatrix} e^t \\ e^{2t} \end{bmatrix}$, $y(t) = \begin{bmatrix} 1 & 1 \end{bmatrix} \begin{bmatrix} e^t \\ e^{2t} \end{bmatrix} = [e^t + e^{2t}]$

So, $y(t) = e^t + e^{2t}$ For $t = \log_e 2$, $y(t) = e^{\log 2} + e^{\log 4} = 2 + 4 = 6$

Example 124. A small ball of mass 1 kg moving with a velocity of 12 m/s undergoes a direct central impact with a stationary ball of mass 2 kg. The impact is perfectly elastic. The speed (in m/s) of 2 kg mass ball after the impact will be _ [ME 2015]

Answer: 8

$m_1 = 1, m_2 = 2, u_1 = 12, u_1 = 12, u_2 = 0, v_1 =$ vel. of first ball after impact,

$v_2 =$ vel. of second ball after impact,

$e = 1$ $m_1u_1 + m_2u_2 = m_1v_1 + m_2v_2 \Rightarrow 12 = v_1 + 2v_2 ---(i)$

$v_2 - v_1 = e(u_1 - u_2) = 12 ----(ii)$ Solving (i) & $(ii), v_2 = 8$

Example 125. Consider an ant crawling along the curve $(x-2)^2 + y^2 = 4,$ where x and y are in meters. The ant starts from the point (4, 0) and moves counter −clockwise with a speed of 1.57 meters per second. The time taken by the ant to reach the point (2, 2) in seconds___. [ME 2015] Answer: 2

The points (4, 0) and (2, 2) lie on the circle and their distance along the circumference $= \dfrac{1}{4} (2\pi \times 2) = \pi$ meter. Time $=\pi/ 1.57 =3.14/1.57=2$ sec

EXERCISE

Question of One mark:

1. Which one of the following functions is strictly bounded?

(A) $1/x^2$ (B) e^x (C) x^2 (D) e^{-x^2} [EC 2007]

2. The infinite series $1 + x + \dfrac{x^2}{2!} + \dfrac{x^3}{3!} + --$ corresponds to

(A) $\sec x$ (B) e^x (C) $\cos x$ (D) $1 + \sin^2 x$ [CS/IT 2012]

3. The $\lim\limits_{x\to 0} \dfrac{\sin[2x/3]}{x}$ is

(A)   2/3 (B)   1 (C)   3/2 (D) ∞ [CS/IT 2010]

4. $\lim\limits_{x\to \pi/4} \dfrac{\sin^2(x - \pi/4)}{(x - \pi/4)}$ equals

(A)   0 (B)   1/2 (C)   1   (D)   2   [CS/IT 2000]

5. The value of the function $\lim_{x \to 0} \dfrac{x^3 + x^2}{2x^3 - 7x^2}$ is

(A)　0　(B)　- 1/7　(C)　1/7　(D)　∞　[CS/IT 2004]

Question of Two marks:

6. A function $f(x)$ is continuous in the interval. It is known that
$f(0) = f(2) = -1$ and $f(1) = 1$.
Which of the following statements must be true?

(A)　There exists a y in the interval (0, 1) such that $f(y) = f(y+1)$

(B)　For every y in the interval (0, 1), $f(y) = f(2-y)$

(C)　The maximum value of the function in the interval (0, 2) is 1

(D)　There exists a y in the interval (0, 1) such that
$f(y) = -f(2-y)$　[CS/IT 2014]

7. The function $f(x) = 2x^3 - 3x^2 - 36x + 2$ has its maxima at

(A)　$x = -2$ only　　　(B)　$x = 0$ only

(C)　$x = 3$ only　　　(D)　$x = -2$ and $x = 3$ both　[CS/IT 2000]

8. The period of a simple pendulum is $T = 2\pi\sqrt{l/g}$. The maximum error in T due to the possible error up to 1% in l and 2.5% in g is

(A)　1.5% (B)　1.75% (C)　2% (D)　2.25%

9. A rail engine accelerates from its stationary position for 8 seconds and travels a distance of 280 m. According to Mean Value Theorem, the speedometer at a certain time must read exactly

(A)　0 (B)　8 kmph (C)　75 kmph (D)　126 kmph [CS/IT 2010]

10. By Maclaurin's expansion $e^{\sin x} =$

(A)　$1 - x + x^2/2 - -$　　　(B)　$1 - x - x^2/2 - -$

(C)　$1 + x + x^2/2 - -$　　　(D)　$1 + x - x^2/2 - - -$

Answers:

1. (B)　2. (B)　3. (A)　4. (A)　5. (B)
6. (A)　7. (A)　8. (B)　9. (D)　10. (C)

Chapter - 7

INDEFINITE AND DEFINITE INTEGRALS

Indefinite Integral: Let $f(x)$ be a function of x. If there exists another function of x, say $F(x)$, such that $\dfrac{d}{dx}F(x) = f(x)$, then $F(x)$ is defined as an *indefinite integral* of $f(x)$ w.r.t. x. Symbolically, $F(x) = \int f(x)dx$

As $\dfrac{d}{dx}F(x) = \dfrac{d}{dx}[F(x) + c]$, c being an arbitrary constant.

So, $\int f(x)dx = F(x) + c$. For this c, the integral is called *Indefinite*.

Definite Integral: Let $f(x)$ be a function of x and $x = a$ & $x = b$ be two given values of x. Let there exists another function of x, say, $F(x)$, such that $\dfrac{d}{dx}F(x) = f(x)$. Then the change in value of $F(x)$ as x changes from a to b [i.e., $F(b) - F(a)$] is defined as the *definite integral* of $f(x)$ w. r.t. x from $x = a$ to $x = b$. It is unique.

Symbolically, $F(b) - F(a) = \int\limits_a^b f(x)dx$

Formulae and Results:

1. $\lim\limits_{n \to \infty} \dfrac{1}{n}\sum\limits_{r=0}^{n-1} f(r/n) = \lim\limits_{h \to 0} h\sum\limits_{r=1}^{n} f(rh) = \int\limits_0^1 f(x)dx$ where $nh = 1 \Rightarrow h = \dfrac{1}{n}$

   [Put $r/n = r h = x$]

2. If $f(x) = \phi'(x)$, then $\int\limits_a^b f(x)dx = \phi(b) - \phi(a)$

188

3.(i) $\int_a^b f(x)dx = \int_a^b f(z)dz$ 　　　(ii) $\int_a^b f(x)dx = - \int_b^a f(x)dx$

(iii) $\int_a^b f(x)dx = \int_a^c f(x)dx + \int_c^b f(x)dx$, (a < c < b)

(iv) $\int_0^a f(x)dx = \int_0^a f(a-x)dx$

(v) $\int_0^{na} f(x)dx = n\int_0^a f(x)dx$ if $f(a+x) = f(x)$

(vi) $\int_0^{2a} f(x)dx = \begin{cases} 2\int_0^a f(x)dx, & \text{if } f(2a-x) = f(x) \\ 0, & \text{if } f(2a-x) = -f(x) \end{cases}$

(vii) $\int_{-a}^a f(x)dx = \begin{cases} 2\int_0^a f(x)dx, & \text{if } f(-x) = f(x) \\ 0, & \text{if } f(-x) = -f(x) \end{cases}$

(viii) $\int_a^b f(x)dx = \int_a^b f(a+b-x)dx$

4. If f(x) be continuous in a≤ x ≤ b, then at every point x of [a, b], F(x) =

$\int_0^x f(x)dx$, a ≤ x ≤ b possesses a derivative

and $\dfrac{d}{dx} F(x) = \dfrac{d}{dx}\int_a^x f(x)dx = f(x)$.

5. *Gamma* functions:

$\Gamma(n) = \int_0^\infty e^{-x}x^{n-1}dx$, $n > 0 : \Gamma(n+1) = n\Gamma(n)$ $\Gamma(n+1) = n!$, if n is a

positive

integer. $\Gamma(1) = 1$ & $\Gamma(1/2)) = \sqrt{\pi}$

$\Gamma(m).\Gamma(1-m) = \pi \cos ec(m\pi), 0 < m < 1$

6. *Beta* functions:

i) $\beta(m,n) = \int_0^1 x^{m-1}(1-x)^{n-1}dx, \ m > 0, n > 0$

ii) $\beta(m,n) = \beta(n,m)$ & $\beta(m,n) = \dfrac{\Gamma(m).\Gamma(n)}{\Gamma(m+n)}$

$\int_0^{\pi/2} \cos^p\theta \sin^q\theta \, d\theta = \dfrac{\Gamma\left(\dfrac{p+1}{2}\right).\Gamma\left(\dfrac{q+1}{2}\right)}{2.\Gamma\left(\dfrac{p+q+2}{2}\right)}, \ p > -1, q > -1$

7. $\int_0^\infty \dfrac{\sin t}{t} dt = \dfrac{\pi}{2}$

8. Area bounded by the curve $y = f(x)$, the x-axis and two fixed ordinates $x = a$ & $x = b$ is given by $A = \int_a^b y\,dx = \int_a^b f(x)dx.$

9. Area bounded by the curve $x = f(y)$, the y-axis and two fixed abscissa $y = c$ & $y = d$ is given by $A = \int_c^d x\,dy = \int_c^d f(y)dy.$

10. Area bounded by two given curves $y = f_1(x), y = f_2(x)$ and two fixed ordinates $x = a$ & $x = b$ is given by $\left| \int_a^b f_1(x)dx - \int_a^b f_2(x)dx \right|$

Question of one mark:

Example 1. Evaluate $\displaystyle\int \dfrac{dx}{e^x - 1}$ (Note: C is a constant of integration.)

(A) $\dfrac{e^x}{e^x - 1} + C$

(B) $\dfrac{\ln\left(e^x - 1\right)}{e^x} + C$

(C) $\ln\left(\dfrac{e^x}{e^x - 1}\right) + C$

(D) $\ln\left(1 - e^{-x}\right) + C$ [CH 2013]

Answer (D)

$$I = \int \frac{dx}{e^x - 1} = \int \frac{e^{-x}}{1 - e^{-x}} dx \left[1 - e^{-x} = z \Rightarrow e^{-x} dx = dz \right]$$

$$I = \int \frac{dz}{z} = \ln z + C = \ln\left(1 - e^{-x}\right) + C$$

Example 2. If $\int \frac{\sin^2 x}{\cos^6 x} dx = A \tan^3 x + B \tan^5 x + C$, then $(A, B) =$

(A) $\left(\dfrac{1}{3}, \dfrac{1}{5}\right)$ (B) $\left(\dfrac{1}{5}, \dfrac{1}{3}\right)$ (C) $\left(\dfrac{1}{2}, \dfrac{1}{6}\right)$ (D) $\left(\dfrac{1}{6}, \dfrac{1}{2}\right)$

Answer (A)

$\tan x = z, \sec^2 x dx = dz$

$$I = \int \tan^2 x \sec^4 x dx = \int z^2 \left(1 + z^2\right) dz = \int \left((z^2 + z^4\right) dz$$

$$= \frac{z^3}{3} + \frac{z^5}{5} + c = \frac{1}{3}\tan^3 x + \frac{1}{5}\tan^5 x + C \Rightarrow (A, B) = \left(\frac{1}{3}, \frac{1}{5}\right)$$

Example 3. The value of $\int\limits_0^2 |1 - x| dx =$

(A)   0   (B)   1   (C)   2   (D)   1/2

Answer: (B)

The given integral

$$= \int\limits_0^1 |1 - x| dx + \int\limits_1^2 |1 - x| dx = \int\limits_0^1 (1 - x) dx + \int\limits_1^2 (x - 1) dx$$

$$= \left[x - \frac{x^2}{2} \right]_0^1 + \left[\frac{x^2}{2} - x \right]_1^2 = \left(1 - \frac{1}{2}\right) + \left(2 - 2 - \frac{1}{2} + 1\right) = 1$$

[As $(1 - x) > 0$ in [0, 1], so $|1 - x| = (1 - x)$ and

$(1 - x) < 0$ in [1, 2], so $|1 - x| = -(1 - x) = x - 1$]

Example 4. The value of $\int\limits_{-1}^1 x \, |x| \, dx$ is

(A)   0   (B)   1   (C)   2   (D)   1/2

191

Answer: (A)

Let $f(x) = x \mid x \mid \Rightarrow f(-x) = (-x) \mid -x \mid = -x \mid x \mid = -f(x)$

So, $f(x)$ is an odd function of x. $\displaystyle\int_{-1}^{1} f(x)dx = \int_{-1}^{1} x \mid x \mid dx = 0$

Example 5. Let x be a continuous variable defined in the interval $(-\infty, \infty)$, and $f(x) = e^{-x-e^{-x}}$. The integral $g(x) = \int f(x)dx$ is equal to

(A) $e^{e^{-x}}$ (B) $e^{e^{-x}}$ (C) $e^{e^{x}}$ (D) e^{-x} [CE 2017]

Answer: (B)

$e^{-x} = z, -e^{-x}dx = dz$

$g(x) = \int f(x)dx = \int e^{-x}.e^{-e^{-x}} = \int (-dz)e^{-z} = e^{-z} = e^{-e^{-x}}$

Example 6. $\displaystyle\int_{1/\pi}^{2/\pi} \frac{\cos\left(\dfrac{1}{x}\right)}{x^2} dx = \underline{\quad\quad}$ [CS/IT 2015]

Answer: - 1

$\dfrac{1}{x} = z, -\dfrac{1}{x^2}dx = dz, \ x = \dfrac{1}{\pi} \Rightarrow z = \pi \ \& \ x = \dfrac{2}{\pi} \Rightarrow z = \dfrac{\pi}{2}$

$I = -\displaystyle\int_{\pi}^{\pi/2} \cos z \, dz = -[\sin z]_{\pi}^{\pi/2} = -(1-0) = -1$

Example 7. If $f(x) = \begin{cases} 2x, 0 \le x \le 2 \\ x^2, 2 \le x \le 3 \end{cases}$ then the value of $\displaystyle\int_{0}^{3} f(x)dx$ is

(A)   41/3   (B)   47/3   (C)   37/3   (D)   31/3

Answer: (D)

$\displaystyle\int_{0}^{3} f(x)dx = \int_{0}^{2} f(x)dx + \int_{2}^{3} f(x)dx = \int_{0}^{2} 2xdx + \int_{2}^{3} x^2 dx$

$$= \left[x^2 \right]_0^2 + \left[\frac{x^3}{3} \right]_2^3 = 4 + \frac{1}{3}(27-8) = \frac{31}{3}$$

Example 8. The value of $\int\limits_{-a}^{a} x\sqrt{a^2 - x^2}\, dx$ is

(A)   0   (B)   a   (C) a^2 (D) a^3

Answer: (A)

Let $f(x) = x\sqrt{a^2 - x^2}$ then $f(-x) = -x\sqrt{a^2 - (-x)^2}$

$= -x\sqrt{a^2 - x^2} = -f(x)$, so $f(x)$ is an odd function of

x.i.e., $\int\limits_{-a}^{a} f(x)dx = 0 \Rightarrow \int\limits_{-a}^{a} x\sqrt{a^2 - x^2}\, dx = 0.$

Example 9. If $i = \sqrt{-1}$, what will be the evaluation of the definite

integral $\int\limits_{0}^{\pi/2} \frac{\cos x + i\sin x}{\cos x - i\sin x}\, dx$?   [CS/IT 2011]

(A)   0   (B)   2   (C) $-i$ (D) i

Answer: (D)

$$\int\limits_{0}^{\pi/2} \frac{\cos x + i\sin x}{\cos x - i\sin x}\, dx = \int\limits_{0}^{\pi/2} \frac{e^{ix}}{e^{-ix}}\, dx = \int\limits_{0}^{\pi/2} e^{2ix}\, dx = \left[\frac{e^{2ix}}{2i} \right]_{0}^{\pi/2}$$

$$= \frac{1}{2i}\left[e^{i\pi} - 1 \right] = \frac{1}{2i}\left[\cos\pi + i\sin\pi - 1 \right] = \frac{1}{2i}\left[-1 - 1 \right] = -\frac{1}{i} = \frac{i^2}{i} = i$$

Example 10. If a is a constant then the value of the integral

$$a^2 \int\limits_{0}^{\infty} xe^{-ax}\, dx \text{ is}$$

(A) $\dfrac{1}{a}$ (B) a (C)   1   (D)   0 [CE 2012]

Answer (C)

193

$$I = a^2 \int_0^\infty x e^{-ax} dx = a^2 \left[x . \frac{e^{-ax}}{-a} - \int 1 . \frac{e^{-ax}}{-a} dx \right]_0^\infty$$

$$= a^2 . \frac{1}{a} \left[\int e^{-ax} dx \right]_0^\infty = a . \left[\frac{e^{-ax}}{-a} \right]_0^\infty = -\frac{a}{a} . (0 - 1) = 1$$

Example 11. The value of the integral $\int_0^2 \frac{(x-1)^2 \sin(x-1)}{(x-1)^2 + \cos(x-1)} dx$ is

(A) 3   (B) 0   (C) - 1   (D) - 2   [ME 2014]

Answer: (B)

Let

$$x - 1 = z \Rightarrow dx = dz, \ x = 0 \Rightarrow z = -1 \ \& \ x = 2 \Rightarrow z = 1$$

$$I = \int_{-1}^1 \frac{z^2 \sin z}{z^2 + \cos z} dx$$

$$f(z) = \frac{z^2 \sin z}{z^2 + \cos z} \Rightarrow f(-z) = \frac{(-z)^2 \sin(-z)}{(-z)^2 + \cos(-z)} = -\frac{z^2 \sin z}{z^2 + \cos z} = -f(z)$$

$f(z)$ is an odd function of z. $I = \int_{-1}^1 f(z) dz = 0$

Example 12. The value of $\int_{-\pi/2}^{\pi/2} \sqrt{\frac{1}{2}(1 - \cos x)} dx$ is

(A)   π   (B)   π/2   (C)   π/4   (D)   0

Answer: (D)

$$\sqrt{\frac{1}{2}(1 - \cos x)} = \sqrt{\frac{1}{2} . 2 \sin^2 x/2} = \sin x/2$$

$I = \int_{-\pi/2}^{\pi/2} \sin \frac{x}{2} dx = 0$ as $\sin \frac{x}{2}$ is an odd function of x.

Example 13. The value of $\int_{-\pi}^{\pi} \frac{x e^{-x^2}}{1 + x^2} dx$ is

194

(A)   π/2   (B)   π/4   (C)   0   (D)   1

Answer: (C)

Let $f(x) = \dfrac{xe^{-x^2}}{1+x^2}$ then $f(-x) = \dfrac{-xe^{-x^2}}{1+x^2} = -f(x)$

So $f(x)$ is an odd function of x.   i.e., $\displaystyle\int_{-\pi}^{\pi} f(x)dx = 0 \Rightarrow \int_{-\pi}^{\pi} \dfrac{xe^{-x^2}}{1+x^2}dx = 0$

Example 14. If $f(x)$ is an even function and a is a positive real number,

then $\displaystyle\int_{-a}^{a} f(x)dx$ is

(A)   0   (B) a (C) $2a$ (D) $2\displaystyle\int_{0}^{a} f(x)dx$ [ME 2011]

Answer (D)
It follows from the property 3.(vii) of the definite integration.

Example 15. If $f^*(x)$ is the complex conjugate of $f(x) = \cos x + i\sin x$

then for real a & b, $\displaystyle\int_{a}^{b} f^*(x).f(x)dx$ is ALWAYS

(A) positive (B) negative (C) real (D) imaginary [CH 2014]

Answer (C)
$f^*(x) = \cos x - i\sin x \Rightarrow f^*(x).f(x) = \cos^2 x + \sin^2 x = 1$

$\displaystyle\int_{a}^{b} f^*(x).f(x)dx = \int_{a}^{b} dx = b - a$ which is always real as a & b are real.

Example 16. The value of the quantity $P,$, where $P = \displaystyle\int_{0}^{1} xe^x dx$

(A)   0   (B)   1   (C)   e   (D)   1/e   [EE 2010]
Answer (B)
Using the rule of the integration by parts with
$u = x$ & $v = e^x$

$$P = \int_0^1 xe^x\,dx = \left[x.e^x - \int 1.e^x dx\right]_0^1 = \left[xe^x - e^x\right]_0^1 = (e-e)-(0-1) = 1$$

Example 17. The value of $\displaystyle\int_0^{\log 2} \frac{e^x}{1+e^x}dx$ is

(A) $\log\dfrac{3}{2}$ (B) $\log\dfrac{4}{3}$ (C) $\log 2$ (D) $\quad 1$

Answer: (A)

Let $I = \displaystyle\int_0^{\log 2} \frac{e^x}{1+e^x}dx$

$1+e^x = z, e^x dx = dz : x = 0 \Rightarrow z = 2, x = \log 2 \Rightarrow z = 1+e^{\log 2} = 1+2 = 3$

$$I = \int_2^3 \frac{dz}{z} = \left[\log z\right]_2^3 = \log 3 - \log 2 = \log\frac{3}{2}$$

Example 18. The value of $\displaystyle\int_0^{\pi/4} x\cos x^2\,dx$ correct to three decimal places

(assuming that π = 3.14) is _____.   [CS/IT 2018]

Answer: 0.27 – 0.30

$$\int_0^{\pi/4} x\cos x^2\,dx = \frac{1}{2}\int_0^{\pi/4} \cos x^2 d(x^2) = \frac{1}{2}\sin x^2\Big|_0^{\pi/4} = \frac{1}{2}\sin\left(\frac{\pi^2}{16}\right)$$

$$= \frac{1}{2}\sin\left(\frac{\pi^2}{16}\times\frac{180}{\pi}\right) = \frac{1}{2}\sin(35.325)^0 = \frac{1}{2}\times 0.5782 = 0.29$$

Example 19. The value of $\displaystyle\int_0^{\pi/2} \sin^6 x\cos^3 x\,dx$ is

(A) $\quad 5/81$ (B) $\quad 9/64$ (C) $\quad 4/21$ (D) $\quad 2/63$

Answer: (D)

$$I = \frac{\Gamma(7/2).\Gamma 2}{2.\Gamma(11/2)} \text{ [By formula 6.ii)]} = \frac{\Gamma(7/2).1}{2.(9/2)(7/2).\Gamma(7/2)} = \frac{2}{63}$$

196

Example 20. Which of the following integrals is unbounded?

(A) $\int_0^{\pi/4} \tan x\, dx$ (B) $\int_0^\infty \dfrac{dx}{x^2+1}$ (C) $\int_0^\infty xe^{-x}\, dx$ (D) $\int_0^1 \dfrac{dx}{1-x}$ [ME 2008]

Answer (D)

At $x=1$, $f(x)=\dfrac{1}{1-x}$ has an infinite discontinuity.

So, $\int_0^1 \dfrac{dx}{1-x}$ is unbounded. All other are bounded.

Example 21. Which of the following integrals is unbounded?

(A) $\int_0^\infty e^{-x}\, dx$ (B) $\int_0^1 \dfrac{dx}{x^{2/3}}$ (C) $\int_0^\infty \dfrac{x\, dx}{x^2+4}\, dx$ (D) $\int_0^\infty \dfrac{\sin^2 x}{x^2}\, dx$

Answer (C)

$\int_0^\infty e^{-x}\, dx = \left[e^{-x}\right]_0^\infty = 1$ $\qquad \int_0^1 \dfrac{dx}{x^{2/3}} = \left[3x^{1/3}\right]_0^1 = 3$

$\int_0^\infty \dfrac{x\, dx}{x^2+4}\, dx = \dfrac{1}{2}\left[\log(x^2+4)\right]_0^\infty$ -which is unbounded.

As $\dfrac{\sin^2 x}{x^2} \le \dfrac{1}{x^2}$ & $\int_0^\infty \dfrac{1}{x^2}dx$ is bounded so $\int_0^\infty \dfrac{\sin^2 x}{x^2}dx$ is bounded.

Example 22. The improper integral $\int_0^\infty \cos tx\, dx$ is

(A) 0 (B) 1 (C) $\dfrac{1}{t}$ (D) unbounded

Answer (D)

$\int_0^\infty \cos tx\, dx = \lim_{X\to\infty}\left[\dfrac{\sin tx}{t}\right]_0^X$ which is unbounded.

Example 23. The value of $\int_0^\infty \dfrac{dx}{x^2+2x+2}$ is

(A) 0 (B) $\dfrac{\pi}{2}$ (C) $\dfrac{\pi}{4}$ (D) $\log\dfrac{1}{2}$

Answer (C)

$$\int_0^\infty \frac{dx}{x^2+2x+2} = \int_0^\infty \frac{dx}{(x+1)^2+1} = \left[\tan^{-1}(x+1)\right]_0^\infty$$

$$= \tan^{-1}(\infty) - \tan^{-1} 1 = \frac{\pi}{2} - \frac{\pi}{4} = \frac{\pi}{4}$$

Example 24. $\Gamma\left(\dfrac{1}{3}\right)\Gamma\left(\dfrac{2}{3}\right)$ equals

(A) $\dfrac{2}{\sqrt{3}}\pi$ (B) $\dfrac{\sqrt{3}}{2}\pi$ (C) $\dfrac{1}{\sqrt{3}}\pi$ (D) $\dfrac{1}{3}\pi$

Answer (A)

$$\Gamma\left(\frac{1}{3}\right)\Gamma\left(\frac{2}{3}\right) = \Gamma\left(\frac{1}{3}\right).\Gamma\left(1-\frac{1}{3}\right) = \pi\cos ec\left(\frac{1}{3}\pi\right) = \pi.\frac{2}{\sqrt{3}} = \frac{2}{\sqrt{3}}\pi$$

[Taking m =1/3 in Result 5]

Example 25. The value of $\displaystyle\int_0^\infty e^{-x^2}\,dx$ is

(A)   0   (B) $\dfrac{1}{2}\sqrt{\pi}$ (C) $\dfrac{1}{4}\sqrt{\pi}$ (D)   e

Answer (B)

$$x^2 = z,\ 2xdx = dz \Rightarrow dx = \frac{dz}{2x} = \frac{dz}{2\sqrt{z}}$$

$$I = \int_0^\infty e^{-z}\frac{dz}{2\sqrt{z}} = \frac{1}{2}\int_0^\infty e^{-z}z^{1/2-1}dz = \frac{1}{2}\Gamma(1/2) = \frac{1}{2}\sqrt{\pi}$$

Example 26. Let $f(x) = \displaystyle\int_1^x \sqrt{2-t^2}\,dt$, then the real roots of the equation

$x^2 - f'(x) = 0$ are:

(A)   ± 1   (B) $\pm\dfrac{1}{\sqrt{2}}$ (C) $\pm\dfrac{1}{2}$ (D)   0 and 1

Answer: (A)

$$f(x) = \int_1^x \sqrt{2-t^2}\,dt \Rightarrow f'(x) = \sqrt{2-x^2}.1 - \sqrt{2-1}.0 = \sqrt{2-x^2}$$

$$x^2 = f'(x) \Rightarrow x^2 = \sqrt{2-x^2} \Rightarrow x^4 = 2 - x^2 \Rightarrow x^4 + x^2 - 2 = 0$$
$$\Rightarrow \left(x^2 + 2\right)\left(x^2 - 1\right) = 0 \text{ The real roots are given by } x^2 - 1 = 0 \Rightarrow x = \pm 1$$

Question of two marks

Example 27. Assuming $i = \sqrt{-1}$ and t is a real number, $\displaystyle\int_0^{\pi/3} e^{it}\, dt$ is

(A) $\dfrac{\sqrt{3}}{2} + i.\dfrac{1}{2}$ (B) $\dfrac{\sqrt{3}}{2} - i.\dfrac{1}{2}$ (C) $\dfrac{1}{2} + i.\dfrac{\sqrt{3}}{2}$ (D) $\dfrac{1}{2} + i.\left(1 - \dfrac{\sqrt{3}}{2}\right)$ [ME 2006]

Answer (A)

$$\int_0^{\pi/3} e^{it}\, dt = \frac{1}{i}\left[e^{it}\right]_0^{\pi/3} = \frac{1}{i}\left[e^{i\pi/3} - 1\right] = \frac{1}{i}\left[\cos\frac{\pi}{3} + i\sin\frac{\pi}{3} - 1\right]$$
$$= \frac{1}{i}\left[\frac{1}{2} + i.\frac{\sqrt{3}}{2} - 1\right] = \frac{1}{i}\left[i.\frac{\sqrt{3}}{2} - \frac{1}{2}\right] = \frac{\sqrt{3}}{2} - \frac{1}{i}.\frac{1}{2} = \frac{\sqrt{3}}{2} + \frac{i^2}{i}.\frac{1}{2} = \frac{\sqrt{3}}{2} + i.\frac{1}{2}$$

Example 28. If $\displaystyle\int_0^{\pi/2} \cos^m x \sin^m x\, dx = \lambda \int_0^{\pi/2} \sin^m x\, dx$, then $\lambda =$

(A) 2^{2m} (B) 2^{-2m} (C) 2^m (D) 2^{-m}

Answer: (D)

$$I = \int_0^{\pi/2} \cos^m x \sin^m x\, dx = 2^{-m} \int_0^{\pi/2} (2\cos x \sin x)^m\, dx = 2^{-m} \int_0^{\pi/2} (\sin 2x)^m\, dx$$
$$\left[2x = z, \Rightarrow 2dx = dz \Rightarrow dx = \frac{1}{2}dz,\ x = 0 \Rightarrow z = 0\ \&\ x = \frac{\pi}{2} \Rightarrow z = \pi\right]$$
$$I = 2^{-m}\int_0^{\pi} \sin^m z\left(\frac{1}{2}dz\right) = 2^{-m}.2\int_0^{\pi/2} \sin^m z\left(\frac{1}{2}dz\right) \text{ [by prop. 3(v) with}$$
n=2] $$I = 2^{-m} \int_0^{\pi/2} \sin^m z\, dz = 2^{-m} \int_0^{\pi/2} \sin^m x\, dx \Rightarrow \lambda = 2^{-m}$$

199

Example 29. If $y = \int\limits_0^x \frac{1-t}{1+t} dt \, (x > 0)$, then at x = 1, the value of $\frac{dy}{dx}$

(A)   0   (B)   1   (C)   2   (D)   1/2

Answer: (A)

$$y = \int\limits_0^x \frac{-(1+t)+2}{1+t} dt = -\int\limits_0^x dt + 2\int\limits_0^x \frac{dt}{1+t} = -x + 2\log(1+x)$$

Then $\frac{dy}{dx} = -1 + \frac{2}{1+x}$. At $x = 1$, $\frac{dy}{dx} = -1 + 1 = 0$

Example 30. The value of the integral $\int\limits_{-20\pi}^{20\pi} |\cos x| \, dx$ is

(A)   40π   (B)   0   (C)   80   (D)   40

Answer (C)

$I = \int\limits_{-20\pi}^{20\pi} |\cos x| \, dx = 2 \int\limits_0^{20\pi} |\cos x| \, dx$ as $|\cos x|$ is an even function of x.

$$I = \int\limits_{-20\pi}^{20\pi} |\cos x| \, dx = 2 \times 20 \int\limits_0^{\pi} |\cos x| \, dx = 40 \left[\int\limits_0^{\pi/2} |\cos x| \, dx + \int\limits_{\pi/2}^{\pi} |\cos x| \, dx \right]$$

$$= 40 \left[\int\limits_0^{\pi/2} \cos x \, dx + \int\limits_{\pi/2}^{\pi} (-\cos x) dx \right] \text{ as}$$

$\cos x > 0$ in $\left(0, \frac{\pi}{2}\right)$ & $\cos x < 0$ in $\left(\frac{\pi}{2}, \pi\right)$

$I = 40 \left\{ [\sin x]_0^{\pi/2} - [\sin x]_{\pi/2}^{\pi} \right\} = 40(1+1) = 80$

Example 31. If $\int\limits_0^{2\pi} |x \sin x| \, dx = k\pi$ then the value of k is equal

to _____.   [CS/IT 2014]

Answer: 4

In $0 < x < \pi$, $x \sin x > 0 \Rightarrow |x \sin x| = x \sin x$

In $\pi < x < 2\pi$, $x \sin x < 0 \Rightarrow |x \sin x| = -x \sin x$

$$I = \int\limits_{0}^{2\pi} |x\sin x|\,dx = \int\limits_{0}^{\pi} |x\sin x|\,dx + \int\limits_{\pi}^{2\pi} |x\sin x|\,dx$$

$$\int\limits_{0}^{\pi} x\sin x\,dx + \int\limits_{\pi}^{2\pi}(-x\sin x)dx = \left[x(-\cos x) - \int\limits_{0}^{\pi}(-\cos x)dx \right]_{0}^{\pi}$$

$$-\left[x(-\cos x) - \int\limits_{\pi}^{2\pi}(-\cos x)dx \right]_{\pi}^{2\pi}$$

$$= \pi(-\cos\pi) + \left[\sin x\right]_{0}^{\pi} + \left[x\cos x - \int\cos x\,dx\right]_{\pi}^{2\pi}$$

$$= \pi + 0 + (2\pi\cos 2\pi - \pi\cos\pi) - \left[\sin x\right]_{\pi}^{2\pi} = \pi + 2\pi + \pi - 0 + 4\pi \text{ So,}$$

$$k\pi = 4\pi \Rightarrow k = 4$$

Example 32. The value of the integral given below is $\int\limits_{0}^{\pi} x^2 \cos x\,dx$

(A) -2π (B) π (C) $-\pi$ (D) 2π [CS/IT 2014]

Answer (A)

$$\int\limits_{0}^{\pi} x^2 \cos x\,dx = \left[x^2 \sin x - 2\int x\sin x\,dx\right]_{0}^{\pi} = 0 - 2\left[-x\cos x - \int 1.(-\cos x)dx\right]_{0}^{\pi}$$

$$= -2\{[(-\pi.(-1))] + \left[\sin x\right]_{0}^{\pi}\} = -2\pi$$

Example 33. The value of $\int\limits_{1}^{\sqrt{e}} x\log x\,dx$ is

(A)   e/4   (B)   1/4   (C)   1/2   (D)   e/2

Answer: (B)

Using the rule of the integration of parts and taking

$$u = \log x, v = x, \int\limits_{1}^{\sqrt{e}} x\log x\,dx =$$

$$\left[\log x \int x\,dx - \int \frac{1}{x}\frac{x^2}{2}\,dx\right]_{1}^{\sqrt{e}} = \left[\log x.\frac{x^2}{2} - \frac{1}{2}.\frac{x^2}{2}\right]_{1}^{\sqrt{e}}$$

$$= \left(\frac{e}{2} \cdot \log \sqrt{e} - \frac{1}{4} e \right) - \left(0 - \frac{1}{4} \right) = \frac{e}{4} \log e - \frac{e}{4} + \frac{1}{4} = \frac{1}{4}$$

Example 34. The value of the definite integral $\int_1^e \sqrt{x} \ln(x) dx$ is

(A) $\frac{4}{9} \sqrt{e^3} + \frac{2}{9}$ (B) $\frac{2}{9} \sqrt{e^3} - \frac{4}{9}$

(C) $\frac{2}{9} \sqrt{e^3} + \frac{4}{9}$ (D) $\frac{4}{9} \sqrt{e^3} - \frac{2}{9}$ [ME 2013]

Answer (C)

Now, $\int \ln(x) x^{1/2} dx = \left[\ln(x) \cdot \frac{2}{3} x^{3/2} - \int \frac{1}{x} \cdot \frac{2}{3} x^{3/2} dx \right]$

$= \frac{2}{3} x^{3/2} \ln(x) - \frac{2}{3} \int x^{1/2} dx = \frac{2}{3} x^{3/2} \ln(x) - \frac{2}{3} \cdot \frac{2}{3} x^{3/2}$

$\frac{2}{3} x^{3/2} \ln(x) - \frac{4}{9} x^{3/2} \int_1^e \sqrt{x} \ln(x) dx = \left[\frac{2}{3} \sqrt{x^3} \ln(x) - \frac{4}{9} x^{3/2} \right]_1^e$

$= \frac{2}{3} \sqrt{e^3} - \frac{4}{9} \sqrt{e^3} - 0 + \frac{4}{9}$ [As $\ln e = 1$ & $\ln 1 = 0$]

Example 35. $\int_0^{\pi/4} \frac{1 - \tan x}{1 + \tan x} dx$ evaluates to

(A) 0   (B) 1   (C) $\ln 2$ (D) $\frac{1}{2} \ln 2$ [CS/IT 2009]

Answer (D)

$I = \int_0^{\pi/4} \frac{1 - \tan x}{1 + \tan x} dx = \int_0^{\pi/4} \tan \left(\frac{\pi}{4} - x \right) dx$

$\frac{\pi}{4} - x = z, \ -dx = dz \quad x = \frac{\pi}{4} \rightarrow z = 0, \ x = 0 \rightarrow z = \frac{\pi}{4}$

$I = - \int_{\pi/4}^0 \tan z dz = \left[\ln(\sec z) \right]_0^{\pi/4} = \ln \left(\sec \frac{\pi}{4} \right) = \ln \sqrt{2} = \frac{1}{2} \ln 2$

Example 36. The value of $\displaystyle\int_{0}^{\pi/2} \sin\phi\cos\phi\sqrt{a^2\sin^2\phi + b^2\cos^2\phi}\,d\phi$ is

(A) $\dfrac{a^2 + ab + b^2}{3(a+b)}$

(B) $\dfrac{a^2 - ab + b^2}{3(a-b)}$

(C) $\dfrac{a^2 + ab + b^2}{2(a+b)}$

(D) $\dfrac{a^2 - ab + b^2}{2(a-b)}$

Answer: (A)

$a^2\sin^2\phi + b^2\cos^2\phi = z^2, 2(a^2 - b^2)\sin\phi\cos\phi\,d\phi = 2z\,dz$

When $\phi = 0$, z = b and when ϕ = $\pi/2$, z =

a $\displaystyle\int_{0}^{\pi/2} \sin\phi\cos\phi\sqrt{a^2\sin^2\phi + b^2\cos^2\phi}\,d\phi$ =

$\dfrac{1}{a^2 - b^2}\displaystyle\int_{b}^{a} z.(zdz) = \dfrac{1}{a^2 - b^2}\left[\dfrac{z^3}{3}\right]_{b}^{a} = \dfrac{1}{3(a^2 - b^2)}(a^3 - b^3)$

$= \dfrac{1}{3(a-b)(a+b)}(a-b)(a^2 + ab + b^2) = \dfrac{a^2 + ab + b^2}{3(a+b)}$

Example 37. The value of $\displaystyle\int_{0}^{\log 2} \dfrac{e^x}{1+e^x}dx$ is

(A) $\log\dfrac{3}{2}$ (B) $\log\dfrac{4}{3}$ (C) $\log 2$ (D) 1

Answer: (A)

$I = \displaystyle\int_{0}^{\log 2} \dfrac{e^x}{1+e^x}dx \ \ 1+e^x = z, e^x dx = dz : x = 0 \Rightarrow z = 2$

$x = \log 2 \Rightarrow z = 1 + e^{\log 2} = 1 + 2 = 3$

$I = \displaystyle\int_{2}^{3}\dfrac{dz}{z} = \left[\log z\right]_{2}^{3} = \log 3 - \log 2 = \log\dfrac{3}{2}$

Example 38. The value of $\displaystyle\int_{0}^{\pi/2}\dfrac{\cos x\,dx}{(1+\sin x)(2+\sin x)}$ is

(A) $\log\dfrac{3}{2}$ (B) $\log\dfrac{5}{2}$ (C) $\log\dfrac{1}{2}$ (D) $\log\dfrac{4}{3}$

203

Answer: (D)

Let $\sin x = z$, $\cos x \, dx = dz$, $x = 0$, $z = 0$ and $x = \pi/2$, $z =$

$$1\int_0^1 \frac{dz}{(1+z)(2+z)} = \int_0^1 \frac{(2+z)-(1+z)}{(1+z)(2+z)} dz = \int_0^1 \frac{dz}{1+z} - \int_0^1 \frac{dz}{2+z}$$

$$= \left[\log(1+z)\right]_0^1 - \left[\log(2+z)\right]_0^1 = [\log 2 - \log 1] - [\log 3 - \log 2]$$

$$= \log 2 - 0 - \log 3 + \log 2 = 2\log 2 - \log 3 = \log 4 - \log 3 = \log \frac{4}{3}$$

Example 39. The value of $\displaystyle\int_0^{\pi/2} \frac{\sqrt{\sin x}}{\sqrt{\sin x} + \sqrt{\cos x}} dx$ is

(A) $\pi/2$ (B) $\pi/4$ (C) 0 (D) 1

Answer: **(B)**

$$I = \int_0^{\pi/2} \frac{\sqrt{\sin x}}{\sqrt{\sin x} + \sqrt{\cos x}} dx \; --- (i)$$

$$I = \int_0^{\pi/2} \frac{\sqrt{\sin\left(\dfrac{\pi}{2} - x\right)}}{\sqrt{\sin\left(\dfrac{\pi}{2} - x\right)} + \sqrt{\cos\left(\dfrac{\pi}{2} - x\right)}} dx = \int_0^{\pi/2} \frac{\sqrt{\cos x}}{\sqrt{\cos x} + \sqrt{\sin x}} dx \; --(ii)$$

Adding (i) & (ii), $2I = \displaystyle\int_0^{\pi/2} \frac{\sqrt{\sin x} + \sqrt{\cos x}}{\sqrt{\cos x} + \sqrt{\sin x}} dx = \int_0^{\pi/2} dx = \frac{\pi}{2} \Rightarrow I = \frac{\pi}{4}$

Example 40. What is the value of the definite integral

$$\int_0^a \frac{\sqrt{x}}{\sqrt{x} + \sqrt{a-x}} dx \, ?$$

(A) 0 (B) $a/2$ (C) a (D) $2a$ [CE 2011]

Answer (B)

Let $f(x) = \dfrac{\sqrt{x}}{\sqrt{x} + \sqrt{a-x}}$, then

$$f(a-x) = \frac{\sqrt{a-x}}{\sqrt{a-x} + \sqrt{x}} \qquad I = \int_0^a \frac{\sqrt{x}}{\sqrt{x} + \sqrt{a-x}} dx \; ----(i)$$

204

$$I = \int_0^a \frac{\sqrt{a-x}}{\sqrt{a-x}+\sqrt{x}}dx - - - -(ii) \quad \left[As \int_0^a f(x)dx = \int_0^a f(a-x)dx \right]$$

Adding (i) & (ii), $2I = \int_0^a \frac{\sqrt{x}+\sqrt{a-x}}{\sqrt{x}+\sqrt{a-x}}dx = \int_0^a dx = a \Rightarrow I = \frac{a}{2}$

Example 41. The value of $\int_0^\pi x\sin^2 x dx$ is

(A) $\dfrac{\pi^2}{8}$ (B) $\dfrac{\pi^2}{2}$ (C) $\dfrac{\pi^2}{4}$ (D) π^2

Answer: (C)

$$I = \int_0^\pi x\sin^2 x dx \text{ --------- (i)}$$

$$I = \int_0^\pi (\pi-x)\sin^2(\pi-x)dx = \int_0^\pi (\pi-x)\sin^2 x dx \quad [As \ \sin(\pi-x) = \sin x\ ]$$

$$= \pi\int_0^\pi \sin^2 x dx - \int_0^\pi x\sin^2 x dx = \pi\int_0^\pi \sin^2 x dx - I \Rightarrow 2I = \pi\int_0^\pi \sin^2 x$$

$$\int_0^\pi x\sin^2 x dx = \pi\int_0^\pi \sin^2 x dx - I \Rightarrow 2I = \pi\int_0^\pi \sin^2 x dx$$

$$= \pi\frac{1}{2}\int_0^\pi (1-\cos 2x)dx \Rightarrow 2I = \frac{\pi}{2}\left[x - \frac{\sin 2x}{2} \right]_0^\pi = \frac{\pi}{2}.\pi = \frac{\pi^2}{2} \ or, I = \frac{\pi^2}{4}$$

Example 42. What is the value of the definite integral

$$\int_{\pi/6}^{\pi/3} \frac{1}{1+\sqrt{\tan x}}dx ?$$

(A)   0   (B) $2\pi/3$ (C) $\pi/2$ (D) $\pi/12$

Answer (D)

$$\text{Let } I = \int_{\pi/6}^{\pi/3} \frac{1}{1+\sqrt{\tan x}} dx = \int_{\pi/6}^{\pi/3} \frac{\sqrt{\cos x}}{\sqrt{\cos x}+\sqrt{\sin x}} dx ---(i)$$

$$I = \int_{\pi/6}^{\pi/3} \frac{\sqrt{\sin x}}{\sqrt{\sin x}+\sqrt{\cos x}} dx ---(ii) \text{ [By prop. 3(viii)as } a+b=\pi/2]$$

$$(i)+(ii), \ 2I = \int_{\pi/6}^{\pi/3} dx = \frac{\pi}{3}-\frac{\pi}{6} = \frac{\pi}{6} \Rightarrow I = \frac{\pi}{12}$$

Example 43. If $\int_{a}^{b} \frac{x^n}{x^n+(16-x)^n} dx = 6, \ a,b =$

(A) 12, 4 (B) 2, 14 (C) 4, 12 (D) 14, 2

Answer (B)

Choosing $a+b=16 ---(i)$

$$I = \int_{a}^{b} \frac{x^n}{x^n+(a+b-x)^n} dx \Rightarrow I = \int_{a}^{b} \frac{(a+b-x)^n}{(a+b-x)^n+x^n} dx$$

[By prop. 3(viii)as $a+b=16$]

$$2I = \int_{a}^{b} dx \Rightarrow I = \frac{b-a}{2} = 6 \ (given) \Rightarrow b-a = 12 ----(ii)$$

Solving, $(i) \ \& \ (ii), \ a = 2, b = 14$

Example 44. The value of $\int_{0}^{1} \cot^{-1}(1-x+x^2)dx$ is

(A) $2\int_{0}^{1} \cot^{-1} x dx$ (B) $\int_{0}^{1} \cot^{-1} x dx$ (C) $\int_{0}^{1} \tan^{-1} x dx$ (D) $2\int_{0}^{1} \tan^{-1} x dx$

Answer: (D)

$$I = \int_{0}^{1} \cot^{-1}(1-x+x^2)dx = \int_{0}^{1} \tan^{-1}\left(\frac{1}{1-x+x^2}\right)dx$$

$$= \int_{0}^{1} \tan^{-1}\left(\frac{x+(1-x)}{1-x(1-x)}\right)dx = \int_{0}^{1} (\tan^{-1} x)dx + \int_{0}^{1} \tan^{-1}(1-x)dx$$

$$= \int_0^1 (\tan^{-1} x)dx + \int_0^1 \tan^{-1}[1-(1-x)]dx = \int_0^1 (\tan^{-1} x)dx + \int_0^1 (\tan^{-1} x)dx$$

$$= 2\int_0^1 (\tan^{-1} x)dx \qquad [\text{As } \int_0^a f(x)dx = \int_0^a f(a-x)dx \ \& \ a = 1]$$

Example 45. The value of $\displaystyle\int_a^b \frac{dx}{\sqrt{(x-a)(b-x)}}, 0 < a < b$ is

(A) $\dfrac{\pi}{8}$ (B) $\dfrac{\pi}{2}$ (C) $\dfrac{\pi}{4}$ (D) π

Answer: (D)

$x = a\cos^2 t + b\sin^2 t \Rightarrow dx = 2(b-a)\sin t \cos t\, dt$

$x - a = a\cos^2 t + b\sin^2 t - a = (b-a)\sin^2 t : x = a \Rightarrow t = 0$

$b - x = b - (a\cos^2 t + b\sin^2 t) = (b-a)\cos^2 t : x = b \Rightarrow t = \pi/2$

$$I = \int_0^{\pi/2} \frac{2(b-a)\sin t \cos t\, dt}{\sqrt{(b-a)\sin^2 t(b-a)\cos^2 t}} = 2\int_0^{\pi/2} dt = 2\left[\frac{\pi}{2} - 0\right] = \pi$$

Example 46. The integral $\displaystyle\frac{1}{2\pi}\int_0^{2\pi} \sin(t-\tau)\cos\tau\, d\tau$ equals

(A) $\sin t \cos t$ (B) 0   (C) $\dfrac{1}{2}\cos t$ (D) $\dfrac{1}{2}\sin t$ [EE 2007]

Answer (D)

$$\int_0^{2\pi} \sin(t-\tau)\cos\tau\, d\tau = \int_0^{2\pi} \left(\sin t \cos^2 \tau - \cos t \sin\tau \cos\tau\right) d\tau$$

$$= \frac{1}{2}\left\{\sin t \int_0^{2\pi} (1+\cos 2\tau)d\tau - \cos t \int_0^{2\pi} \sin 2\tau\, d\tau\right\}$$

$$= \frac{1}{2}\left\{\sin t \left[\tau + \frac{\sin 2\tau}{2}\right]_0^{2\pi} - \cos t \left[-\frac{\cos 2\tau}{2}\right]_0^{2\pi}\right\}$$

$$= \frac{1}{2}\left\{ \sin t.(2\pi - 0) + \frac{\cos t}{2}.(1-1) \right\} = \pi \sin t$$

$$\frac{1}{2\pi} \int_0^{2\pi} \sin(t-\tau)\cos \tau \, d\tau = \frac{1}{2\pi}(\pi \sin t) = \frac{1}{2}\sin t$$

Example 47. The value of $\displaystyle\int_2^3 \frac{dx}{(x-1)\sqrt{x^2-2x}}$ is

(A) $\pi/8$ (B) $\pi/4$ (C) $\pi/3$ (D) $\pi/2$

Answer:(C)

$$I = \int_2^3 \frac{dx}{(x-1)\sqrt{x^2-2x}} = \int_2^3 \frac{dx}{(x-1)\sqrt{(x-1)^2-1}}$$

$$x - 1 = z, dx = dz : x = 2 \Rightarrow z = 1 \ \& \ x = 3 \Rightarrow z = 2$$

$$I = \int_1^2 \frac{dz}{z\sqrt{z^2-1}} = \left[\sec^{-1} z\right]_1^2 = \sec^{-1} 2 - \sec^{-1} 1 = \frac{\pi}{3} - 0 = \frac{\pi}{3}$$

Example 48. Consider the following definite integral $\displaystyle\int_0^1 \frac{\left(\sin^{-1} x\right)^2}{\sqrt{1-x^2}} dx.$

The value of the integral is

(A) $\dfrac{\pi^3}{24}$ (B) $\dfrac{\pi^3}{12}$ (C) $\dfrac{\pi^3}{48}$ (D) $\dfrac{\pi^3}{64}$ [CE 2017]

Answer: (A)

$$\sin^{-1} x = z, \frac{1}{\sqrt{1-x^2}} dx = dz, x = 0 \Rightarrow z = 0 \ \& \ x = 1 \Rightarrow z = \frac{\pi}{2}$$

$$\int_0^1 \frac{\left(\sin^{-1} x\right)^2}{\sqrt{1-x^2}} dx. = \int_0^{\pi/2} z^2 dz = \frac{z^3}{3}\Bigg]_0^{\pi/2} = \frac{\pi^3}{24}$$

Example 49. The solution of $\displaystyle\int_0^{\pi/6} \cos^4 3\theta \sin^3 6\theta \, d\theta$ is

(A) 0 (B) $\dfrac{1}{15}$ (C) 1 (D) $\dfrac{8}{3}$ [CE /BT 2013]

208

Answer (B)

Let $3\theta = z \Rightarrow 3d\theta = dz, \theta = 0 \Rightarrow z = 0$ & $\theta = \pi/6 \Rightarrow z = \pi/2$

$$I = \frac{1}{3}\int_0^{\pi/2}\cos^4 z \sin^3 2z\, dz = \frac{1}{3}\int_0^{\pi/2}\cos^4 z\,(2\sin z\cos z)^3\, dz$$

$$= \frac{8}{3}\int_0^{\pi/2}\cos^7 z \sin^3 z\, dz = \frac{8}{3}\frac{\Gamma 4.\Gamma 2}{2\Gamma 6} \quad \text{[Result 6 with p =3, q = 6]}$$

$$= \frac{4}{3}\frac{3!\times 1}{5!} = \frac{4}{3}.\frac{6}{120} = \frac{1}{15}$$

Example 50. The value of $\displaystyle\int_0^\infty \frac{1}{1+x^2}\,dx + \int_0^\infty \frac{\sin x}{x}\,dx$ is

(A) $\dfrac{\pi}{2}$ (B) π (C) $\dfrac{3\pi}{2}$ (D) 1 [CE 2016]

Answer (B)

$$\int_0^\infty \frac{1}{1+x^2}\,dx + \int_0^\infty \frac{\sin x}{x}\,dx = \tan^{-1}x]_0^\infty + \frac{\pi}{2} = \frac{\pi}{2} + \frac{\pi}{2} = \pi$$

Example 51. The value of the integral

$$\int_{-\infty}^\infty 12\cos(2\pi t)\frac{\sin(4\pi t)}{4\pi t}\,dt \text{ is -\_\_\_\_.} \qquad\qquad \text{[EC 2015]}$$

Answer: 3

$$f(t) = 12\cos(2\pi t)\frac{\sin(4\pi t)}{4\pi t} \Rightarrow f(-t) = 12\cos(-2\pi t)\frac{-\sin(4\pi t)}{-4\pi t}$$

$$12\cos(2\pi t)\frac{\sin(4\pi t)}{4\pi t} = f(t)\text{ So, } f(t)\text{ is an even function of}$$

$$t.\ I = \int_{-\infty}^\infty 12\cos(2\pi t)\frac{\sin(4\pi t)}{4\pi t}\,dt = 2.12\int_0^\infty \cos(2\pi t)\frac{\sin(4\pi t)}{4\pi t}\,dt$$

$$\left[2\pi t = z \Rightarrow dt = \frac{dz}{2\pi} \right]$$

$$I = 24\int_0^\infty \cos z.\frac{\sin 2z}{2z}\frac{dz}{2\pi} = \frac{3}{\pi}\int_0^\infty \frac{2\sin 2z \cos z}{z}dz = \frac{3}{\pi}\int_0^\infty \frac{\sin 3z + \sin z}{z}dz$$

$$= \frac{3}{\pi}\left[\int_0^\infty \frac{\sin 3z}{3z}d(3z) + \int_0^\infty \frac{\sin z}{z}dz \right] = \frac{3}{\pi}\left(\frac{\pi}{2} + \frac{\pi}{2} \right) = \frac{3}{\pi}.\pi = 3$$

Example 52. The value of $\displaystyle\int_0^\infty \frac{dx}{(x+1)(x+2)}$ is

(A)   0   (B) $\log 2$ (C) $\log\dfrac{1}{2}$ (D)   2

Answer (B)

$$\int_0^\infty \frac{dx}{(x+1)(x+2)} = \int_0^\infty \left(\frac{1}{x+1} - \frac{1}{x+2} \right)dx = \left[\log\frac{x+1}{x+2} \right]_0^\infty$$

$$= \lim_{x\to\infty}\log\left[\frac{1+1/x}{1+2/x} \right] - \left[\log\frac{x+1}{x+2} \right]_{x=0} = \log 1 - \log\frac{1}{2} = 0 - \log\frac{1}{2} = \log 2$$

Example 53. The solution of $\displaystyle\int_0^1 x^3\left(1 - x^2\right)^{5/2} dx$ is

(A) $\dfrac{2}{63}$ (B) $\dfrac{2}{15}$ (C) $\dfrac{3}{56}$ (D) $\dfrac{1}{32}$

Answer (A)

$$x = \sin\theta,\ dx = \cos\theta\, d\theta\ \ x = 0 \Rightarrow \theta = 0\ \&\ x = 1 \Rightarrow \theta = \frac{\pi}{2}$$

$$I = \int_0^{\pi/2} \sin^3\theta\cos^5\theta.\cos\theta\, d\theta = \int_0^{\pi/2} \sin^3\theta\cos^6\theta.d\theta$$

$$= \frac{\Gamma\left(\frac{3+1}{2}\right).\Gamma\left(\frac{6+1}{2}\right)}{2.\Gamma\left(\frac{3+6+2}{2}\right)}$$ [Result 6 with p =3, q = 6]

$$= \frac{\Gamma(2).\Gamma\left(\frac{7}{2}\right)}{2.\Gamma\left(\frac{11}{2}\right)} = \frac{1.\Gamma\left(\frac{7}{2}\right)}{2.\Gamma\left(\frac{9}{2}+1\right)} = \frac{1.\Gamma\left(\frac{7}{2}\right)}{2.\frac{9}{2}.\frac{7}{2}\Gamma\left(\frac{7}{2}\right)} = \frac{2}{63}$$

Example 54. Consider the function $g(t) = e^{-t}\sin(2\pi t)u(t)$ where $u(t)$ is the unit step function. The area under $g(t)$ is _____.[EC 2015]
Answer: 0.14 − 0.16

$$u(t) = \begin{cases} 1, t \geq 0 \\ 0, t < 0 \end{cases}$$

Area= $\displaystyle\int_{-\infty}^{\infty} g(t)dt = \int_{-\infty}^{0} e^{-t}\sin(2\pi t)u(t)dt + \int_{0}^{\infty} e^{-t}\sin(2\pi t)u(t)dt$

$$= 0 + \int_{0}^{\infty} e^{-t}\sin(2\pi t)dt = \left[\frac{e^{-t}}{1+4\pi^2}\{-\sin(2\pi t) - 2\pi\cos(2\pi t\}\right]_{0}^{\infty}$$

$$= 0 - \left[\frac{1}{1+4\pi^2}(0-2\pi)\right] = \frac{2\pi}{1+4\pi^2} = \frac{6.28}{40.44} = 0.155$$

Example 55. The value of $\displaystyle\int_{0}^{3}\int_{0}^{x}(6-x-y)dxdy$ is

(A) 13.5   (B) 27.0   (C) 40.5   (D) 54.0   [CS/IT 2008]
Answer (A)

$$I = \int_{x=0}^{3}\int_{y=0}^{x}[(6-x-y)dy]dx = \int_{0}^{3}\left[6y-xy-\frac{y^2}{2}\right]_{y=0}^{x}dx$$

$$= \int_{0}^{3}\left(6x-3\frac{x^2}{2}\right)dx = \left[3x^2-\frac{x^3}{2}\right]_{0}^{3} = 13.5$$

Example 56. The value of the integral $\displaystyle\int_{0}^{2}\int_{0}^{x} e^{x+y}dy\,dx$ is

(A) $\dfrac{1}{2}(e-1)$ (B) $\dfrac{1}{2}\left(e^2-1\right)^2$ (C) $\dfrac{1}{2}\left(e^2-e\right)$ (D) $\dfrac{1}{2}\left(e-\dfrac{1}{e}\right)^2$ [ME 2014]

Answer: (B)

$$\int_0^2\int_0^x e^{x+y}\,dy\,dx = \int_0^2\left[e^{x+y}\right]_{y=0}^x dx = \int_0^2\left(e^{2x}-e^x\right)dx = \left[\dfrac{e^{2x}}{2}-e^x\right]_0^2$$

$$=\left[\left(\dfrac{e^4}{2}-e^2\right)-\left(\dfrac{1}{2}-1\right)\right]=\dfrac{e^4}{2}-e^2+\dfrac{1}{2}=\dfrac{1}{2}\left(e^4-2e^2+1\right)=\dfrac{1}{2}\left(e^2-1\right)^2$$

Example 57. The values of the integrals $\displaystyle\int_0^1\left(\int_0^1\dfrac{x-y}{(x+y)^3}\,dy\right)dx$ and

$\displaystyle\int_0^1\left(\int_0^1\dfrac{x-y}{(x+y)^3}\,dx\right)dy$ are

(A) same and equal to 0.5   (B) same and equal to - 0.5
(C) 0.5 and $-$ 0.5, respectively   (D) - 0.5 and 0.5, respectively   [EC 2017]
Answer: (C)

Now, $\displaystyle\int_0^1\dfrac{x-y}{(x+y)^3}\,dy = \int_0^1\dfrac{2x-(x+y)}{(x+y)^3}\,dy = \int_0^1\left[\dfrac{2x}{(x+y)^3}-\dfrac{1}{(x+y)^2}\right]dy$

$$=\left[-\dfrac{2x}{2(x+y)^2}+\dfrac{1}{x+y}\right]_{y=0}^1 = -\dfrac{x}{(x+1)^2}+\dfrac{1}{x+1}+\dfrac{1}{x}-\dfrac{1}{x} = \dfrac{1}{(x+1)^2}$$

$$\int_0^1\left(\int_0^1\dfrac{x-y}{(x+y)^3}\,dy\right)dx = \int_0^1\dfrac{dx}{(x+1)^2} = -\left[\dfrac{1}{x+1}\right]_0^1 = 1-\dfrac{1}{2} = 0.5$$

As the integrand is discontinuous at (0, 0), the two integrals are unequal.

Example 58. Let $I = \displaystyle\int_C\left(x^2dx + xy\,dy\right)$ where x, y are real and C be the

line segment from point A: (1, 0) to point B: (0, 1). The value of I is __.
Answer: - 1/6

Equ. of $AB : \dfrac{x-1}{1-0} = \dfrac{y-0}{0-1} \Rightarrow y = -x+1, dy = -dx, x$ varies from 1 to 0.

$$I = \int_{x=1}^0\left[x^2dx + x(-x+1)(-dx)\right] = \int_1^0\left(x^2+x^2-x\right)dx = \left[\dfrac{2x^3}{3}-\dfrac{x^2}{2}\right]_1^0 = -\dfrac{1}{6}$$

Example 59. Let $I = \int_C (2zdx + 2ydy + 2xdz)$ where x, y, z are real

and C be the straight line segment from point A: (0, 2, 1) to point

B: (4, 1, -1). The value of I is _____. [EC 2017]

Answer: - 11

Equation of $AB : \dfrac{x-0}{4-0} = \dfrac{y-2}{1-2} = \dfrac{z-1}{-1-1} = t \Rightarrow x = 4t, y = 2-t$

$z = 1-2t \Rightarrow dx = 4dt, dy = -dt, dz = -2dt$ At $A, t = 0$ & at

$B, t = 1$

$I = \int_0^1 [8(1-2t) - 2(2-t) - 4.4t]dt = \int_0^1 (4-30t)dt = \left[4t - 15t^2\right]_0^1 = -11$

Example 60. To evaluate the double integral $\int_0^8 \left(\int_{y/2}^{(y/2)+1} \dfrac{2x-y}{2} dx \right) dy$, we

make the substitution $u = \dfrac{2x-y}{2}$ & $v = \dfrac{y}{2}$.. The integral will reduce to

(A) $\int_0^4 \left(\int_0^2 2udu \right) dv$

(B) $\int_0^4 \left(\int_0^1 2udu \right) dv$

(C) $\int_0^4 \left(\int_0^1 udu \right) dv$

(D) $\int_0^4 \left(\int_0^2 udu \right) dv$ [EE 2014]

Answer: (B)

$x = \dfrac{y}{2} \Rightarrow u = 0$ & $x = \dfrac{y}{2} + 1 \Rightarrow u = 1$ $y = 0 \Rightarrow v = 0$ & $y = 8 \Rightarrow v = 4$

Also, $u + v = x$ & $2v = y$

$J = \begin{vmatrix} \partial x / \partial u & \partial y / \partial u \\ \partial x / \partial v & \partial y / \partial v \end{vmatrix} = \begin{vmatrix} 1 & 0 \\ 1 & 2 \end{vmatrix} = 2$

$\int_0^8 \left(\int_{y/2}^{(y/2)+1} \dfrac{2x-y}{2} dx \right) dy = \int_{v=0}^4 \left(\int_{u=0}^1 u| J | du \right) dv = \int_0^4 \left(\int_0^1 2udu \right) dv$

Example 61. The area enclosed between the straight line $y = x$ and

the parabola $y = x^2$ in the x-y plane is (in sq. units)

213

(A)   1/6   (B)   1/4   (C)   1/3   (D) 1/2 [ME 2012]

Answer (A)

Solving the two equations, the points of intersection of the two curves are (0, 0) and (1, 1).

$$\text{Area} = \left| \int_0^1 y_1 \, dx - \int_0^1 y_2 \, dx \right| = \left| \int_0^1 x \, dx - \int_0^1 x^2 \, dx \right| = \frac{1}{2} - \frac{1}{3} = \frac{1}{6} \text{ sq. units}$$

Example 62. The area enclosed between the straight line $y = x$ and the parabola $y^2 = 4x$ in the x-y plane is (in sq. units)

(A)   1/3   (B)   2/3   (C)   4/3   (D) 8/3

Answer (D)

Solving the two equations, the points of intersection of the two curves are (0, 0) and (4, 4).

$$\text{Area} = \left| \int_0^4 y_1 \, dx - \int_0^4 y_2 \, dx \right| = \left| \int_0^4 x \, dx - 2\int_0^4 \sqrt{x} \, dx \right| = \left| \left[\frac{x^2}{2} - 2.\frac{2}{3} x^{3/2} \right]_0^4 \right|$$

$$= \left| 8 - \frac{4}{3}.8 \right| = \frac{8}{3}$$

Example 63. The area enclosed between the curves $y^2 = 4x$ & $x^2 = 4y$ is

(A)   16/3   (B)   8   (C)   32/3   (D)   16   [ME 2009]

Answer (A)

Solving the two equations, the points of intersection of the two curves are (0, 0) and (4, 4). Area

$$= \left| \int_0^4 y_1 \, dx - \int_0^4 y_2 \, dx \right| = \left| \int_0^4 2x^{1/2} \, dx - \frac{1}{4}\int_0^4 x^2 \, dx \right|$$

$$= \left| 2.\frac{2}{3} x^{3/2} - \frac{1}{4}\frac{x^3}{3} \right|_0^4 = \frac{4}{3} \times 8 - \frac{1}{12} \times 64 = \frac{32}{3} - \frac{16}{3} = \frac{16}{3} \text{ sq. units}$$

Example 64. The area of the region bounded by the parabola $y = x^2 + 1$ and the straight line $x + y = 3$ is

(A)   59/6   (B)   9/2   (C)   10/3   (D)   7/6 [CE 2016]

Answer (B)

214

Solving the equations $x = -2, 1$

$$\text{Area} = \left| \int_{-2}^{1} (x^2 + 1)dx - \int_{-2}^{1} (3 - x)dx \right| = \left| \left[\frac{x^3}{3} + x \right]_{-2}^{1} + \left[\frac{x^2}{2} - 3x \right]_{-2}^{1} \right|$$

$$= \left| \left(\frac{1}{3} + 1 \right) - \left(-\frac{8}{3} - 2 \right) + \left(\frac{1}{2} - 3 \right) - (2 + 6) \right| = \left| \frac{4}{3} + \frac{8}{3} + 2 - \frac{5}{2} - 8 \right| = \frac{9}{2}$$

Example 65. The length of the curve $y = \frac{2}{3} x^{3/2}$ between $x = 0 \,\&\, x = 1$ is

(A)   0.27   (B)   0.67   (C)   1   (D) 1.22   [ME 2008]

Answer (D)

$$y = \frac{2}{3} x^{3/2} \Rightarrow \frac{dy}{dx} = x^{1/2}, \quad \frac{ds}{dx} = \sqrt{1 + \left(\frac{dy}{dx} \right)^2} = \sqrt{1 + x}$$

$$ds = (1 + x)^{1/2} dx \Rightarrow \int_0^s ds = \int_0^1 (1 + x)^{1/2} dx = \frac{2}{3} \left[(1 + x)^{3/2} \right]_0^1$$

$$= \frac{2}{3} \left(2\sqrt{2} - 1 \right) = 1.22$$

Example 66. The length of the curve $y = \log \sec x$ between $x = 0$

and $x = \frac{\pi}{3}$ is

(A) $\log(2 - \sqrt{3})$ (B) $\log(2 + \sqrt{3})$ (C) $\log(\sqrt{3} - 1)$ (D) $\log(\sqrt{3} + 1)$

Answer (B)

$$y = \log \sec x \Rightarrow \frac{dy}{dx} = \frac{1}{\sec x}.(\sec x \tan x) = \tan x$$

$$\frac{ds}{dx} = \sqrt{1 + \left(\frac{dy}{dx} \right)^2} = \sqrt{1 + \tan^2 x} = \sec x \Rightarrow ds = \sec x dx$$

Integrating, $s = \int_0^{\pi/3} \sec x dx = \left[\log(\sec x + \tan x) \right]_0^{\pi/3} = \log(2 + \sqrt{3})$

EXERCISE

Question of one mark:

1. $\int_{-1}^{1} \{x + |x|\} dx =$

(A)   0   (B)   1   (C)   2   (D)   1/2

2. $\int_{0}^{\pi/2} \dfrac{\sin x - \cos x}{1 + \sin x \cos x} dx =$

(A)   0   (B)   1   (C)   2   (D)   1/2

3. $\int_{0}^{\pi} \dfrac{x \tan x}{\sec x + \tan x} dx =$

(A) π(π-2)/8   (B)   π(π+2)/8   (C) π(π-3)/16   (D) π (π-2)/2

4. $\int_{0}^{a} \dfrac{dx}{x + \sqrt{a^2 - x^2}} =$

(A) $\dfrac{\pi}{8}$ (B) $\dfrac{\pi}{2}$ (C) $\dfrac{\pi}{4}$ (D) π

5. $\int_{0}^{\pi} x \sin x \cos^4 x \, dx =$

(A) $\dfrac{\pi}{5}$ (B) $\dfrac{\pi}{6}$ (C) $\dfrac{\pi}{4}$ (D) $\dfrac{\pi}{3}$

Question of two marks:

6. $\int_{0}^{1} x^2 (1 - x^2)^{3/2} dx =$

(A)   8/225   (B)   16/315   (C) 4/125   (D)   2/75

7. $\int\limits_{0}^{\pi/2} \sin 2x \log \tan x\, dx =$

(A) 0 (B) 1 (C) 2 (D) 1/2

8. $\int\limits_{0}^{\pi/2} (a^2 \cos^2 x + b^2 \sin^2 x)\, dx =$

(A) $(a^2 + b^2)\dfrac{\pi}{8}$ (B) $(a^2 + b^2)\dfrac{\pi}{6}$ (C) $(a^2 + b^2)\dfrac{\pi}{4}$ (D) $(a^2 + b^2)\dfrac{\pi}{2}$

9. $\int\limits_{0}^{\pi/2} \dfrac{x}{\sin x + \cos x}\, dx =$

(A) $\dfrac{\pi}{\sqrt{2}}\log(\sqrt{2}+1)$ (B) $\dfrac{\pi}{2\sqrt{2}}\log(\sqrt{2}+1)$ (C) $\pi\log(\sqrt{2}+1)$ (D) $\dfrac{\pi}{2}\log(\sqrt{2}+1)$

10. The value of $\int\limits_{C} \left[(xy + y^2)\, dx + x^2 dy \right]$ (where C is the closed curve of the region bounded by $y = x$ & $y = x^2$) is _____.

11. A three dimensional R of finite volume is described by $x^2 + y^2 \le z^3, 0 \le z \le 1$, where x, y, z are real. The volume of R (up to two decimal places) is_____. [EC 2017]

Answers:

1. (B) 2. (A) 3. (D) 4. (C) 5. (A) 6. (B) 7. (A) 8. (C) 9. (B) 10. -1/20
11. 0.70 - 0.85

Formulae:

i) $P(A)$ Probability of *occurrence* of the event A and

$P(\overline{A})$ or $P(A')$ or $P(A^c)$ = Probability of *non- occurrence* of the event A.

ii) $0 \le P(A) \le 1$ and $P(A) + P(\overline{A}) = 1$

iii) For a certain event A, $P(A) = 1$ and for an impossible event

$A, P(A) = 0$

iv) *Classical or Mathematical formula of Probability*;

$P(A) = \dfrac{m}{n}$, m = number of outcomes favourable to the event A and

n = total number of mutually exclusive, exhaustive and equally likely outcomes of the experiment.

v) If odds in favour of A be $\dfrac{u}{v}$ then $P(A) = \dfrac{u}{u+v}$, $P(A^c) = \dfrac{v}{u+v}$

If odds against A be $\dfrac{u}{v}$ then $P(A) = \dfrac{v}{u+v}$, $P(A^c) = \dfrac{u}{u+v}$

If $P(A) = \dfrac{m}{n}$, then odds in favour of $A = \dfrac{m}{n-m}$ and

against $A = \dfrac{n-m}{m}$

vi) For any two events A and $B, P(A \cup B)$ or $P(A + B)$ = Probability of occurrence of at least one of the events A **and** B

and $P(A \cap B)$ = Probability of joint occurrence of the events A **and** B.

vii)*Theorem of Total Probability:* For any two events A **and** B,

$P(A \cup B) = P(A) + P(B) - P(A \cap B)$ \hfill For

two mutually exclusive events A **and** B, $P(A \cup B) = P(A) + P(B)$

viii) If the n events $A_1, A_2, - - -, A_n$ are mutually exclusive, exhaustive

and equally likely events, then $P(A_1) = P(A_2) = - - - = P(A_n) = \dfrac{1}{n}$

ix) *Conditional Probability:* Probability of occurrence of the event B given that the event A has actually occurred = $P(B/A)$

x) *Theorem of Compound Probability:* $P(B/A) = \dfrac{P(AB)}{P(A)}$, $P(A) \neq 0$ and

$P(A/B) = \dfrac{P(AB)}{P(B)}$, $P(B) \neq 0$

xi) *Independent Events:* If A and B are independent events,
$P(B/A) = P(B) \Rightarrow P(A \cap B) = P(A).P(B)$
If A and B are independent events, then
a) $\overline{A}$ and $\overline{B}$ b) A and $\overline{B}$ c) $\overline{A}$ and B are also *independent*.

xii) *Baye's theorem:*

$$P(B_i / A) = \frac{P(B_i)P(A/B_i)}{P(B_1)P(A/B_1) + ... + P(B_n)P(A/B_n)}, \quad i = 1,2,---,n$$

WORKED OUT EXAMPLES:

Question of one mark:

Example 1. If A and B are any two events such that
$P(A \cap B) = \dfrac{1}{2}, P(\overline{A} \cap \overline{B}) = \dfrac{1}{3}, P(A) = P(B) = p$, then $p =$

(A) $\dfrac{5}{12}$ (B) $\dfrac{7}{12}$ (C) $\dfrac{5}{6}$ (D) $\dfrac{1}{6}$

Answer: (B)

$P(\overline{A} \cap \overline{B}) = \dfrac{1}{3} \Rightarrow P(A \cup B)' = \dfrac{1}{3} \Rightarrow 1 - P(A \cup B) = \dfrac{1}{3} \Rightarrow P(A \cup B) = \dfrac{2}{3}$

Or, $P(A) + P(B) - P(A \cap B) = \dfrac{2}{3} \Rightarrow 2p = \dfrac{2}{3} + \dfrac{1}{2} = \dfrac{7}{6} \Rightarrow p = \dfrac{7}{12}$

Example 2. If A and B be any two events and
$P(A \cup B) = \dfrac{1}{2}, P(A) = \dfrac{2}{5}, P(B) = \dfrac{1}{4}$, then the value of $P(\overline{A} \cup \overline{B})$ is

(A) 17/20 (B) 3/20 (C) 7/10 (D) 3/10

Answer: (A)

$P(A \cap B) = P(A) + P(B) - P(A + B) = \dfrac{2}{5} + \dfrac{1}{4} - \dfrac{1}{2} = \dfrac{3}{20}$,

Now, $P(\overline{A} \cup \overline{B}) = P(A \cap B)'$ [By DE-Morgan's Rule]

$P(\overline{A} \cup \overline{B}) = 1 - P(A \cap B) = 1 - \dfrac{3}{20} = \dfrac{17}{20}$

Example 3. If A and B be any two events and

$P(A \cup B) = \dfrac{1}{2}, P(A) = \dfrac{1}{4}, P(B) = \dfrac{2}{5},$ then the events are

(A)   equally likely   (B)   mutually exclusive
(C)   independent   (D)   none of these

Answer: (D)

As $P(A) \neq P(B),$ so the events are not equally.

As $P(A \cap B) \neq 0$ the events are not mutually exclusive.

As $P(A \cap B) \neq P(A).P(B)$

So, the events are not independent, and the correct option is (D).

Example 4. Suppose A and B are two independent events with probabilities $P(A) \neq 0$ and $P(B) \neq 0.$ Let $\overline{A}$ & $\overline{B}$ be their complements. Which one of the following statements is FALSE?

(A) $P(A \cap B) = P(A).P(B)$ (B) $P(A/B) = P(A)$
(C) $P(A \cup B) = P(A) + P(B)$ (D) $P(\overline{A} \cap \overline{B}) = P(\overline{A}).P(\overline{B})$

Answer (C)

Option(C) should be $P(A \cup B) = P(A) + P(B) - P(A \cap B)$

Example 5. The probability that at least one of the events A and B occurs is 0.6. If A and B occur simultaneously with probability 0.2, then $P(\overline{A}) + P(\overline{B}) =$

(A)   0.4   (B)   0.8   (C)   1.2   (D)   1.4

Answer: (C)

Given, $P(A \cup B) = 0.6, P(A \cap B) = 0.2$

$P(A \cup B) = P(A) + P(B) - P(A \cap B) \Rightarrow 0.6 = \{1 - P(\overline{A})\} + \{1 - P(\overline{B})\} - 0.2$

or, $P(\overline{A}) + P(\overline{B}) = 1 + 1 - 0.2 - 0.6 = 1.2$

Example 6. Let P (E) denote the probability of the event E. Given P(A) = 1, $P(B) = \dfrac{1}{2}$, the values of P(A|B) and P(B|A) respectively are

(A) $\dfrac{1}{4}, \dfrac{1}{2}$ (B) $\dfrac{1}{2}, \dfrac{1}{4}$ (C) $\dfrac{1}{2}, 1$ (D) $1, \dfrac{1}{2}$ [CS/IT 2003]

Answer (D)

As $P(A) = 1$ so A is the sample space. So, $A \cap B = B$

$$P(A/B) = \frac{P(A \cap B)}{P(B)} = \frac{P(B)}{P(B)} = 1 \;\&\; P(B/A) = \frac{P(A \cap B)}{P(A)} = \frac{P(B)}{P(A)} = \frac{1}{2}$$

Example 7. If A and B are independent events and $P(A) = \dfrac{2}{3}, P(B) = \dfrac{3}{5}$, then the value of $P(\overline{A}/B)$ is

(A)   1/2   (B)   1/3   (C)   1/4   (D)   1/5

Answer: (B)

As A and B are independent events, P(A∩B) = P(A).P(B) = 2/5

By the law of Compound Probability, $P(\overline{A}/B) = 1 - P(A/B)$

$$= 1 - \frac{P(A \cap B)}{P(B)} = 1 - \frac{2/5}{3/5} = \frac{1}{3}$$

Example 8. If $P(X) = \dfrac{1}{4}, P(Y) = \dfrac{1}{3}, P(X \cap Y) = \dfrac{1}{12}$, the value of $P(Y/X)$ is

(A) $\dfrac{1}{4}$ (B) $\dfrac{4}{25}$ (C) $\dfrac{1}{3}$ (D) $\dfrac{29}{50}$ [ME 2015]

Answer: (C)

$$P(Y/X) = \frac{P(X \cap Y)}{P(X)} = \frac{1/12}{1/4} = \frac{1}{3}$$

Example 9. 1 hour rainfall of 10 cm has return period of 50 year. The probability that 1 hour of rainfall 10 cm or more will occur in each of two successive years is

(A)   0.04   (B)   0.2   (C)   0.2   (D) 0.0004   [CE 2013]

Answer: (D)

Probability of occurrence of 1 hour of rainfall of 10 cm $= \dfrac{1}{50} = .02$

Probability of occurrence of 1 hour of rainfall of 10 cm in two successive years $= (0.02)^2 = 0.004$

Example 10. An unbiased coin is tossed five times. The outcome of each toss is either a head or a tail. The probability of getting at least one head is

(A) $\dfrac{1}{32}$ (B) $\dfrac{13}{32}$ (C) $\dfrac{16}{32}$ (D) $\dfrac{31}{32}$ [ME 2011]

Answer: (D)

P(one Tail) $= \dfrac{1}{2}$ P(five Tails) $= \left(\dfrac{1}{2}\right)^5 = \dfrac{1}{32}$

P(at least one head) $= 1 - $ P (no heads) $= 1 -$ P (5 tails) $= 1 - \dfrac{1}{32} = \dfrac{31}{32}$

Example 11. Two coins are tossed simultaneously. The probability (up to two decimal points accuracy) of getting at least one head is _____.   [ME 2017]

Answer: 0.75

The sample space is {H H, H T, TH, TT}. $n = 4, m = 3$

Probability $= \dfrac{m}{n} = \dfrac{3}{4} = 0.75$

Example 12. The probability of getting a "head" in a single toss of a biased coin is 0.3. The coin is tossed repeatedly till a "head" is obtained. If the tosses are independent, then the probability of getting "head" for the first time in the fifth toss is ___. [EC 2016]

Answer: 0.072

Probability of getting a "tail" = 1 − 0.3 = 0.7, Here the results of 4 tosses will be "tail" and the result 5th toss will be "head".

Required probability $= (0.7)^4 \times 0.3 = 0.2401 \times 0.3 = 0.072$

Example 13. A fair coin is tossed till a head appears for the first time. The probability that the number of required tosses is odd, is

(A) $\dfrac{1}{3}$ (B) $\dfrac{1}{2}$ (C) $\dfrac{2}{3}$ (D) $\dfrac{3}{4}$ [EC 2012]

Answer: (C)

H will appear first time in 1st or 3rd or 5th ,----tosses.

The sample space: {H, TTH, TTTTH,----}

Probability $= = \dfrac{1}{2} + \dfrac{1}{8} + \dfrac{1}{32} = ---- = \dfrac{1/2}{1-1/4} = \dfrac{2}{3}$

Example 14. If two fair coins are flipped and at least one of the outcomes is known to be a head, what is the probability that both outcomes are heads?

(A) $\dfrac{1}{3}$ (B) $\dfrac{1}{4}$ (C) $\dfrac{1}{2}$ (D) $\dfrac{2}{3}$ [CS/IT 2011]

Answer: (A)

Sample space = {HH, HT, TH} Total no. of sample points = $n = 3$
No. of sample points favorable to the event = $m = 1$ (Only HH)

Required probability $= \dfrac{m}{n} = \dfrac{1}{3}$

Example 15. A fair coin is tossed n times. What is the probability that the difference between the number of heads and tails is $(n-3)$?

(A) $\dfrac{1}{3}$ (B) $\dfrac{1}{4}$ (C) $\dfrac{1}{2}$ (D) 0

Answer: (D)

If E be the required event and x & y be the no. of heads and tails, then

$x + y = n,\ x - y = n - 3$ Solving, $x = \dfrac{1}{2}(2n-3),\ y = \dfrac{3}{2}$

But x & y should be positive integers $\Rightarrow$ E is an impossible event.
So, P(E) = 0

Example 16. Consider the following experiment:
Step 1. Flip a fair coin twice.
Step 2. If the outcomes are (Tails, Heads), then output Y and stop.
Step 3. If the outcomes are either (Heads, Heads) or (Heads, Tails), then

output N and stop.

Step 4. If the outcomes are (Tails, Tails) then go to Step 1.

The probability that the output of the experiment is Y is (up to two decimal places) _____ [CS/IT 2017]

Answer: 0.33 (2m)

Sample space = {(H, H), (H, T), (T, H), (T, T)} where H→Head, T→Tail.

Let Z be the event of getting (T, T). $P(Y) = \dfrac{1}{4}$, $P(N) = \dfrac{1}{2}$, $P(Z) = \dfrac{1}{4}$,

Required event = Y OR (Z and Y) OR ----

Required Probability = P(Y) + P(Z).P(Y) + --

$$= \frac{1}{4} + \frac{1}{4} \cdot \frac{1}{4} + ---- = \frac{1}{4}\left[\frac{1}{1-1/4}\right] = \frac{1}{4} \cdot \frac{4}{3} = \frac{1}{3} = 0.33$$

Example 17. If two events A & B are independent and $\overline{A}$ & $\overline{B}$ are respectively complements of A & B respectively, then which one is NOT true?

(A) $\overline{A}$ & $\overline{B}$ are independent (B) $\overline{A}$ & B are independent

(C) A & $\overline{B}$ are independent (D) None of these

Answer (D)

Example 18. A box contains 10 screws 3 of which are defective. Two screws are drawn at random with replacement. The probability that none of the screws is defective will be

(A)   100% (B) 50% (C) 49% (D) None of these [CS 2003]

Answer: (C)

Probability of drawing one non-defective screw out of 10 screws $\dfrac{7}{10}$.

Probability of drawing two non-defective screws one by one with

replacement $= \left(\dfrac{7}{10} \times \dfrac{7}{10} \times 100\right)\% = 49\%$

Example 19. A two faced fair coin has its faces designated as head (H) and tail (T). This coin is tossed three times in succession to record the following outcomes: H, H, H. If the coin is tossed once more time, the probability (up to one decimal place) of obtaining H again, given the previous realizations of H, H and H, would be _____. [CE 2017]

Answer: 0.5

$$P(3H) = \frac{1}{2} \times \frac{1}{2} \times \frac{1}{2} = \frac{1}{8}$$

Required prob. $= P(H/3H) = \dfrac{P(H \cap 3H)}{P(3H)} = \dfrac{P(H).P(3H)}{P(3H)} = P(H) = 0.5$

Example 20. Two newspapers X and Y are published in a certain city. It is estimated from a survey that 16% read X, 14% read Y and 5% both. The probability that a randomly selected person read only Y is

(A)   0.06   (B)   0.07   (C)   0.08   (D)   0.09

Answer: (D)

Let A and B be the events of reading the newspapers X & Y respectively.

$P(A) = 0.16$ $P(B) = 0.14$ $P(AB) = 0.05$

The probability that a randomly selected person read only Y

$$= P(\overline{A}B) = P(B) - P(AB) = 0.14 - 0.05 = 0.09 \quad [(\overline{A}B) \cup (AB) = B]$$

Example 21. Three vendors were asked to supply a very high precision component. The respective probabilities of their meeting the strict design specifications are 0.8, 0.7 and 0.5. Each vendor supplies one component. The probability that out of total three components supplied by the vendors, at least one will meet the design specification is____[ME 2015]

Answer: 0.97

Probability of no component meeting the specification

= (1-0.8)(1-0.7)(1-0.5)= 0.2 ×0.3 × 0.5 = 0.03

Probability of at least one component meeting the specification

=1-.03=0.97

Example 22. A box contains 2 washers, 3 nuts and 4 bolts. Items are drawn from the box at random one at a time without replacement. The probability of drawing 2 washers first followed by 3 nuts and subsequently the 4 bolts is

(A) $\dfrac{2}{315}$ (B) $\dfrac{1}{630}$ (C) $\dfrac{1}{1260}$ (D)$\dfrac{1}{252}$ [ME 2010]

Answer: (C)

Total number of items = 2 + 3 + 4 = 9

Required probability = $\left(\dfrac{2}{9}\cdot\dfrac{1}{8}\right)\times\left(\dfrac{3}{7}\cdot\dfrac{2}{6}\cdot\dfrac{1}{5}\right)\times\left(\dfrac{4}{4}\cdot\dfrac{3}{3}\cdot\dfrac{2}{2}\cdot\dfrac{1}{1}\right)=\dfrac{1}{1260}$

Example 23. What is the chance that a leap year, selected at random, will contain 53 Sundays?

(A) $\dfrac{2}{7}$ (B) $\dfrac{3}{7}$ (C) $\dfrac{1}{7}$ (D) $\dfrac{5}{7}$ [EE 2014]

Answer: (A)

A leap year has 366 days-52 complete weeks and 2 more days.

52 complete weeks have 52 Sundays. For 53 Sundays, out of the rest 2 days, one should be a Sunday.

Out of 7 pairs (Saturday, Sunday), (Sunday, Monday), (Monday, Tuesday) etc. the 2 pairs (Saturday, Sunday), (Sunday, Monday) are favourable.

$\therefore$ Probability of 53 Sundays $=\dfrac{2}{7}$.

Example 24. The probability that a k digit number does NOT contain the digits 0, 5 or 9 is

(A) 0.3^k (B) 0.6^k (C) 0.7^k (D) 0.9^k [EE 2017]

Answer: (C)

The probability that 0, 5, 9 being not in a number $1-\dfrac{3}{10}=0.7$

For the k digit number, probability = 0.7^k

Example 25. A product is an assembly of 5 different components. The product can be sequentially assembled in two possible ways. If the 5 components are placed in a box and these are drawn at random from the box, then the probability of getting a correct box is

(A) $\dfrac{2}{5!}$ (B) $\dfrac{2}{5}$ (C) $\dfrac{2}{(5-2)!}$ (D) $\dfrac{2}{(5-3)!}$ [PI 2015]

Answer: (A)

Probability $=\dfrac{^2C_1}{5!}=\dfrac{2}{5!}$

Example 26. A determinant is chosen at random from the set of all determinants of order 2 with elements 0 or 1 only. The probability that the value of the determinant chosen being positive is

(A) $\dfrac{1}{8}$ (B) $\dfrac{3}{16}$ (C) $\dfrac{5}{16}$ (D) $\dfrac{7}{16}$

Answer: (B)

Since each of the 4 places of the 2nd order determinant can be filled by either 0 or 1, $n = 2^4 = 16$

The favourable cases are $\begin{vmatrix} 1 & 0 \\ 0 & 1 \end{vmatrix}, \begin{vmatrix} 1 & 0 \\ 1 & 1 \end{vmatrix}, \begin{vmatrix} 1 & 1 \\ 0 & 1 \end{vmatrix}$ as only they are positive.

$m = 3$ Probability $= \dfrac{m}{n} = \dfrac{3}{16}$

Example 27. A fair dice is rolled twice. The probability that an odd number will follow an even number is

(A) $\dfrac{1}{2}$ (B) $\dfrac{1}{6}$ (C) $\dfrac{1}{3}$ (D) $\dfrac{1}{4}$ [EC 2005]

Answer (D)

When a fair dice is rolled twice, the total number of

outcomes $= n = 6^2 = 36$

The favourable outcomes:

{(2, 1), (2, 3), (2, 5), (4, 1), (4, 3), (4, 5), (6,1), (6, 3), (6, 5)}

Number of favourable outcomes $= m = 9$

Probability $= \dfrac{m}{n} = \dfrac{9}{36} = \dfrac{1}{4}$

Example 28. Three fair cubical dice are thrown simultaneously. The probability that all three dice have the same number of dots in the faces showing up is (up to third decimal places) _____.   [EC 2017]

Answer: 0.0278

Total number of cases $= n = 6^3$ Number of favourable cases $= m = 6$

Probability $= \dfrac{m}{n} = \dfrac{6}{6^3} = \dfrac{1}{36} = 0.0278$

Example 29. An urn contains 5 red balls and 5 black balls. In the first draw, one ball is picked at random and discarded without noticing its colour. The probability of drawing a red ball in the second draw is

(A) $\dfrac{1}{2}$ (B) $\dfrac{4}{9}$ (C) $\dfrac{5}{9}$ (D) $\dfrac{6}{9}$ [EE 2017]

Answer: (A)

Events are {1 R from 5 R & 5 B balls and 1 R from 4 R & 5 B balls}

OR {1 B from 5 R & 5 B balls and 1 R from 5 R & 4 B balls}

Prob. $= \dfrac{5}{10} \times \dfrac{4}{9} + \dfrac{5}{10} \times \dfrac{5}{9} = \dfrac{1}{2}\left(\dfrac{4}{9} + \dfrac{5}{9}\right) = \dfrac{1}{2}$

Example 30. Two coins R & S are tossed. The 4 joint events $H_R H_s, T_R T_S, H_R T_s, H_S T_R$ have probabilities 0.28, 0.18, 0.30, 0.24, respectively, where H represents head and T represents tail. Which one of the following is TRUE?

(A) The coin tosses are independent

(B) R is fair, S it not.

(C) S is fair, R it not.

(D) The coin tosses are dependent   [EE 2015]

Answer (D)

If coin tosses are independent corresponding probabilities will be

$\dfrac{1}{2}\cdot\dfrac{1}{2}, \dfrac{1}{2}\cdot\dfrac{1}{2}, \dfrac{1}{2}\cdot\dfrac{1}{2}, \dfrac{1}{2}\cdot\dfrac{1}{2} = \dfrac{1}{4}, \dfrac{1}{4}, \dfrac{1}{4}, \dfrac{1}{4}$

But the given probabilities are 0.28, 0.18, 0.3, 0.24 respectively and we cannot decide whether R is fair or S is fair.

So, the coin tosses are dependent.

Example 31. Two urns contain respectively 2 white, 1 black balls and 1 white 5 black balls. One ball is transferred from the 1st to the 2nd urn and then a ball is drawn from the 2nd urn. The probability that the ball drawn is white:

(A) $\dfrac{16}{21}$ (B) $\dfrac{5}{21}$ (C) $\dfrac{3}{7}$ (D) $\dfrac{4}{7}$

Answer (B)

Probability = P (1 white ball from urn I and 1white ball from urn II) +

P (1 black ball from urn I and 1white ball from urn II) $= \dfrac{2}{3} \times \dfrac{2}{7} + \dfrac{1}{3} \times \dfrac{1}{7} = \dfrac{5}{21}$

Example 32. Three identical dice are rolled. The probability that the same number will appear on each of them is

(A) $\dfrac{1}{6}$ (B) $\dfrac{1}{36}$ (C) $\dfrac{1}{18}$ (D) $\dfrac{3}{28}$

Answer (B)

$n = 6^3 = 216$

Favourable cases are

{(1, 1,1), (2, 2, 2), (3, 3, 3), $, 4, 4), (5, 5, 5), (6, 6, 6)}

Number of favourable cases $= m = 6$ Probability $= \dfrac{m}{n} = \dfrac{6}{36} = \dfrac{1}{6}$

Example 33. A deck of 5 cards (each carrying a distinct number from 1 to 5) is shuffled thoroughly. Two cards are then removed one at a time from the deck. What is the probability that the two cards are selected with the number on the first card being one higher than the number on the second card?

(A) $\dfrac{1}{5}$ (B) $\dfrac{4}{25}$ (C) $\dfrac{1}{4}$ (D) $\dfrac{2}{5}$ [CS/IT 2011]

Answer: (A)

From 5 cards (number 1-5), 2 cards may be drawn one by one in
5 × 4 = 20 ways   or, n = 20
The favorable numbers in the card will be
(2, 1), (3, 2), (4, 3), (5, 4)   or, m = 4

Required probability $= \dfrac{m}{n} = \dfrac{4}{20} = \dfrac{1}{5}$

Example 34. An examination consists of two papers, Paper 1 and Paper 2. The probability of failing in Paper 1 is 0.3 and that in Paper 2 is 0.2. Given that a student has failed in Paper 2, the probability of failing in Paper 1 is 0.6. The probability of a student failing in both the papers is:

(A) 0.5   (B) 0.18   (C) 0.12   (D) 0.06   [EC 2007]

Answer: (C)

Let A and B be the events denoting failing in Paper 1 and failing in Paper 2

$P(A) = 0.3, P(B) = 0.2, P(A/B) = 0.6$

Probability of failing in both the papers
$$= P(A \cap B) = P(B).P(A/B) = 0.2 \times 0.6 = 0.12$$

Example 35. An anti-aircraft gun can take a maximum four shots at an enemy plane moving away from it. The probabilities of hitting the plane at first, second, third and fourth shot are $\dfrac{1}{2}, \dfrac{1}{3}, \dfrac{1}{4}$ & $\dfrac{1}{5}$.

The probability that the gun hits the plane is

(A) $\dfrac{1}{4}$ (B) $\dfrac{1}{5}$ (C) $\dfrac{4}{5}$ (D) $\dfrac{2}{3}$

Answer: (C)

The gun cannot hit the plane if none of the shots fail to hit and the

probability $\left(1 - \dfrac{1}{2}\right)\left(1 - \dfrac{1}{3}\right)\left(1 - \dfrac{1}{4}\right)\left(1 - \dfrac{1}{5}\right) = \dfrac{1}{5}$

The probability that the gun hits the plane $= 1 - \dfrac{1}{5} = \dfrac{4}{5}$

Example 36. A box contains 25 parts of which 10 are defective. Two parts are drawn in a random manner from the box. The probability of both the parts being good is

(A) $\dfrac{7}{20}$ (B) $\dfrac{42}{125}$ (C) $\dfrac{25}{29}$ (D) $\dfrac{5}{9}$ [ME 2014]

Answer: (A)

$$n = {}^{25}C_2 = \dfrac{25 \times 24}{2}, \quad m = {}^{15}C_2 = \dfrac{15 \times 14}{2}, \quad Prob. = \dfrac{15 \times 14}{25 \times 24} = \dfrac{7}{20}$$

Example 37. The probability that a screw manufactured by a company is defective is 0.1. The company sells screws in packets containing 5 screws and gives a guarantee of replacement if one or more screws in the packet are found to be defective. The probability that a packet would have to be replaced is _____

Answer: 0.4095

The probability that no screw in the packet is

defective $= (0.9)^5 = 0.5905$

Probability of at least one defective screw in the packet
= 1 - 0.5905 = 0.4095   Required probability = 0.4095

Example **38.** In a certain class 25% of the students failed in Mathematics, 15% failed in Chemistry and 10% in both. A student is selected at random. The probability that the student has passed in Mathematics if he failed in Chemistry

(A) $\dfrac{3}{5}$ (B) $\dfrac{3}{8}$ (C) $\dfrac{1}{3}$ (D) $\dfrac{1}{4}$

Answer: (C)

Let A and B denote the events –' the student failed in Mathematics' and' the student failed in Chemistry' respectively. Then,

$$P(A) = \frac{25}{100}, P(B) = \frac{15}{100} \ \& \ P(A \cap B) = \frac{10}{100}$$

Probability = $P(\overline{A}/B) = 1 - P(A/B) = 1 - \dfrac{P(A \cap B)}{P(B)} = 1 - \dfrac{10}{15} = \dfrac{1}{3}$

Example **39.** A person moving through a tuberculosis prone zone has a 50% probability of becoming infected. However, only 30% of infected people develop the disease. What percentage of people moving through a tuberculosis prone zone remains infected but does not show symptoms of disease?

(A)   15   (B)   33   (C)   35   (D)   37   [ME 2016]

Answer: (C)

Let A be the event of being infected and B be the event of developing the disease. $P(A) = 0.5, P(B/A) = 0.3 \Rightarrow P(AB) = P(A).P(B/A) = 0.15$

Required probability

$= P(A\overline{B}) = P(A) - P(AB) = 0.5 - 0.15 = 0.35 = 35\%$

Example **40.** A person draws a card from a pack of playing cards, replaces it and shuffles the pack. He continues doing this until he shows a spade. The chance that he will fail the first two times is

(A) $\dfrac{9}{64}$ (B) $\dfrac{1}{64}$ (C) $\dfrac{1}{16}$ (D) $\dfrac{9}{16}$

Answer: (A)

He will not get sped in the first two chances but will get sped in the third chance. Out of 52 cards, 39 cards are non-sped and 13 are sped.

Probability $= \dfrac{39}{52} \times \dfrac{39}{52} \times \dfrac{13}{52} = \dfrac{3}{4} \times \dfrac{3}{4} \times \dfrac{1}{4} = \dfrac{9}{64}$

Example 41. A coin is tossed thrice. Let X be the event that head occurs in each of the first two tosses. Let Y be the event that a tail occurs in the third toss. Let Z be the event that two tails occur in the three tosses. Based on the above information, which one of the above statements is TRUE?

(A) X and Y are not independent (B) Y and Z are dependent
(C) Y and Z are independent (D) X and Z are independent   [ME 2015]

Answer: (B)

For 3 tosses of a coin, the total number of outcomes $= 2^3 = 8$

$$X : HHT, HHH \Rightarrow P(X) = \frac{2}{8} = \frac{1}{4}, \; Y : HHT, TTT, HTT, THT \Rightarrow P(Y) = \frac{4}{8} = \frac{1}{2}$$

$$Z : TTH, THT, HTT \Rightarrow P(Z) = \frac{3}{8}, \; X \cap Y = HHT \Rightarrow P(X \cap Y) = \frac{1}{8}$$

$$Y \cap Z = HTT, THT \Rightarrow P(Y \cap Z) = \frac{2}{8} = \frac{1}{4}, \; X \cap Z = \phi \Rightarrow P(X \cap Z) = 0$$

$P(X \cap Y) = P(X).P(Y)$ Option (A) is wrong.

$P(Y \cap Z) \neq P(Y).P(Z)$ Option (B) is TRUE.

Example 42. There are 25 calculators in a box. Two of them are defective. Suppose 5 calculators are picked for inspection (i,e, each has the same chance of being selected), what is the probability that only one of the defective calculators will be included in inspection?

(A) $\dfrac{1}{2}$ (B) $\dfrac{1}{3}$ (C) $\dfrac{1}{4}$ (D) $\dfrac{1}{5}$ [CS/IT 2006]

Answer (B)

$$n = {}^{25}C_5 = \frac{25!}{20!5!}, \; m = {}^{23}C_4 \times {}^2C_1 = \frac{23!}{19!4!} \times 2$$

Probability $= 2 \times \dfrac{23!}{19!4!} \times \dfrac{20!5!}{25!} = 2 \times \dfrac{20 \times 5}{24 \times 25} = \dfrac{1}{3}$

Example 43. Two players, A and B, alternately keep rolling a fair dice. The person to get a six first wins the game. Given that player A starts the game, the probability that A wins the game is

(A) $\dfrac{5}{11}$ (B) $\dfrac{1}{2}$ (C) $\dfrac{7}{13}$ (D) $\dfrac{6}{11}$ [EE 2015]

Answer: (D)

Prob. $= P\left[(A) \cup (\overline{A}\,\overline{B}A) \cup --\right] = P(A) + P(\overline{A})P(\overline{B}).P(A) + --$

$$= \dfrac{1}{6} + \left(\dfrac{5}{6}\right)^2 \times \dfrac{1}{6} + -- = \dfrac{1}{6}\left[1 + \left(\dfrac{5}{6}\right)^2 + ---\right] = \dfrac{1}{6} \times \dfrac{1}{1 - 25/36} = \dfrac{6}{11}$$

Example 44. A fair dice is tossed two times. The probability that the second toss results in a value that is higher than the first toss is

(A) $\dfrac{2}{36}$ (B) $\dfrac{2}{6}$ (C) $\dfrac{5}{12}$ (D) $\dfrac{1}{2}$ [EC 2011]

Answer: (C)

Total number of outcomes $= n = 6^2 = 36 = 36$

Favourable outcomes:

{(1, 2), (1, 3), (1, 4), (1, 5), (1, 6)

(2, 3), (2, 4), (2, 5), (2,6)

(3, 4), (3, 5), (3, 6)

(4, 5), (4, 6)

(5, 6)}

Total number of favorable causes $= m = 5+4+3+2+1=15$

Then probability $= \dfrac{15}{36} = \dfrac{5}{12}$

Example 45. A pair of dice is thrown at random. If the two numbers appearing be different, then the probability that the sum is 6:

(A) $\dfrac{2}{15}$ (B) $\dfrac{3}{5}$ (C) $\dfrac{5}{12}$ (D) $\dfrac{7}{36}$

Answer: (A)

In a single throw of two dice, the number of mutually exclusive, equally likely and exhaustive outcomes $= 6^2 = 36$

The number of outcomes in which the numbers appearing are different $=$ $36 - 6 = 30$ [Exclude (1, 1), (2, 2), (3, 3), (4, 4), (5, 5), (6, 6)] $\Rightarrow n = 30$

Let A be the event that 'the two numbers are different and their sum is 6'

A = {(1, 5), (5, 1), (2, 4), (4, 2)} $\Rightarrow m = 4$. So, $P(A) = \dfrac{m}{n} = \dfrac{4}{30} = \dfrac{2}{15}$

Example 46. Four fair six- sided dice are rolled. The probability that the sum being 22 is $x/1296$. The value of x is ____.

(A)   7   (B)   8   (C)   9   (D)   10 [CS/IT 2014]

Answer: (D)

In throwing 4 dice, the total number of outcomes $= n = 6^4 = 1296$

The sum of numbers appeared will be 22 and it can happen in the following two mutually exclusive ways:

(i) 3 will show 6 and 1 will show 4 i.e. (6, 6, 6, 4). No. Of ways $= \dfrac{4!}{3!} = 4$

(ii) 2 will show 6 and 2 will show 5 i.e. (6, 6, 5, 5). No. Of ways $= \dfrac{4!}{2!2!} = 6$

Total no. Of favourable cases $= m = 6 + 4 = 10$

Probability $= \dfrac{m}{n} = \dfrac{10}{1296} \Rightarrow x = 10$

Example 47. In rolling of two fair dice, the outcome of an experiment is considered as the sum of the numbers appearing on the dice. The probability is highest for the outcome of ____.   [CH 2014]

Answer: 7

In throwing 2 dice, the total number of outcomes $= n = 6^2 = 36$

The outcomes are $\begin{Bmatrix} (6,1),(6,2),(6,3),(6,4),(6,5),(6,6) \\ (5,1),----------(5,6) \\ (4,1),----------(4,6) \\ (3,1),----------(3,6) \\ (2,1),----------(2,6) \\ (1,1),----------(1,6) \end{Bmatrix}$

The sum of numbers appeared will be 7 for the following 6 outcomes {(6, 1), (1, 6), (5, 2), (2, 5), (4, 3), (3, 4)} and in each case the probability of occurrence is 1/6 that is the total probability is 6/36.

The sum of numbers appeared will be2, 3, 4, 5, 6, 8, 9, 10, 11, 12 for lesser number of outcomes. In each case the probability will be < 6/36.

So, the probability is highest for the outcome of 7.

Example 48. Let S be a sample space and two mutually exclusive events A and B be such that $A \cup B = S$. If $P()$ denotes the probability of the event, then the maximum value of $P(A).P(B)$ is _[EC 2014, CS/IT 2014]

Answer 0.25

$A \cup B = S \Rightarrow P(A \cup B) = P(S) = 1 \Rightarrow P(A) + P(B) = 1$

If $P(A) = x, P(B) = 1 - x$ Let $y = P(A).P(B) = x(1-x) = x - x^2$

$\Rightarrow \dfrac{dy}{dx} = 1 - 2x$ So, $\dfrac{dy}{dx} = 0 \Rightarrow x = \dfrac{1}{2}$ and $\dfrac{d^2y}{dx^2} = -2 < 0.$ y is maximum

at $x = \dfrac{1}{2}.$ $[P(A)P(B)]_{max} = y_{max} = \dfrac{1}{2}\left(1 - \dfrac{1}{2}\right) = \dfrac{1}{4} = 0.25$

Example 49. The chance of a student passing an exam is 20%. The chance of a student passing the exam and getting above 90% marks in it is 5%. Given that a student passes the examination, the probability that the student gets above 90% marks is

(A) $\dfrac{1}{18}$ (B) $\dfrac{1}{4}$ (C) $\dfrac{2}{9}$ (D) $\dfrac{5}{18}$ [ME 2015]

Answer: (B)

Let A be the event the student passes and B be the event that the student gets above 90% marks $P(A) = \dfrac{20}{100} = \dfrac{1}{5}$ & $P(A \cap B) = \dfrac{5}{100} = \dfrac{1}{20}$

It is given that the student passes, the probability that the student has obtained more than 90% marks $P(B/A) = \dfrac{P(A \cap B)}{P(A)} = \dfrac{1/20}{1/5} = \dfrac{1}{4}$

Example 50. The probability that a student knows the correct answer to a multiple choice question is 2/3. If the student does not know the answer, then the student guesses the answer. The probability of the guessed answer being correct is 1/4. Given that the student has answered the question correctly, the conditional probability that the student knows the correct answer is

(A) $\dfrac{2}{3}$ (B) $\dfrac{3}{4}$ (C) $\dfrac{5}{6}$ (D) $\dfrac{8}{9}$ [ME 2013]

Answer: (D)

Let A be the event that the student answers the question correctly, B_1 be the event that the student knows the correct answer and B_2 be the event that the student guesses the correct

answer. $P(B_1) = \dfrac{2}{3}, P(B_2) = \dfrac{1}{3}, \quad P(A/B_1) =$ Probability of giving the

correct answer when he knows the correct answer = 1

Probability of giving the correct answer when he guesses the correct answer = 1/4

$$P(A) = P(B_1).P(A/B_1) + P(B_2).P(A/B_2) = \frac{2}{3} \times 1 + \frac{1}{3} \times \frac{1}{4} = \frac{3}{4}$$

$$P(A \cap B_1) = P(B_1).P(A/B_1) = \frac{2}{3} \times 1 = \frac{2}{3}$$

Required probability $= P(B_1/A) = \dfrac{P(A \cap B_1)}{P(A)} = \dfrac{2}{3} \times \dfrac{4}{3} = \dfrac{8}{9}$

Example 51. Two people, P and Q, decide to independently roll two identical dice each with six faces, numbered 1 to 6. The person with the lower number wins. In case of a tie, they roll the dice repeatedly until there is no tie. Define a trial as the throw of the dice by P and Q. Assume that all six numbers on each dice is equi-probable and that all trials are independent. The probability (rounded to three decimal places) that one of them wins in the third trial is _____.   [CS/IT 2018]

Answer: 0.023

P(tie in any trial) = P(P = 1 & Q = 1) + P(P = 2 & Q = 2) + --+P(P = 6 & Q =

$$6) = \frac{1}{6} \times \frac{1}{6} + \frac{1}{6} \times \frac{1}{6} + ---- + \frac{1}{6} \times \frac{1}{6} = \frac{6}{36} = \frac{1}{6}$$

P(one of them wins) $= 1 - \dfrac{1}{6} = \dfrac{5}{6}$

P(one of them wins in the third trial) =P(tie in the 1st trial) × P(tie in the

2nd trial) × P(one of them wins in the 3rd trial) $= \dfrac{1}{6} \times \dfrac{1}{6} \times \dfrac{5}{6} = \dfrac{5}{216} = 0.023$

Example 52. Two people, P and Q, decide to independently roll two identical dice each with six faces, numbered 1 to 6. The person with the lower number wins. In case of a tie, they roll the dice repeatedly until there is no tie. Define a trial as the throw of the dice by P and Q. Assume that all six numbers on each dice is equi-probable and that all trials are independent. The probability (rounded to three decimal places) that one of them wins in the third trial is _____. [CS/IT 2018]

Answer: 0.023

P(tie in any trial) = P(P = 1 & Q = 1) + P(P = 2 & Q = 2) + -------

$$+ \text{ P(P = 6 \& Q = 6)} = \frac{1}{6} \times \frac{1}{6} + \frac{1}{6} \times \frac{1}{6} + ---- + \frac{1}{6} \times \frac{1}{6} = \frac{6}{36} = \frac{1}{6}$$

$$P(\text{one of them wins}) = 1 - \frac{1}{6} = \frac{5}{6}$$

P(one of them wins in the third trial) =P(tie in the 1st trial) × P(tie in the

2nd trial) × P(one of them wins in the 3rd trial) $= \frac{1}{6} \times \frac{1}{6} \times \frac{5}{6} = \frac{5}{216} = 0.023$

Example 53. P and Q are considering to apply for a job. The probability

that P applies for the job is $\frac{1}{4}$, the probability that P applies for the job

given that Q applies for the job is $\frac{1}{2}$, and the probability that Q applies

for the job given that P applies for the job is $\frac{1}{3}$. Then the probability that

P does not apply for the job given that Q does not apply for the job is

(A) $\frac{4}{5}$ (B) $\frac{5}{6}$ (C) $\frac{7}{8}$ (D) $\frac{11}{12}$ [CS/IT 2017]

Answer: (A)

$$p(P) = \frac{1}{4}, \quad p(P/Q) = \frac{1}{2}, \quad p(Q/P) = \frac{1}{3} \text{ and}$$

$$p(P \cap Q) = p(Q/P).P(P) = \frac{1}{12}$$

Again $p(P \cap Q) = p(P/Q).P(Q) \Rightarrow \frac{1}{12} = \frac{1}{2}.P(Q) \Rightarrow P(Q) = \frac{1}{6}$ and

$$p(P \cup Q) = p(P) + p(Q) - p(P \cap Q) = \frac{1}{4} + \frac{1}{6} - \frac{1}{12} = \frac{1}{3}$$

Required probability $= p(\overline{P}/\overline{Q}) = \dfrac{p(\overline{P} \cap \overline{Q})}{p(\overline{Q})} = \dfrac{p(P \cup Q)^c}{p(\overline{Q})}$

$$= \dfrac{1 - p(P \cup Q)}{1 - p(Q)} = \dfrac{1 - 1/3}{1 - 1/6} = \dfrac{4}{5}$$

***Example* 54.** A group consists of equal number of men and women. Of this group 20% of the men and 50% of the women are unemployed. If a person is selected at random from this group, the probability of the selected person being employed is _ [ME 2014]

Answer: 0.65

Let M and W be the events of being 'man' and 'woman'. U be the event of being 'unemployed'. P(M) = P(W) = 0.5 P(U/M) = 0.2, P(U/W) = 0.5
The probability that the selected person will be unemployed
= P(U) = P(M).P(U/M) + P(W).P(U/W) = 0.5(0.2 + 0.5) = 0.35
The probability that the selected person will be employed
= 1 − P(U) = 0.65

***Example* 55.** Out of all the 2-digit integers between 1 and 100, a 2-digit number has to be selected at random. What is the probability that the selected number is not divisible by 7?

(A) $\dfrac{13}{90}$ (B) $\dfrac{12}{90}$ (C) $\dfrac{78}{90}$ (D) $\dfrac{77}{90}$ [CS/IT 2013, ME 2013]

Answer: (D)
Number of 2-digit integers between 1 and 100 = 100 − (9 + 1) = 90
Number of 2-digit integers between 1 and 100 and divisible by 7 is 13

Probability that the selected number is divisible by 7 is $\dfrac{13}{90}$

Probability that the selected number is not divisible by 7 is $1 - \dfrac{13}{90} = \dfrac{77}{90}$

***Example* 56.** An integer is selected at random from the first 100 positive integers. The probability that the integer is divisible by 6 or 8 is

(A) $\dfrac{4}{25}$ (B) $\dfrac{6}{25}$ (C) $\dfrac{2}{25}$ (D) $\dfrac{1}{5}$

Answer: (B)

Among the first 100 positive integers, the number divisible by 6 is 16, the number divisible by 8 is 12 and the number divisible by both 6 and 8 (i.e. by their LCM 24) is 4.

If A be the event that the number is divisible by 6, then $P(A) = \dfrac{16}{100}$

and B be the event that the number is divisible by 8, then $P(B) = \dfrac{12}{100}$

Also, $A \cap B$ is the event that the number is divisible by both 6 & 8,

then $P(A \cap B) = \dfrac{4}{100}$

Required probability $= P(A) + P(B) - P(A \cap B)$

$= \dfrac{16}{100} + \dfrac{12}{100} - \dfrac{4}{100} = \dfrac{6}{25}$

Example 57. A box contains 4 red balls and 6 black balls. Three balls are selected randomly from the box one after another without replacement. The probability that the selected set contains one red ball and two black balls is

(A) $\dfrac{1}{20}$ (B) $\dfrac{1}{12}$ (C) $\dfrac{3}{10}$ (D) $\dfrac{1}{2}$ [ME 2012]

Answer: (D)

Total number of balls = 10

The balls will be in the order (RBB) or (BRB) or (BBR)

Probability of selecting balls in the order of (RBB) $= \dfrac{4}{10} \cdot \dfrac{6}{9} \cdot \dfrac{5}{8} = \dfrac{120}{720} = \dfrac{1}{6}$

Probability of selecting balls in the order of (BRB) $= \dfrac{6}{10} \cdot \dfrac{4}{9} \cdot \dfrac{5}{8} = \dfrac{120}{720} = \dfrac{1}{6}$

Probability of selecting balls in the order of (BBR) $= \dfrac{6}{10} \cdot \dfrac{5}{9} \cdot \dfrac{4}{8} = \dfrac{120}{720} = \dfrac{1}{6}$

Required probability = P(RBB) + P(BRB) + P(BBR) $= \dfrac{1}{6} + \dfrac{1}{6} + \dfrac{1}{6} = \dfrac{1}{2}$

Example 58. An urn contains 5 red and 7 green balls. A ball is drawn at random and its colour is noted. The ball is placed back into the urn along with another ball of the same colour. The probability of getting a red ball in the next draw is

(A) $\dfrac{65}{156}$ (B) $\dfrac{67}{156}$ (C) $\dfrac{79}{156}$ (D) $\dfrac{89}{156}$ [IN 2016]

Answer (A)

Two cases may arise:

i) First red ball is drawn and then after adding one more red ball, a red ball is drawn. Probability $\dfrac{5}{12} \times \dfrac{6}{13} = \dfrac{30}{156}$

ii) First green ball is drawn and then after adding one more green ball, a red ball is drawn. Probability $\dfrac{7}{12} \times \dfrac{5}{13} = \dfrac{35}{156}$

Required probability $= \dfrac{30}{156} + \dfrac{35}{156} = \dfrac{65}{156}$

Example 59. Three identical boxes I, II, III contain respectively 4 white and 3 red balls, 3 white and 7 red balls, 2 white and 3 red balls. A box is chosen at random and a ball is drawn out of it. If the ball is found to be white, the probability that the box II is selected:

(A) $\dfrac{23}{89}$ (B) $\dfrac{23}{79}$ (C) $\dfrac{21}{79}$ (D) $\dfrac{21}{89}$

Answer (D)

Let the events of the selection of the boxes I, II, III are A, B, C.

$P(A) = P(B) = P(C) = \dfrac{1}{3}$. Let the event of drawing a white ball is W.

$P(W/A) = \dfrac{4}{7}$, $P(W/B) = \dfrac{3}{10}$, $P(W/C) = \dfrac{2}{5}$

Required probability =

$P(B/W) = \dfrac{P(B).P(W/B)}{P(A).P(W/A) + P(B).P(W/B) + P(C).P(W/C)} = \dfrac{21}{89}$

Example 60. Three identical urns are termed as a, b and c. The table below lists the number of red, black or white balls the urns contain.

Urn name	Number of balls		
	White	black	red
a	1	2	3
b	2	1	1
c	4	5	3

One of the urns is selected at random and two balls are drawn one after another. These turn out to be white and red balls. The probability that they come from urn b or c is

(A) $\dfrac{55}{118}$ (B) $\dfrac{85}{118}$ (C) $\dfrac{35}{118}$ (D) $\dfrac{75}{118}$

Answer (B)

$P(a) = P(b) = P(c) = \dfrac{1}{3}$

Let D be the event that the two balls are white and red. $P(D/a) = \dfrac{1}{6} \times \dfrac{3}{5} + \dfrac{3}{6} \times \dfrac{1}{5} = \dfrac{1}{5}$

$P(D/b) = \dfrac{2}{4} \times \dfrac{1}{3} + \dfrac{1}{4} \times \dfrac{2}{3} = \dfrac{1}{3}$ $P(D/c) = \dfrac{4}{12} \times \dfrac{3}{11} + \dfrac{3}{12} \times \dfrac{4}{11} = \dfrac{2}{11}$

$P(b/D) = \dfrac{P(b).P(D/b)}{P(a).P(D/a) + P(b).P(D/b) + P(C).P(D/c)} = \dfrac{55}{118}$

Similarly, $P(c/D) = \dfrac{30}{118}$

Required probability $= P(b/D) + P(c/D) = \dfrac{85}{118}$

Example 61. Ten boys among whom A and B are arranged randomly in a ring. The probability that A and B will not be together is

(A) $\dfrac{4}{9}$ (B) $\dfrac{5}{9}$ (C) $\dfrac{2}{3}$ (D) $\dfrac{7}{9}$

Answer: (D)

10 boys may be arranged in a ring in $\dfrac{1}{2}(10-1)! = \dfrac{1}{2}(9!)$ ways.

[See *formula* of *Permutation*]

Keeping A & B together, 10 boys may be arranged in the ring in

$$\frac{1}{2}(9-1)!\times2!=\frac{1}{2}(8\,!)\times2!\,\text{ways.}$$

The probability that A and B will be together is $\dfrac{8!\times2!}{9!}=\dfrac{2}{9}$

The probability that A and B will not be together is $=1-\dfrac{2}{9}=\dfrac{7}{9}$

Example 62. A box contains six pair of shoes. Four shoes are randomly selected. The probability of having exactly one complete pair is

(A) $\dfrac{16}{33}$ (B) $\dfrac{14}{33}$ (C) $\dfrac{13}{33}$ (D) $\dfrac{10}{33}$

Answer: (A)

Four shoes can be selected from 12 shoes in $^{12}C_4$ = 45 × 11 ways. One pair among six pairs are taken and the remaining two must be different. It can be selected in $^6C_1\times^5C_2\times2^2$ = 6 × 10 × 4 ways.

Required probability = $\dfrac{6\times10\times4}{45\times11}=\dfrac{16}{33}$

Example 63. A box contains 10 pairs of shoes. Eight shoes are randomly selected. The probability that there is no complete pair is

(A) $\dfrac{^{10}C_6\times2^6}{^{20}C_8}$ (B) $\dfrac{^{10}C_8\times2^6}{^{20}C_8}$ (C) $\dfrac{^{10}C_6\times2^8}{^{20}C_8}$ (D) $\dfrac{^{10}C_8\times2^8}{^{20}C_8}$

Answer: (D)

8 shoes can be taken from 20 shoes (10 pairs) in $^{20}C_8$ ways. $n=^{20}C_8$.

Here 8 of the 10 pairs are taken such that at most one shoe belongs to the same pair. It can be done in $m = ^{10}C_8\times2^8$ ways.

P (no complete pair) = $\dfrac{m}{n}=\dfrac{^{10}C_8\times2^8}{^{20}C_8}$

Example 64. A fair coin is tossed till a head appears for the first time. The probability that the number of tosses is odd, is

(A) $\dfrac{1}{3}$ (B) $\dfrac{1}{2}$ (C) $\dfrac{2}{3}$ (D) $\dfrac{3}{4}$ [EE 2012]

242

Answer (C)

Probability of getting a head = Probability of getting a tail $= \dfrac{1}{2}$

Event of getting a head in odd number of tosses

= (H) or (TTH) or (TTTTH) or ------

Probability of getting a head in odd number of tosses

= P(H) + P(TTH) + P(TTTTH) + --[As the events are mutually exclusive]

= P(H) + P(T).P(T).P(H) + P(T).P(T).P(T).P(T).P(H) + --

[As the events are independent]

$$= \frac{1}{2} + \frac{1}{2}.\frac{1}{2}.\frac{1}{2} + \frac{1}{2}.\frac{1}{2}.\frac{1}{2}.\frac{1}{2}.\frac{1}{2} + ---- = \frac{1}{2}\left(1 + \frac{1}{4} + \frac{1}{4^2} + ---\right) = \frac{1}{2}.\frac{1}{1-1/4} = \frac{2}{3}$$

Example 65. In an examination, 10 True/False type of questions are given. To pass the examination, an examinee is to give at least 8 correct answers. The probability of his passing the examination is

(A) $\dfrac{5}{128}$ (B) $\dfrac{7}{128}$ (C) $\dfrac{11}{64}$ (D) $\dfrac{9}{64}$

Answer: (B)

Answer to each question may be of 2 types: T/F.

For each answer of the first question there may be 2 answers of the

second. So, the answers of the first 2 questions may be given in 2^2 ways.

The answers of the 10 questions may be given in $n = 2^{10} = 1024$ ways.

Let A be the event of 'at least 8 correct answers'.

This may be given in $^{10}C_8 + {}^{10}C_9 + {}^{10}C_{10} = 45 + 10 + 1 = 56$ ways.   The

number of cases favourable to A is $m = 56$

Required probability = $P(A) = \dfrac{m}{n} = \dfrac{56}{1024} = \dfrac{7}{128}$

Example 66. A box contains 10 items of which 2 are defective. 3 items are drawn at random. The probability of drawing at least one defective item is

(A) $\dfrac{8}{15}$ (B) $\dfrac{1}{25}$ (C) $\dfrac{2}{15}$ (D) $\dfrac{4}{15}$

Answer: (A)

243

3 items can be drawn out of 10 items in $^{10}C_3 = 120$ ways. $n = 120$

Let A be the event of 'drawing at least one defective item'.

Number of ways of drawing 1 defective out of 2 and 2 non-defective out of 8 is $^2C_1 \times {^8C_2} = 56$

Number of ways of drawing 2 defective out of 2 and 1 non-defective out of 8 non-defective is $^2C_2 \times {^8C_1} = 8$ So, $m = 56 + 8 = 64$.

Required probability = $P(A) = \dfrac{m}{n} = \dfrac{64}{120} = \dfrac{8}{15}$

Example 67. A box contains 20 defective items and 80 non-defective items. If two items are selected at random without replacement, what will be the probability that both items are defective?

(A) $\dfrac{1}{5}$ (B) $\dfrac{1}{25}$ (C) $\dfrac{20}{99}$ (D) $\dfrac{19}{495}$ [ME 2006]

Answer (D)

Initially there are 100 items of which 20 are defective.

Probability of drawing I defective item $= \dfrac{^{20}C_1}{^{100}C_1} = \dfrac{20}{100} = \dfrac{1}{5}$

Now, 99 items of which 19 are defective.

Probability of drawing I defective item $= \dfrac{^{19}C_1}{^{99}C_1} = \dfrac{19}{99}$

Required probability $= \dfrac{1}{5} \times \dfrac{19}{99} = \dfrac{19}{495}$

Example 68. The probabilities that a particular sum will be solved by A and B are 0.5 and 0.6 respectively. If the events of A solving and B solving the sum are independent, then the probability that the sum will be solved is:

(A)   0.4   (B)   0.3   (C)   0.8   (D)   0.6

Answer (C)

Let A and B denote respectively the events of solving the problem by

244

A & by B.

$P(A) = 0.5$, $P(\overline{A}) = 1 - 0.5 = 0.5$ $P(B) = 0.6$, $P(\overline{B}) = 1 - 0.6 = 0.4$

As A and B are independent events, $\overline{A}$ and $\overline{B}$ are also independent.

So, $P(\overline{A} \cap \overline{B}) = P(\overline{A}).P(\overline{B}) = 0.5 \times 0.4 = 0.2$

The problem will not be solved (i.e. Both A and B will not be able to solve the problem) $= P(\overline{A} \cap \overline{B}) = 0.2$

Probability that the problem will be solved $= 1 - P(\overline{A} \cap \overline{B}) = 1 - 0.2 = 0.8$

Example 69. The probability that a teacher will take a surprise test during any class meeting is $\dfrac{1}{5}$. If a student is absent on two days, the probability that he will miss at least one test is

(A) $\dfrac{2}{25}$ (B) $\dfrac{4}{25}$ (C) $\dfrac{8}{25}$ (D) $\dfrac{9}{25}$

Answer: (D)

Probability of a surprise test on a day $= \dfrac{1}{5}$.

Probability of not a surprise test on a day $= \dfrac{4}{5}$.

The student will miss at least one test if one of three mutually exclusive events occur

i) test on the first day and not a test on the second day, probability

$= \dfrac{1}{5} \times \dfrac{4}{5} = \dfrac{4}{25}$ [As the cases are independent]

ii) test on the second day and not a test on the first day, probability

$= \dfrac{4}{5} \times \dfrac{1}{5} = \dfrac{4}{25}$ [As the cases are independent]

iii) test on both the days, probability $= \dfrac{1}{5} \times \dfrac{1}{5} = \dfrac{1}{25}$ [Same reason]

As the cases are mutually exclusive, probability $= \dfrac{4}{25} + \dfrac{4}{25} + \dfrac{1}{25} = \dfrac{9}{25}$

Example 70. A candidate is selected for interview for three posts. The number of candidates for the first, second and third posts are

3, 4 and 2 respectively. The probability that the candidate will get at least one post is:

(A) $\dfrac{3}{4}$ (B) $\dfrac{2}{3}$ (C) $\dfrac{1}{3}$ (D) $\dfrac{1}{4}$

Answer: (A)

Let $A, B \,\&\, C$ be the events of getting the first, second and third posts respectively.

$$P(A) = \dfrac{1}{3} \Rightarrow P(\overline{A}) = \dfrac{2}{3}, P(B) = \dfrac{1}{4} \Rightarrow P(\overline{B}) = \dfrac{3}{4}, P(C) = \dfrac{1}{2} \Rightarrow P(\overline{C}) = \dfrac{1}{2}$$

Probability of not getting post $= P(\overline{A}).P(\overline{B}).P(\overline{C}) = \dfrac{2}{3} \times \dfrac{3}{4} \times \dfrac{1}{2} = \dfrac{1}{4}$

Probability of getting at least one of the posts

$$= 1 - P(\overline{A}).P(\overline{B}).P(\overline{C}) = 1 - \dfrac{1}{4} = \dfrac{3}{4}$$

***Example* 71.** The integers x and y are chosen at random with replacement from the nine natural numbers 1, 2, 3 ...9. The probability of $|x^2 - y^2|$ being even is

(A) $\dfrac{23}{81}$ (B) $\dfrac{21}{125}$ (C) $\dfrac{41}{81}$ (D) $\dfrac{43}{125}$

Answer: (C)

As x and y are chosen at random from the integers 1, 2, ... ,9 with

replacement, number of possible selections $= n^2 = 9^2 = 81$

Also, $|x^2 - y^2|$ will be even if x and y are both even or both odd.

The possible ways in which x and y are both odd will be $5^2 = 25$

and x and y will be both even $4^2 = 16$

So, the number of favourable cases $= m = 25 + 16 = 41$

Required probability $= \dfrac{m}{n} = \dfrac{41}{81}$

***Example* 72.** Four fair six sided dice are rolled. The probability that the

sum of the results being 22 is $\dfrac{X}{1296}$.

The value of X is ___.   [CS/IT 2014]

Answer: 10

When 4 fair dice are rolled, the number of cases $= n = 6^4 = 1296$

Favourable cases will be

i) Getting 6 at 2 dice and 5 at 2 dice, no. of such cases $= \dfrac{4!}{2!2!} = 6$

ii) Getting 6 at 3 dice and 4 at 1 dice, no. of such cases $= \dfrac{4!}{3!} = 4$ Total no.

of favourable cases $= 6 + 4 = 10 = X$

Example 73. A pair of dice is rolled. What is the probability that the sum is neither 7 nor 11?

(A) $\dfrac{4}{9}$ (B) $\dfrac{5}{9}$ (C) $\dfrac{7}{9}$ (D) $\dfrac{2}{9}$

Answer (C)

Let A be the event that the sum is 7 and B be the event that the sum is 11

In throwing a pair of dice, the sample space contains $6^2 = 36$

sample points i.e. $n = 36$

The sum of points on two dice will be 7, the cases favourable to A are

$\{(6, 1), (1, 6), (5, 2), (2, 5), (3, 4), (4, 3)\}$ i.e. $m = 6$ $P(A) = \dfrac{m}{n} = \dfrac{6}{36} = \dfrac{1}{6}$

$P(A) = \dfrac{m}{n} = \dfrac{6}{36} = \dfrac{1}{6}.$

The sum of points on two dice will be 11, the cases favourable to are

$\{(6, 5), (5, 6)\}$ i.e. $m = 2$ $P(B) = \dfrac{m}{n} = \dfrac{2}{36} = \dfrac{1}{18}.$

As A and B are mutually exclusive events, $P(A \cup B) = P(A) + P(B)$

$= \dfrac{1}{6} + \dfrac{1}{18} = \dfrac{2}{9}$ [Occurrence of A & B simultaneously is not possible].

Probability that the sum is neither 7 nor

$11 = P(A \cup B)' = 1 - P(A \cup B) = 1 - \dfrac{2}{9} = \dfrac{7}{9}$

Example 74. If λdt denotes the probability of a radioactive atom decaying in an infinitely small time dt, the probability that the atom is not decayed in time t will be _____.

Answer: $e^{-\lambda t}$

Since the probability of decaying $= \lambda\,dt$, probability of not decaying

$= 1 - \lambda\,dt$ and $t = \lim\limits_{n\to\infty}(n.dt)$

Required probability $= \lim\limits_{n\to\infty}(1 - \lambda dt)^n = \lim\limits_{n\to\infty}\left(1 - \lambda\dfrac{t}{n}\right)^n = e^{-\lambda t}$

Example 75. An urn contains 5 red and 7 green balls. A ball is drawn at random and its colour is noted. The ball is placed back in the urn with another ball of the same colour. The probability of drawing a red ball in the next draw is

(A) $\dfrac{65}{156}$ (B) $\dfrac{67}{156}$ (C) $\dfrac{79}{156}$ (D) $\dfrac{89}{156}$ [IN 2016]

Answer: (A)

(1 R ball out of 5 R & 7 G balls and then 1 R ball from 6 R & 7 G balls)

OR (1 G ball out of 5 R & 7 G balls and then 1 R ball from 5 R & 8 G balls)

Prob. $= \dfrac{5}{12}\times\dfrac{6}{13} + \dfrac{7}{12}\times\dfrac{5}{13} = \dfrac{65}{156}$

Example 76. Three cards were drawn from a pack of 52 cards. The probability that they are a king, a queen and a jack is

(A) $\dfrac{16}{5525}$ (B) $\dfrac{64}{2197}$ (C) $\dfrac{3}{13}$ (D) $\dfrac{8}{16575}$ [ME 2016]

Answer: (A)

The no. of ways in which three cards may be drawn from a pack of 52

cards is $n = {}^{52}C_3 = \dfrac{52!}{49!\,3!} = \dfrac{52\times51\times50}{6} = 26\times17\times50$

The no. of ways in which one K card from four K cards etc., may be drawn is $m = 4\times4\times4$

Probability $= \dfrac{m}{n} = \dfrac{4\times4\times4}{26\times17\times50} = \dfrac{16}{13\times17\times25} = \dfrac{16}{5525}$

Example 77. A card is drawn from a full pack of playing cards. What is the probability that it is either a spade or an ace?

(A) $\dfrac{2}{13}$ (B) $\dfrac{3}{13}$ (C) $\dfrac{4}{13}$ (D) $\dfrac{5}{13}$

Answer: (C)

One card is drawn out of 52 cards in $^{52}C_1 = 52$ ways. So, the total number of mutually exclusive, equally likely cases for the event $= n = 52$.

Let A be the event that the card drawn is 'either a spade or an ace'.

There are 13 spades and 3 aces (other than that of spade).

Total number of cards = 13 + 3 = 16. 1 card may be drawn out of 16 cards in $^{16}C_1 = 16$ ways.

So, the number of cases favourable to the A is $m = 16$

The required Probability $P(A) = \dfrac{m}{n} = \dfrac{16}{52} = \dfrac{4}{13}$

Example 78. A's skill is to B is as 1:3, to C as 3:2 and to D as 4:3. The chance that A in three trials, one with each person, will succeed twice at least is

(A) $\dfrac{9}{28}$ (B) $\dfrac{11}{28}$ (C) $\dfrac{13}{28}$ (D) $\dfrac{15}{28}$

Answer: (C)

$A = \dfrac{1}{4}, B = \dfrac{3}{4}; \quad A = \dfrac{3}{5}, C = \dfrac{2}{5}; \quad A = \dfrac{4}{7}, D = \dfrac{3}{7};$ The favourable cases are

i) A wins with B, C, D or

ii) A wins with C & D but fails with B or

iii) A wins with B & D but fails with C or

iv) A wins with B & C but fails with D-all the cases are mutually exclusive.

Probability $= \dfrac{1}{4} \cdot \dfrac{3}{5} \cdot \dfrac{4}{7} + \dfrac{3}{5} \cdot \dfrac{4}{7} \cdot \dfrac{3}{4} + \dfrac{1}{4} \cdot \dfrac{4}{7} \cdot \dfrac{2}{5} + \dfrac{1}{4} \cdot \dfrac{3}{5} \cdot \dfrac{3}{7} = \dfrac{13}{28}$

Example 79. The probability that A speaks the truth is $\dfrac{3}{4}$ and that B speaks the truth is $\dfrac{4}{5}$. The probability that each contradicts other in stating a fact:

(A) $\dfrac{13}{20}$ (B) $\dfrac{7}{20}$ (C) $\dfrac{11}{15}$ (D) $\dfrac{4}{15}$

Answer (B)

Let X be the event that A speaks the truth, $P(X) = \dfrac{3}{4} \Rightarrow P(\overline{X}) = \dfrac{1}{4}$

and Y be the event that B speaks the truth, $P(Y) = \dfrac{4}{5} \Rightarrow P(\overline{Y}) = \dfrac{1}{5}$ 4/5,

A and B will contradict each other if one of the *mutually excnts* occurs:

(a) A speaks the truth and B does not i.e., $X \cap \overline{Y}$ or

(b) B speaks the truth and A does not i.e., $\overline{X} \cap Y$

The required probability = $P[(X \cap \overline{Y}) \cup (\overline{X} \cap Y)]$

$= P(X \cap \overline{Y}) + P(\overline{X} \cap Y) = P(X).P(\overline{Y}) + P(\overline{X}).P(Y)$

[As A speaking the truth and B speaking the truth are independent events, so $X \& Y$ are independent events and so are $X \& \overline{Y}$ and $\overline{X} \& Y$]

$= \dfrac{3}{4}.\dfrac{1}{5} + \dfrac{1}{4}.\dfrac{4}{5} = \dfrac{7}{20}$

Example 80. Let $A \& B$ be two independent events such that the

probability is $\dfrac{1}{8}$ that they will occur simultaneously and $\dfrac{3}{8}$ that neither of

them will occur. Then $P(A)$ equals

(A) $\quad \dfrac{1}{4}$ (B) $\quad \dfrac{1}{5}$ (C) $\quad \dfrac{4}{5}$ (D) $\quad \dfrac{2}{3}$

Answer: (A)

Let $P(A) = x$ As $A \& B$ are independent,

$P(A).P(B) = P(A \cap B) = \dfrac{1}{8} \Rightarrow P(B) = \dfrac{1}{8x}$

$\dfrac{3}{8} = P(A \cup B)' = 1 - P(A \cup B) = 1 - [P(A) + P(B) - P(A \cap B)]$

or, $P(A) + P(B) - P(A \cap B) = 1 - \dfrac{3}{8} = \dfrac{5}{8}$

or, $x + \dfrac{1}{8x} - \dfrac{1}{8} = \dfrac{5}{8} \Rightarrow x + \dfrac{1}{8x} = \dfrac{5}{8} + \dfrac{1}{8} = \dfrac{3}{4} \Rightarrow x = \dfrac{1}{4} \Rightarrow P(A) = \dfrac{1}{4}$

250

Example 81. The minimum number of times a die has to be thrown such that the probability of no six is less than $\dfrac{1}{2}$:

(A)　　2　　(B)　　3　　　　(C)　4　　(D)　　5

Answer (C)

In a single throw of a dice, the probability of getting 6 is $\dfrac{1}{6}$ and not getting 6 is $1 - \dfrac{1}{6} = \dfrac{5}{6}$ Let the number of throws = n.

The probability of not getting 6 in n throws $= \left(\dfrac{5}{6}\right)^n$

By the given condition, $\left(\dfrac{5}{6}\right)^n < \dfrac{1}{2}$

For $n = 1$, $\dfrac{5}{6} > \dfrac{1}{2}$; For $n = 2$, $\left(\dfrac{5}{6}\right)^2 = \dfrac{25}{36} > \dfrac{1}{2}$;

For $n = 3$, $\left(\dfrac{5}{6}\right)^3 = \dfrac{125}{216} > \dfrac{1}{2}$; For $n = 4$, $\left(\dfrac{5}{6}\right)^4 = \dfrac{625}{1296} < \dfrac{1}{2} \Rightarrow n = 4$

Example 82. The minimum number of tosses required of an unbiased coin so that the probability of at least one success being 0.8 is

(A)　　2　　(B)　　3　　(C)　　4　　(D)　　5

Answer: (B)

Suppose the required number of tosses $= n$

P(at least 1 head in n trials) = 1 - P(not a single head in n trials) $1 - \left(\dfrac{1}{2}\right)^n \geq 0.8 \Rightarrow -\left(\dfrac{1}{2}\right)^n \geq 0.8 - 1 = -0.2 = -\dfrac{1}{5}$

or, $\left(\dfrac{1}{2}\right)^n \leq \dfrac{1}{5} \Rightarrow 2^n \geq 5$　The least value of n is 3

Example 83. A five figure number is formed by the digits 0, 1, 2, 3, 4 (without repetition). The probability that the number formed is an even number:

(A)   5/16   (B) 3/8   (C) 11/16   (D)   5/8

Answer (D)

The number of five figure numbers formed by the digits 0, 1, 2, 3, 4 (without repetition) = $n = 5! - 4! = 96$

[The numbers with 0 at the beginning are to be excluded]

The number will be an even number if it contains 0, 2 or 4 at the end.

The number of five figure numbers with 0 at the end = $4! = 24$

The number of five figure numbers with 2 at the end = $4! - 3! = 18$

The number of five figure numbers with 4 at the end = $4! - 3! = 18$[The numbers with 0 at the beginning are to be excluded in the last two cases]

The number of even five figure numbers $m = 24 + 18 + 18 = 60$

Required probability = $\dfrac{m}{n} = \dfrac{60}{96} = \dfrac{5}{8}$

Example 84. A room has 3 lamp sockets. From a collection of 10 bulbs of which 6 are bad, a person selects 3 at random & puts them in the sockets. The probability that there will be light is

(A) $\dfrac{5}{8}$ (B) $\dfrac{5}{6}$ (C) $\dfrac{3}{8}$ (D) $\dfrac{2}{3}$

Answer: (B)

From 10 bulbs 3 bulbs may be selected in $^{10}C_3 = 120$ ways.

There will be light if one, two or three good bulbs are selected from 4 good bulbs and the rest are bad bulbs. This can be done in $^4C_1 \times {}^6C_2 + {}^4C_2 \times {}^6C_1 + {}^4C_3 \times {}^6C_0 = 60 + 36 + 4 = 100$ ways.

Required probability = $\dfrac{100}{120} = \dfrac{5}{6}$

Example 85. The natural numbers 1, 2, 3, ... n are arranged in random order. The probability that the three numbers 1, 2, 3 will remain together is

(A) $\dfrac{2}{n}$ (B) $\dfrac{6}{n(n-1)}$ (C) $\dfrac{2}{n(n-1)}$ (D) $\dfrac{6}{n}$

Answer: (B)

The numbers may be arranged in $n!$ ways.

Keeping 1, 2, 3 together, the number of arrangements = $(n-2)!3!$

Required probability = $\dfrac{3!(n-2)!}{n!} = \dfrac{6(n-2)!}{n(n-1)(n-2)!} = \dfrac{6}{n(n-1)}$

Example 86. The probability that a given positive number lying between 1 and 100 (both inclusive) is NOT divisible by 2, 3 or 5 is ___[CS/IT 2014]

Answer: 0.26

Let A, B, C be the events that the number is divisible by 2, 3, 5.

$n(A) = 50, n(B) = 33, n(C) = 20, n(A \cap B) = 16, n(B \cap C) = 6,$
$n(A \cap C) = 10, n(A \cap B \cap C) = 3$

Number of cases that the number is divisible by 2, 3 or 5
$= P(A \cup B \cup C) = P(A) + P(B) + P(C) - P(A \cap B) - P(B \cap C)$
$- P(A \cap C) + P(A \cap B \cap C) = 50 + 33 + 20 - 16 - 6 - 10 + 3 = 74$

Number of cases that the number is not divisible by 2, 3 or 5

$= 100 - 74 = 26$ Required probability $= \dfrac{26}{100} = 0.26$

Example 87. If the letters of the word DIRECTOR are arranged without repetition, the probability that the vowels should be together:

(A) $\dfrac{3}{35}$ (B) $\dfrac{5}{28}$ (C) $\dfrac{3}{28}$ (D) $\dfrac{6}{35}$

Answer: (C)

Total 8 letters with 2 are alike (R). They may be arranged in $n = \dfrac{8!}{2!}$ ways.

There are 3 vowels, keeping together they may be arranged in $m = \dfrac{6!3!}{2!}$ ways.

Required probability = $\dfrac{6!3!}{2!} \times \dfrac{2!}{8!} = \dfrac{6}{7 \times 8} = \dfrac{3}{28}$

Example 88. A hydraulic structure has four gates which operate independently. The probability of failure of each gate is 0.2. Given that gate 1 has failed, the probability that both gates 2 and 3 will fail is (A) 0.240   (B)   0.200   (C) 0.040   (D)   0.008 [CS/IT 2004]

Answer   (C)

Let A, B, C, D be the events of failure of gates 1, 2, 3, 4.

$$P(A) = P(B) = P(C) = P(D) = 0.2$$

Required probability $= P[(B \cap C)/A] = \dfrac{P(B \cap C \cap A)}{P(A)}$

$= \dfrac{P(B).P(C).P(A)}{P(A)} = 0.2 \times 0.2 = 0.040$

Example 89. P and Q are considering to apply for a job. The probability that P applies for the job is $\dfrac{1}{4}$, the probability that P applies for the job given that Q applies for the job is $\dfrac{1}{2}$, and the probability that Q applies for the job given that P applies for the job is $\dfrac{1}{3}$. Then the probability that Q does not apply for the job given that P does not apply for the job is

(A) $\dfrac{4}{5}$ (B) $\dfrac{5}{6}$ (C) $\dfrac{7}{8}$ (D) $\dfrac{11}{12}$ [CS/IT 2017]

Answer   (A)

Let X & Y denote the events that P applies for the job and Q applies for the job. $P(X) = \dfrac{1}{4}, P(X/Y) = \dfrac{1}{2}, P(Y/X) = \dfrac{1}{3}.$

Required probability $= P(\overline{X}/\overline{Y}) = \dfrac{P(\overline{X} \cap \overline{Y})}{P(\overline{Y})} - - - (i)$

$P(X \cap Y) = P(Y/X).P(X) = \dfrac{1}{12}, P(X \cap Y) = P(X/Y).P(Y)$

$\Rightarrow P(Y) = \dfrac{1/12}{1/2} = \dfrac{1}{6} \Rightarrow P(\overline{Y}) = 1 - P(Y) = \dfrac{5}{6}$

$P(X \cup Y) = P(X) + P(Y) - P(X \cap Y) = \dfrac{1}{4} + \dfrac{1}{6} - \dfrac{1}{12} = \dfrac{1}{3}$

$P(\overline{X} \cap \overline{Y}) = P(X \cup Y)' = 1 - P(X \cup Y) = 1 - \dfrac{1}{3} = \dfrac{2}{3}$

From (i), probability $= \dfrac{2/3}{5/6} = \dfrac{4}{5}$

Example 90. Assuming that each child is as likely to be a boy as is to be a girl, then the conditional probability that in a family of two children both

254

are boys, given that at least one of the children is a boy is

(A) $\dfrac{4}{5}$ (B) $\dfrac{1}{3}$ (C) $\dfrac{2}{5}$ (D) $\dfrac{3}{5}$

Answer: (B)

Let A and B denote respectively the events that the older child is a boy and the younger child is a boy. So, $P(A) = P(B) = \dfrac{1}{2}$

$A \cap B$ denotes the events that both the children are boys.

As A and B are independent events,

$$P(A \cap B) = P(A).P(B) = \dfrac{1}{2}.\dfrac{1}{2} = \dfrac{1}{4} \text{ and}$$

$$P(A \cup B) = P(A). + P(B) - P(A \cap B) = \dfrac{1}{2} + \dfrac{1}{2} - \dfrac{1}{4} = \dfrac{3}{4}$$

Required probability =

$$P[(A \cap B)/(A \cup B)] = \dfrac{P[(A \cap B) \cap (A \cup B)]}{P(A \cup B)} = \dfrac{P(A \cap B)}{P(A \cup B)} = \dfrac{1/4}{3/4} = \dfrac{1}{3}$$

Example 91. The probability that a construction job will be finished in time is $\dfrac{17}{20}$, the probability that there will be no strike is $\dfrac{3}{4}$ and the probability that the construction job will be finished in time, assuming that there will be no strike is $\dfrac{14}{15}$. The probability that the construction job will be finished in time and there will be no strike is

(A) $\dfrac{7}{10}$ (B) $\dfrac{3}{10}$ (C) $\dfrac{14}{17}$ (D) $\dfrac{11}{14}$

Answer (A)

Let A be the event that 'the construction job will be finished in time' and B be the event that 'there will be no strike'.

$$P(A) = \dfrac{17}{20}, P(B) = \dfrac{3}{4} \text{ \& } P(A/B) = \dfrac{14}{15}$$

Required probability $= P(A \cap B) = P(B).P(A/B) = \dfrac{3}{4}.\dfrac{14}{15} = \dfrac{7}{10}$

Example 92. The probability that a construction job will be finished in time is 17/20, the probability that there will be no strike is 3/4 and the probability that the construction job will be finished in time, assuming that there will be no strike is 14/15. The probability that there will be no strike given that the job is finished on time is

(A) $\dfrac{7}{10}$ (B) $\dfrac{3}{10}$ (C) $\dfrac{14}{17}$ (D) $\dfrac{11}{14}$

Answer (C)

Let A be the event that 'the construction job will be finished in time' and B be the event that 'there will be no strike'. Then

$P(A) = 17/20$, $P(B) = 3/4$ and $P(A/B) = 14/15$, $P(AB) = P(B).P(A/B) = 7/10$

Required probability = $P(B/A) = \dfrac{P(AB)}{P(A)} = \dfrac{7/10}{17/20} = \dfrac{14}{17}$

Example 93. An urn contains 2 white, 3 black balls and another urn contains 3 white, 4 black balls. One ball is transferred from the first urn to the second urn and then one ball is drawn from the later. The probability of its being a white ball is

(A) $\dfrac{17}{40}$ (B) $\dfrac{9}{40}$ (C) $\dfrac{13}{40}$ (D) $\dfrac{21}{40}$

Answer (A)

Total number of balls in the first urn = 5 and in the second urn = 7

Drawing a white ball from the second urn may be done in either of the 2 following mutually exclusive ways:

i) Ball transferred from the first urn to the second is white and the ball drawn from the second urn is white. If this event be denoted by A, then

$P(A) = \dfrac{2}{5} \times \dfrac{3+1}{7+1} = \dfrac{1}{5}$ [As the events are independent]

ii) Ball transferred from the first urn to the second is black and the ball drawn from the second urn is white. If this event be denoted by B, then

$P(B) = \dfrac{3}{5} \times \dfrac{3}{7+1} = \dfrac{9}{40}$

Required probability $= P(A) + P(B) = \dfrac{1}{5} + \dfrac{9}{40} = \dfrac{17}{40}$

Example 94. In a class 40% students read Mathematics, 25% Biology and

256

15% both Mathematics & Biology. One student is selected at random. The probability that he reads Mathematics if it is known that he reads Biology is

(A)   3/5   (B)   3/8   (C)   4/5   (D)   5/8

Answer: (A)

Let M and B denote respectively the events that the student reads Mathematics and Biology. $P(M) = \dfrac{40}{100}, P(B) = \dfrac{25}{100}, P(M \cap B) = \dfrac{15}{100}$

The probability that he reads Mathematics if it is known that he reads

Biology = $P(M/B) = \dfrac{P(M \cap B)}{P(B)} = \dfrac{15/100}{25/100} = \dfrac{15}{25} = \dfrac{3}{5}$

***Example* 95.** A missile was fired at a plane of which there are two targets T_1 and T_2. The probability of hitting T_1 is p_1 and that of hitting T_2 is p_2. It is known that T_2 was not hit. The probability that T_1 will be hit is

(A) $\dfrac{p_2}{1 - p_1}$ (B) $\dfrac{p_1}{1 - p_2}$ (C) $\dfrac{p_1 p_2}{1 - p_1}$ (D) $\dfrac{p_1 p_2}{1 - p_2}$

Answer: (B)

Let T_1 denotes the event 'T₁ is hit' andT_2 denotes the event 'T₂ is hit'.

By the given condition $P(T_1 \bar{T}_2) = P(T_1)$ [As T₂ was not hit]

Required probability = $P(T_1 / \bar{T}_2) = \dfrac{P(T_1 \bar{T}_2)}{P(\bar{T}_2)} = \dfrac{P(T_1)}{1 - P(T_2)} = \dfrac{p_1}{1 - p_2}$

***Example* 96.** There are four machines and it is known that exactly two of them are faulty. They are tested one by one in a random order till both the faulty machines are identified. Then the probability that only two tests are needed is

(A) $\dfrac{1}{3}$ (B) $\dfrac{1}{6}$ (C) $\dfrac{1}{2}$ (D) $\dfrac{1}{4}$

Answer: (B)

Let A be the event that two defective machines are identified in the first two tests out of four

machines. $n = ^4C_2 = 6$ & $m = ^2C_2 = 1$ $P(A) = \dfrac{m}{n} = \dfrac{1}{6}$

Example 97. If the integers m & n are chosen at random between 1 and 100, then the probability that a number of the form $7^m + 7^n$ is divisible by 5 equals

(A) $\dfrac{1}{4}$ (B) $\dfrac{1}{7}$ (C) $\dfrac{1}{8}$ (D) $\dfrac{1}{49}$

Answer: (C)

$7^m + 7^n$ is divisible by 5 if one ends with 9 and the other with 1 so that the sum will have 0 at the end (unit place).

7^2 ends with 9 and 7^4 ends with 1.

m can be 2, 6, 10, ---, 98 (25 cases) and n can be 4, 8, 12, ---, 100 (25 cases) and vice-versa.

Probability $2 \times \dfrac{25}{100} \times \dfrac{25}{100} = \dfrac{1}{8}$

Example 98. There are two identical urns containing respectively (5 white, 3 red) balls and (4 white, 6 red) balls. An urn is chosen at random and a ball is drawn from it. The probability that the ball drawn is white is

(A) $\dfrac{31}{80}$ (B) $\dfrac{37}{80}$ (C) $\dfrac{41}{80}$ (D) $\dfrac{43}{80}$

Answer (C)

Let A_1, A_2 be the events of selecting the first and the second urn respectively, $P(A_1) = P(A_2) = \dfrac{1}{2}$

Let W be the event of drawing a white ball.

$P(W/A_1)$= Probability of drawing a white ball from the first urn = 5/8 and $P(W/A_2)$= Probability of drawing a white ball from the second urn=4/10.

Probability of drawing a white ball

$= P(W) = P(A_1). P(W/A_1) + P(A_2). P(W/A_2) = \dfrac{1}{2}\left(\dfrac{5}{8} + \dfrac{4}{10}\right) = \dfrac{41}{80}$

Example 99. In a given day in the rainy season, it may rain 70% of the time. If it rains, chance that a village fair will make a loss on that day is 80%. However, if it does not rain, chance that the fair will make a loss on that day is only 10%. If the fair has not made a loss on a given day in the rainy season, what is the probability that it has not rained on that day?

(A)   3/10   (B)   9/11   (C)   14/17   (D)   27/41   [PI 2014]

Answer: (D)

Let R & L be events of rain and

loss. $P(R) = \dfrac{7}{10}, P(\overline{R}) = \dfrac{3}{10}, P(L/R) = \dfrac{8}{10}\ P(L/\overline{R}) = \dfrac{1}{10}$

$P(L) = P(R).P(L/R) + P(\overline{R}).P(L/\overline{R}) = \dfrac{59}{100} \Rightarrow P(\overline{L}) = \dfrac{41}{100}$

$P(R \cap L) = P(R).P(L/R) = \dfrac{56}{100}$

$P(\overline{R} \cap \overline{L}) = 1 - P(R \cup L) = 1 - [P(R) + P(L) - P(R \cap L)]$

$= 1 - \left[\dfrac{7}{10} + \dfrac{59}{100} - \dfrac{56}{100}\right] = \dfrac{27}{100}$

Required prob. $= P(\overline{R}/\overline{L}) = \dfrac{P(\overline{R} \cap \overline{L})}{P(\overline{L})} = \dfrac{27/100}{41/100} = \dfrac{27}{41}$

Example 100. If from each of the three boxes containing 3 white and 1 black, 2 white and 2 black, 1 white and 3 black balls, one ball is drawn at random. The probability that 2 white and 1 black ball will be drawn is

(A) $\dfrac{1}{4}$ (B) $\dfrac{13}{32}$ (C) $\dfrac{1}{32}$ (D) $\dfrac{3}{16}$

Answer: (B)

Selection can be made in the following manner:

i) white, white, black: Probability $= \dfrac{3}{4} \times \dfrac{2}{4} \times \dfrac{3}{4} = \dfrac{18}{64}$

ii) white, black, white: Probability $= \dfrac{3}{4} \times \dfrac{2}{4} \times \dfrac{1}{4} = \dfrac{6}{64}$

iii) black, white, white: Probability $= \dfrac{1}{4} \times \dfrac{2}{4} \times \dfrac{1}{4} = \dfrac{2}{64}$

All the cases are mutually exclusive.

Probability $= \dfrac{18}{64} + \dfrac{6}{64} + \dfrac{2}{64} = \dfrac{26}{64} = \dfrac{13}{32}$

Example 101. A person is assigned three jobs A, B & C. The probability of his doing the jobs A, B, C are p, q & $\frac{1}{2}$. He gets the full payment only if he either does the jobs A & B or A & C or all the three jobs. If the probability of his getting the full payment is $\frac{1}{2}$, then the relation satisfied by p & q is

(A) $p + q = 1$ (B) $pq = 1$ (C) $q(1 + p) = 1$ (D) $p(1 + q) = 1$

Answer: (D)

Probability of getting full payment

$= P(AB\overline{C}) + P(A\overline{B}C) + P(ABC)$ [As all the three cases are mutually exclusive] $= P(A).P(B).P(\overline{C}) + P(A).P(\overline{B}).P(C) + P(A).P(B).P(C)$

[As the cases are

independent] $= p.q.\frac{1}{2} + p.(1 - q).\frac{1}{2} + p.q.\frac{1}{2} = \frac{1}{2}pq + \frac{1}{2}p = \frac{1}{2}$ (given)

So, $p(1 + q) = 1$

Example 102. An urn contains m white and n black balls. A ball is drawn at random and is put back into the urn with k additional balls of the same colour as that of the ball drawn. A ball is again drawn at random. The probability that the ball drawn now being white is

(A) $\dfrac{n}{m + n + k}$ (B) $\dfrac{m}{m + n + k}$ (C) $\dfrac{n}{m + n}$ (D) $\dfrac{m}{m + n}$

Answer: (D)

Case I: 1 white ball drawn from the first urn, k additional white balls put in the second urn & 1 white ball drawn from the second urn

Probability $= = \dfrac{m}{m + n} \times \dfrac{m + k}{m + n + k}$ Case II: 1 black ball drawn from the first urn, k additional black balls put in the second urn & 1 white ball drawn from the second urn Probability $= = \dfrac{n}{m + n} \times \dfrac{m}{m + n + k}$

As the two cases are mutually exclusive,

Required

$$\text{probability} = \frac{m}{m+n} \times \frac{m+k}{m+n+k} + \frac{n}{m+n} \times \frac{m}{m+n+k}$$

$$= \frac{m(m+k)+mn}{(m+n)(m+n+k)} = \frac{m(m+k+n)}{(m+n)(m+n+k)} = \frac{m}{m+n}$$

Example 103. P and Q are considering to apply for a job. The probability that P applies for the job is $\frac{1}{4}$, the probability that P applies for the job given that Q applies for the job is $\frac{1}{2}$, and the probability that Q applies for the job given that P applies for the job is $\frac{1}{3}$. Then the probability that P does not apply for the job given that Q does not apply for the job is

(A) $\quad \frac{4}{5}$ (B) $\quad \frac{5}{6}$ (C) $\quad \frac{7}{8}$ (D) $\quad \frac{11}{12}$ [CS/IT 2017]

Answer: (A)

$p(P) = \frac{1}{4}, \ p(P/Q) = \frac{1}{2}, \ p(Q/P) = \frac{1}{3}$

Now $p(P \cap Q) = p(Q/P).P(P) = \frac{1}{12}$

Again $p(P \cap Q) = p(P/Q).P(Q) \Rightarrow \frac{1}{12} = \frac{1}{2}.P(Q) \Rightarrow P(Q) = \frac{1}{6}$ and

$p(P \cup Q) = p(P) + p(Q) - p(P \cap Q) = \frac{1}{4} + \frac{1}{6} - \frac{1}{12} = \frac{1}{3}$

Required probability $= p(\overline{P}/\overline{Q}) = \dfrac{p(\overline{P} \cap \overline{Q})}{p(\overline{Q})} = \dfrac{p(P \cup Q)^c}{p(\overline{Q})}$

$= \dfrac{1 - p(P \cup Q)}{1 - p(Q)} = \dfrac{1 - 1/3}{1 - 1/6} = \dfrac{4}{5}$

Example 104. Three companies, X, Y and Z supply computers to a university. The percentage of computers supplied by them and the probability of those being defective are tabulated below.

Company	% of computers supplied	Prob. of being defective
X	60%	0.01
Y	30%	0.02
Z	10%	0.03

Given that a computer is defective, the probability that it was supplied by Y is:

(A)   0.1   (B)   0.2   (C)   0.3   (D)   0.4   [EC 2006]

Answer 0.4

If the event of the supplied computer being defective is denoted by D,

$P(X) = 0.6, P(Y) = 0.3, P(Z) = 0.1, P(D/X) = 0.01, P(D/Y) = 0.02, P(D/z) = 0.03$

Probability

$$= P(Y/D) = \frac{P(Y).P(D/Y)}{P(X).P(D/X) + P(Y).P(D/Y) + P(Z).P(D/Z)}$$

$$= \frac{0.3 \times 0.02}{0.6 \times 0.01 + 0.3 \times 0.02 + 0.1 \times 0.03} = 0.4$$

Example 105. The probability that a doctor diagnoses the disease correctly is 80%. The probability that a patient dies after correct diagnosis is 40% and the probability that the patient dies after wrong diagnosis is 70%. The patient died. The probability that the disease was correctly diagnosed is

(A)   31/49   (B)   8/25   (C)   7/10   (D)   16/23

Answer: (D)

If B_1= event of correct diagnosis, then $P(B_1) = 80/100 = 4/5$

B_2 = event of wrong diagnosis, then $P(B_2) = 20/100 = 1/5$

A = event that the patient dies, then $P(A/B_1)=40/100 = 2/5, P(A/B_2) = 7/10$

Required probability = $P(B_1/A) = \dfrac{P(B_1)P(A/B_1)}{P(B_1)P(A/B_1) + P(B_2)P(A/B_2)} = \dfrac{16}{23}$

Example 106. In a factory, two machines M_1 and M_2 manufacture 60% and 40% of the auto components respectively. Out of the total production, 2% of M_1 and 3% of M_2 are found to be defective. If a randomly drawn auto component from the combined lot is found defective, what is the probability that it was manufactured by M_2?

(A)   0.35   (B)   0.45   (C)   0.5   (D)   0.4   [BT 2013]

Answer (B)

$$P(M_1) = \frac{60}{100} = 0.6 \; P(M_2) = \frac{40}{100} = 0.4$$

If D denotes the event of a defective bolt,

$$P(D/M_1) = \frac{2}{100} = 0.02 \; P(D/M_2) = \frac{3}{100} = 0.03$$

Required probability

$$= P(M_2/D) = \frac{P(M_2)P(D/M_2)}{P(M_1)P(D/M_1) + P(M_2)P(D/M_2)}$$

$$= \frac{0.4 \times 0.03}{0.6 \times 0.02 + 0.4 \times 0.03} = 0.5$$

Example 107. In a factory, three machines M_1, M_2 and M_3 manufacture 20% 30% and 50% of the components respectively. Out of the total production, 3% of M_1, 4% of M_2 and 5% of M_3 are found to be defective. If a randomly drawn component from the combined lot is found defective, what is the probability that it was manufactured by M_3?

(A) $\frac{23}{43}$ (B) $\frac{25}{43}$ (C) $\frac{27}{43}$ (D) $\frac{21}{43}$ [BT 2013]

Answer (B)

$$P(M_1) = \frac{20}{100} \; P(M_2) = \frac{30}{100}, \; P(M_3) = \frac{50}{100}$$

If D denotes the event of a defective bolt,

$$P(D/M_1) = \frac{3}{100} \; P(D/M_2) = \frac{4}{100} \; P(D/M_3) = \frac{5}{100}$$

Probability

$$=P(M_3/D)=\frac{P(M_3)P(D/M_3)}{P(M_1)P(D/M_1)+P(M_2)P(D/M_2)+P(M_3)P(D/M_3)}$$

$$=\frac{\dfrac{50}{100}\times\dfrac{5}{100}}{\dfrac{20}{100}\times\dfrac{3}{100}+\dfrac{30}{100}\times\dfrac{4}{100}+\dfrac{50}{100}\times\dfrac{5}{100}}=\frac{25}{43}$$

Example 108. Suppose there are three urns containing 2 white and 3 black balls; 3 white and 2 black balls; 4 white and 1 black balls respectively. There is equal probability of each urn being chosen. One ball is drawn from an urn chosen at random and found to be white. The probability that the ball drawn from the first urn is

(A) $\dfrac{1}{4}$ (B) $\dfrac{1}{5}$ (C) $\dfrac{3}{5}$ (D) $\dfrac{2}{9}$

Answer: (D)

A_i $(i=1,2,3)$ be the event that $i-$th urn is chosen and W be the event that a white ball is drawn. $P(A_1)=P(A_2)=P(A_3)=\dfrac{1}{3}$

$$P(W/A_1)=\frac{2}{5},\ P(W/A_2)=\frac{3}{5},\ P(W/A_3)=\frac{4}{5}$$
$$P(W)=P(A_1).P(W/A_1)+P(A_2).P(W/A_2)+P(A_3).P(W/A_3)$$
$$=\frac{1}{3}\left(\frac{2}{5}+\frac{3}{5}+\frac{4}{5}\right)=\frac{3}{5}$$

The probability that the white ball drawn from the first

urn $=P(A_1/W)=\dfrac{P(A_1).P(W/A_1)}{P(W)}=\dfrac{2}{9}$ [By Bayes' Theorem]

Example 109. Consider a hash table with 100 slots. Collisions are resolved using chaining. Assuming simple uniform hashing, what is the probability that the first 3 slots are unfilled after the first 3 insertions?

(A) $(97\times97\times97)/100^3$ (B) $(99\times98\times97)/100^3$

(C) $(97\times96\times95)/100^3$ (D) $(97\times96\times95)/(3!\times100^3)$ [CS/IT 2014]

Answer (A)

$$P[\text{First insertion where first 3 slots unfilled}] = \frac{^{97}C_1}{^{100}C_1} = \frac{97}{100}$$

$$P[\text{Second insertion where first 3 slots unfilled}] = \frac{^{97}C_1}{^{100}C_1} = \frac{97}{100}$$

$$P[\text{Third insertion where first 3 slots unfilled}] = \frac{^{97}C_1}{^{100}C_1} = \frac{97}{100}$$

$$\text{Required probability} = \frac{97}{100} \times \frac{97}{100} \times \frac{97}{100} = \frac{97 \times 97 \times 97}{100^3}$$

Example 110. Two sets of candidates are competing for the positions of board of directors of a company. The probabilities that the first and the second sets will win are 0.6 and 0.4 respectively. If the first set wins the probability of introducing a new product is 0.8 and the corresponding probability if the second set wins is 0.3.

What is the probability that the new product will be introduced?

(A)   0.40   (B)   0.50   (C)   0.60   (D)   0.70

Answer: (C)

$P(A_1) = $ Probability that the first set wins = 0.6 $P(A_2) = $ Probability that the first set wins = 0.4

Let B be the event of introducing a new product.

If the first set wins the probability of introducing a new product

$= P(B/A_1) = 0.8$

If the second set wins the probability of introducing a new product

$= P(B/A_2) = 0.3$

Probability that a new product is introduced

$$= P(B) = P(B \cap A_1) + P(B \cap A_2) = P(A_1)P(B/A_1) + P(A_2)P(B/A_2)$$
$$= 0.6 \times 0.8 + 0.4 \times 0.3 = 0.60$$

Example 111. For a biased die the probabilities for different faces to turn up are

Face	1	2	3	4	5	6
Probability	0.1	0.32	0.21	0.15	0.05	0.17

The die is tossed and you are told that either face 1 Or 2 has turned up. The probability that it is face 1 is

(A)   13/21   (B)   5/21   (C)   2/35   (D)   11/35

Answer (B)

Let A be the event that either face 1 Or 2 has turned up and B be the event that it is face 1. $P(A) = P(1) + P(2) = 0.1 + 0.32 = 0.42$, $P(B) = 0.1$

As B is a subset of A, $B \cap A = B \Rightarrow P(B \cap A) = P(B) = 0.1$

Required probability $= P(B/A) = \dfrac{P(B \cap A)}{P(A)} = \dfrac{P(B)}{P(A)} = \dfrac{0.1}{0.42} = \dfrac{10}{42} = \dfrac{5}{21}$

Example 112. A biased die is tossed and the respective probabilities for various faces to turn up is

Face	1	2	3	4	5	6
Probability	0.1	0.24	0.19	0.18	0.15	0.14

If an even face has turned up, then the probability that it is face 2 or face 4 is

(A)   0.25   (B)   0.42   (C)   0.75   (D)   0.9

Answer: (C)

Let A be the event that an even face has turned up and B be the event that it is 2 or 4.

$P(A) = P(2) + P(4) + P(6) = 0.24 + 0.18 + 0.14 = 0.56$

$P(B) = 0.24 + 0.18 = 0.42$

As B is a subset of A, $B \cap A = B \Rightarrow P(B \cap A) = P(B) = 0.42$

Required probability $= P(B/A) = \dfrac{P(B \cap A)}{P(A)} = \dfrac{P(B)}{P(A)} = \dfrac{0.42}{0.56} = \dfrac{42}{56} = \dfrac{3}{4} = 0.75$

Example 113. Two independent random variables X and Y are uniformly distributed in the interval [- 1, 1]. The probability that $\max[X, Y]$ is less than 1/2 is

(A)   3/4   (B)   9/16   (C)   1/4   (D)   2/3   [EE 2012]

Answer (B)

As X and Y are uniformly distributed in the interval [- 1, 1], so their pdf are $f(x) = f(y) = \dfrac{1}{1-(-1)} = \dfrac{1}{2}, -1 < x < 1$

$$P\left[\max(X,Y) \leq \dfrac{1}{2}\right] = P\left(X = \dfrac{1}{2}, -1 \leq Y \leq \dfrac{1}{2}\right).P\left(-1 \leq X \leq \dfrac{1}{2}, Y = \dfrac{1}{2}\right)$$

$$= \int_{-1}^{1/2} \dfrac{1}{2} dx \times \int_{-1}^{1/2} \dfrac{1}{2} dy = \dfrac{1}{4} \times \dfrac{3}{2} \times \dfrac{3}{2} = \dfrac{9}{16}$$

EXERCISE

Question of one mark:

1. If A and B be any two events and P(A∪B) = 1/2, P(A) = 1/4, P(B) = 2/5, then the value of $P(\overline{A} \cap B)$ is
(A)   1/10   (B)   1/8   (C)   1/6   (D)   1/4

2. A problem is given to 3 students whose chance of solving it is 1/2, 1/3, 1/4. The probability that the problem will be solved is:
(A)   1/4   (B)   3/4   (C)   1/8   (D)   3/8

3. Two dice are thrown simultaneously. The probability of obtaining at least one 6 is
(A)   1/6   (B)   5/6   (C)   11/36   (D)   25/36

4. The probability that a gentleman will travel by plane is 2/3 and that he will travel by train is 1/5. What is the probability of his travelling by plane or train?
(A)   13/15   (B)   2/15   (C)   7/15   (D)   8/15

5. The natural numbers 1, 2, 3, ... , n are arranged in random order. The probability that the two numbers 1 and 2 will remain together is
(A)   2/n   (B)   1/n   (C)   n/2   (D)   2/(n-1)

Question of two marks:

6. If the letters of the word SOCIETY are arranged without repetition, the probability that the three vowels should be together is
(A)   3/14   (B)   5/28   (C)   2/35   (D)   1/7

7. Assuming that each child is as likely to be a boy as is to be a girl, then the conditional probability that in a family of two children both are boys,

267

given that the older child is a boy is

(A)   3/4   (B)   1/2   (C)   2/3   (D)   4/5

8. The probability that a construction job will be finished in time is 17/20, the probability that there will be no strike is 3/4 and the probability that the construction job will be finished in time, assuming that there will be no strike is 14/15.The probability that there will be strike or the job will not be finished in time is

(A)   7/10   (B)   3/10   (C)   14/17   (D)   11/14

9. In a class 40% students read Mathematics, 25% Biology and 15% both Mathematics & Biology. One student is selected at random. The probability that he reads Biology if it is known that he reads Mathematics

(A)   3/5   (B)   3/8   (C)   4/5 (D)   5/8

10. A card is drawn from a full pack of cards. The probability that it is a 'heart' or a 'queen' is

(A)   2/13 (B)   4/13   (C)   3/52   (D)   5/52

11. Let $A \& B$ be two independent events such that the probability is $\dfrac{1}{8}$ that they will occur simultaneously and $\dfrac{3}{8}$ that neither of them will occur. Then $P(B)$ equals

(A) $\dfrac{1}{4}$ (B) $\dfrac{1}{2}$ (C) $\dfrac{4}{5}$ (D) $\dfrac{2}{3}$

12. Suppose there are three urns containing 2 white and 3 black balls; 3 white and 2 black balls; 4 white and 1 black balls respectively. There is equal probability of each urn being chosen. One ball is drawn from an urn chosen at random. What is the probability that a white ball is drawn?

(A) $\dfrac{1}{4}$ (B) $\dfrac{1}{5}$ (C) $\dfrac{3}{5}$ (D) $\dfrac{2}{9}$

13. Two persons A and B toss a coin 50 times. The probability that both of them get tails at the same time is _____.

14. There are 5 pair of shoes in a shoe rack. Four shoes are drawn one by one at random. The probability of at least one pair of shoes being drawn is

(A)   13/21   (B)   5/28   (C)   2/35   (D)   11/18

15. There are two containers, with one containing 4 red and 3 green balls and the other containing 3 blue and 4 green balls. One ball is drawn at random from each container. The probability that one of the balls is red and the other is blue will be

(A)   1/7   (B)   9/49   (C)   12/49   (D)   3/7 [CS 2011]

Answers:

1. (D) 2. (B) 3. (C) 4. (A) 5. (A) 6. (D) 7. (B) 8. (B) 9. (B) 10. (B)

11. (B) 12. (C) 13. $(1/4)^{50}$ 14. (A) 15. $0.33 - 0.3$

Chapter- 9

PROBABILITY DISRIBUTIONS

Formulae:

1. *Discrete Random Variable X:* Its p. m. f. (probability mass function)

$f(x)$ satisfies the relations $f(x) \geq 0$ and $\displaystyle\sum_{-\infty}^{\infty} f(x) = 1$

The probability that the X will assume the value x is given by $P(X = x) = f(x)$

Mean of $X = \bar{x} = E(X) = \sum f_i x_i$ where f_i is the probability for $X = x_i$

Variance of X = $VAR(X) = \sigma^2 = E(x - \bar{x})^2 = E(x^2) - \bar{x}^2$

s. d. of X = $\sigma = +\sqrt{E(x^2) - \bar{x}^2}$

2. *Continuous Random Variable X:* Its p. d. f. (probability density function)

$f(x)$ satisfies the relations $f(x) \geq 0$ and $\displaystyle\int_{-\infty}^{\infty} f(x)dx = 1$

The probability that X will assume any value in the interval *[a, b]* is given by

$P(a \leq x \leq b) = \displaystyle\int_{a}^{b} f(x)dx$

3. The *Cumulative Distribution Function* (c.d.f.) or Distribution Function F(x) is

given by $F(x) = P(X \leq x) = \displaystyle\int_{-\infty}^{x} f(t)dt$ and $f(x) = \dfrac{d}{dx}F(x);$

$P(c \leq x \leq d) = F(d) - F(c)$

4. *Mean of X* = $\bar{x}$ = *E(X)* = $\displaystyle\int_{-\infty}^{\infty} xf(x)dx$ and

5. *variance* of $x = \sigma^2 = E(x^2) - \bar{x}^2 = \displaystyle\int_{-\infty}^{\infty} x^2 f(x)dx - \bar{x}^2$

6. $E(x - \bar{x}) = 0$, *E(a) = a*, *E(ax) = a E(x)*,
E(ax + b) = a E(x) + b, where *a* and *b* are constants.

7. *var(a) = 0,* *var(ax + b) = a²var(x)*, *var(ax + by) = a²var(x) + b²var(y)*

8. *Binomial Distribution:* The probability of x successes in n independent

trials P(X = x) = f(x) = $^{n}C_x p^x q^{n-x}$, x = 0,1,2, ...,n; p is the probability of success, q is the probability of failure and p + q = 1

Parameters = n, p   Mean = np   Variance = npq

9. *Poisson Distribution :*The probability that the discrete random variable X

assumes the value x is given by $P(X = x) = f(x) = \dfrac{e^{-m}m^x}{x!}, x = 0,1,2,......, \infty$

Parameter = m, Mean = Variance = m

10. Uniform Distribution: Its p. d. f. is given by $f(x) = \dfrac{1}{b-a}, a < x < b$

Parameters = a,b , Mean = $\dfrac{a+b}{2}$ *and Variance =* $\dfrac{(b-a)^2}{12}$

The probability that the continuous random variable X assumes the value x in

the interval $c \le x \le d$ is given by $\displaystyle\int_{c}^{d} f(x)dx$

11. *Normal Distribution:* Its p. d. f. is given by

$f(x) = \dfrac{1}{\sigma\sqrt{2\pi}} e^{-\frac{(x-\mu)^2}{2\sigma^2}}, \quad -\infty < x < \infty$

Parameters = μ,σ , Mean = μ and Variance = σ^2

12. *Standard Normal Distribution:* $z = \dfrac{x-\mu}{\sigma}$ is called standard normal

variate. Its p. d. f. is given by $\phi(z) = \dfrac{1}{\sqrt{2\pi}} e^{-z^2/2}, \quad -\infty < x < \infty$

Mean = 0 and *Variance* = 1. The standard normal variate z is unit free.

$P(z \le c) = \phi(c), \phi(c) + \phi(-c) = 1$

Area between $z = \pm 1$ is 68.27%, Area between $z = \pm 2$ is 95.45%

Area between $z = \pm 3$ is 99.73%

WORKED OUT EXAMPLES:

Question of one mark:

Example 1. If E denotes expectation, the variance of a random variable
X is given by

(A) $E(X^2) - E^2(X)$ (B) $E(X^2) + E^2(X)$

(C) $E(X^2)$ (D) $E^2(X)$ [EC 2007]

Answer: (A)

Example 2. Let X be a real-valued random variable with $E[X]$ and $E[X^2]$ denoting the mean values of X and X^2 respectively. The relation which always holds true is

(A) $(E[X])^2 > E[X^2]$ (B) $E[X^2] \geq (E[X])^2$

(C) $E[X^2] = (E[X])^2$ (D) $E[X^2] > (E[X])^2$ [EC 2014]

Answer (B)

As X be a real-valued random variable, $\text{var}[X] \geq 0$

$\Rightarrow E[X^2] - (E[X])^2 \geq 0 \Rightarrow E[X^2] \geq (E[X])^2$

Example 3. If a random variable X takes the values 1, 2, 3 and 4 such that $2P(X = 1) = 3P(X = 2) = P(X = 3) = 5P(x = 4)$, then the value of $P(X = 1)$ is _____.

Answer: $\dfrac{15}{61}$

$2P(X = 1) = 3P(X = 2) = P(X = 3) = 5P(x = 4) = k$

$\Rightarrow P(X = 1) = \dfrac{k}{2}, P(X = 2) = \dfrac{k}{3}, P(X = 3) = k, P(x = 4) = \dfrac{k}{5}$

So, $\dfrac{k}{2} + \dfrac{k}{3} + k + \dfrac{k}{5} = 1 \Rightarrow \dfrac{61}{30}k = 1 \Rightarrow k = \dfrac{30}{61}, P(X = 1) = \dfrac{k}{2} = \dfrac{15}{61}$

Example 4. A six face fair dice is rolled a large number of times. The mean value of the outcome is -----.   [ME 2017]

Answer: 3.5

Mean = $\sum px = \dfrac{1}{6}(1 + 2 + 3 + 4 + 5 + 6) = 3.5$

Example 5. If f(x) = kx , x = 1, 2, ... , n be the p. m. f. of a random variable, then k =

(A) $\dfrac{n(n+1)}{2}$ (B) $\dfrac{n(n+1)}{6}$ (C) $\dfrac{2}{n(n+1)}$ (D) $\dfrac{6}{n(n+1)}$

Answer: (C)

As f(x)= kx is the p. m. f. so $\sum f(x) = 1$

$\Rightarrow \sum_{x=1}^{n} kx = 1 \Rightarrow k \sum x = 1$

$\Rightarrow k(1 + 2 + - - - + n) = 1 \Rightarrow k.\dfrac{n(n+1)}{2} = 1 \Rightarrow k = \dfrac{2}{n(n+1)}$

Example 6. If $f(x) = kx, x = 1,2,- - -, n$ be the p. m. f. of a random variable X, then the mean of X is

(A) $\dfrac{n+1}{2}$ (B) $\dfrac{n+1}{3}$ (C) $\dfrac{2n+1}{2}$ (D) $\dfrac{2n+1}{3}$

Answer: (D)

From Example 5, $k = \dfrac{2}{n(n+1)}$, Mean of $X = E(X)$

$= \sum xf(x) = k \sum x^2 = \dfrac{2}{n(n+1)} \dfrac{n(n+1)(2n+1)}{6} = \dfrac{2n+1}{3}$

Example 7. A discrete random variable X has the following probability mass functions:

Values of X	0	1	2	3	4	5	6	7
P(X)	0	2k	3k	K	2k	K²	7K²	2K²+k

then k =

(A) $\dfrac{1}{5}$ (B) $\dfrac{1}{10}$ (C) $\dfrac{1}{15}$ (D) $\dfrac{1}{20}$

Answer: (B)

As P(X=x) = f(x) is the p. m. f. of a random variable X,

(i) $f(x) \geq 0$ and (ii) $\sum f(x) = 1$

Using(ii), 0 + 2k + 3k + k + 2k + K² + 7K²+ 2K² +k =1 $\Rightarrow$ 10k² + 10k − 1 = 0

$\Rightarrow$ k = - 1, 1/10

If k = -1, then f(1) = - 2 < 0 which contradicts (i), so k = 1/10

Example 8. A random variable X has the following probability distribution:

273

Values of X	0	1	2	3	4	5	6	7
P(X)	0	2k	3k	K	2k	K²	7K²	2K²+k

Then P(0< X < 5) =

(A) $\dfrac{4}{5}$ (B) $\dfrac{1}{5}$ (C) $\dfrac{2}{5}$ (D) $\dfrac{3}{5}$

Answer: (A)

As in Example 7, k = 1/10 and P(0 < X < 5) = P(1) + P(2) + P(3) + P(4) = 8k = 4/5

Example 9. Consider a random variable X that takes values + 1 and -1 with probability 0.5 each. The values of the cumulative distribution function $F(x)$ at x = -1 and +1 are

(A) 0 and 0.5 (B) 0 and 1 (C) 0.5 and 1 (D) 0.25 and 0.75   [CS/IT 2012]

Answer: (C)

Given $P(X = -1) = P(X = 1) = 0.5$ and

$F(x) = P(X \leq x) \Rightarrow F(-1) = P(X \leq -1) = P(X = -1) = 0.5$
$F(1) = P(X \leq 1) = P(X = -1) + P(X = 1) = 0.5 + 0.5 = 1$

Example 10. A random variable X has the following probability distribution:

X	- 2	- 1	0	1	2	3
P(X)	1/10	K	2/10	2k	3/10	3k

The mean of X is

(A) $\dfrac{1}{15}$ (B) $\dfrac{14}{15}$ (C) $\dfrac{16}{15}$ (D) $\dfrac{19}{15}$

Answer: (C)

$$\dfrac{1}{10} + k + \dfrac{2}{10} + 2k + \dfrac{3}{10} + 3k = 1 \Rightarrow k = \dfrac{1}{15}$$

$$\text{Mean} = (-2).\dfrac{1}{10} + (-1).k + 0 + 1.2k + 2.\dfrac{3}{10} + 3.3k = 10k + \dfrac{2}{5} = \dfrac{16}{15}$$

Example 12. A random variable X has the following probability distribution:

X	1	2	3	4
Probability	2/10	3/10	4/10	1/10

The variance of X is

(A)   17/25   (B)   19/25   (C)   21/25   (D)   23/25

Answer:(C)

$$\overline{X} = E(X) = 1 \times \frac{2}{10} + 2 \times \frac{3}{10} + 3 \times \frac{4}{10} + 4 \times \frac{1}{10} = \frac{12}{5}$$

$$E(X^2) = 1^2 \times \frac{2}{10} + 2^2 \times \frac{3}{10} + 3^2 \times \frac{4}{10} + 4^2 \times \frac{1}{10} = \frac{33}{5}$$

$$Var(X) = E(X^2) - \overline{X}^2 = \frac{21}{25}$$

Example 13. For a discrete random variable X, $ran(X) = \{0, 1, 2, 3\}$ and the cumulative probability $F(X)$ is shown below:

X	0	1	2	3
F(X)	0.5	0.6	0.8	1

The mean value of X is _____.   [BT 2015]

Answer: 1.1

The p.m.f. table of X is

X	0	1	2	3	Total
F(X)	0.5	0.6-0.5 = 0.1	0.8-0.6 = 0.2	1-0.8 =0.2	1

Mean of X = 0 × 0.5 + 1 × 0.1 +2 × 0.2 + 3 × 0.2 = 1.1

Example 14. n unbiased dice are thrown. The mathematical expectation of the sum of points on the n dice is

(A) $\dfrac{n}{2}$ (B) $\dfrac{3n}{2}$ (C) $\dfrac{5n}{2}$ (D) $\dfrac{7n}{2}$

Answer: (D)

For a single die, the numbers appeared on the face with their respective probabilities of occurrence are:

Number appeared (x)	1	2	3	4	5	6
Probability (p)	1/6	1/6	1/6	1/6	1/6	1/6

$$E(X) = = \sum px = 1 \times \frac{1}{6} + 2 \times \frac{1}{6} + 3 \times \frac{1}{6} + 4 \times \frac{1}{6} + 5 \times \frac{1}{6} + 6 \times \frac{1}{6} = \frac{7}{2}.$$

The mathematical expectation of the sum of points on the n dice =

$$\frac{7}{2} \times n = \frac{7n}{2}$$

Example 15 .Consider a dice with the property that the probability of a face with n dots showing up is proportional to n. The probability of the face with three dots showing up is _____.   [EE 2014]

Answer: 1/7

P(n) = k n, n = 1, 2, 3, 4, 5, 6 as the dice has 6 dots only,

P(1) + P(2) + P(3) + P(4) + P(5) + P(6) = 1,

k(1 + 2 + 3 + 4 + 5 + 6) = 1 or, k = 1/21, Required probability = P(3) = k.3 = 1/7

Example 16. Assume that in a traffic junction, the cycle of the traffic signal lights is 2 minutes of green (vehicle does not stop) and 3 minutes of red (vehicle stops). Consider that the arrival time of the vehicles time at the junction is uniformly distributed over a 5 minutes cycle. The expected waiting time (in minutes) for the vehicle at the junction is _ .   [EE 2017]

Answer: 0.9

The pdf of the waiting time (x) is $f(x) = \frac{1}{5}, 0 < x < 3$

Expected waiting time $= \int_0^3 xf(x)dx = \frac{1}{5}\int_0^3 xdx = \frac{1}{5}\left[\frac{x^2}{2}\right]_0^3 = 0.9\, \text{min}.$

Example 17. A probability density function on the interval $[a,1]$ is given by $1/x^2$ and outside this interval the value of the function is 0. The value of a is _____.   [CS/IT 2016]

Answer: 0.5

As $1/x^2$ is the pdf on the interval, $[a,1]$

so $\int_a^1 \frac{1}{x^2}dx = 1 \Rightarrow \left[-\frac{1}{x}\right]_a^1 = 1 \Rightarrow \frac{1}{a}-1 = 1 \Rightarrow \frac{1}{a} = 2 \Rightarrow a = 0.5$

Example 18. For the function $f(x) = a + bx, 0 \le x \le 1$, to be a valid probability density function, which one of the following is correct?

(A) $a = 1, b = 4$ (B) $a = 0.5, b = 1$

(C) $a = 0, b = 1$ (D) $a = 1, b = -1$ [CE 2017]

276

Answer: (B)

$$\int_0^1 (a+bx)dx = 1 \Rightarrow a + \frac{b}{2} = 1$$ The relation is satisfied by (B) only.

Example 19. The variance of the random variable X with probability density function $f(x) = \frac{1}{2}|x|e^{-|x|}$ is _____. [EC 2015]

Answer: 6

$f(x)$ is an even function of x as $f(-x) = f(x)$

$xf(x)$ is an odd function of x and $x^2 f(x)$ is an even function of x.

$$E(x) = \int_{-\infty}^{\infty} xf(x)dx = 0,$$

$$E(x^2) = \int_{-\infty}^{\infty} x^2 f(x)dx = 2.\frac{1}{2}\int_0^{\infty} x^3 e^{-x}dx = \int_0^{\infty} x^{4-1}e^{-x}dx = \Gamma 4 = 3! = 6$$

$$\text{var}(x) = E(x^2) - \{E(x)\}^2 = 6 - 0 = 6$$

Example 20. Probability density function of a random variable X is given below $f(x) = \begin{cases} 0.25, & 1 \le x \le 5 \\ 0, & otherwise \end{cases}$. $P(X \le 4)$ is

(A) $\frac{3}{4}$ (B) $\frac{1}{2}$ (C) $\frac{1}{4}$ (D) $\frac{1}{8}$ [CE 2016]

Answer (A)

$$P(X \le 4) = \int_{-\infty}^4 f(x)dx = \int_{-\infty}^1 f(x)dx + \int_1^4 f(x)dx = 0 + \int_1^4 (0.25)dx$$

$$= 0.25(4-1) = \frac{3}{4}$$

Example 21. The probability density function of evaporation E on any day during a year in a watershed is given by $f(E) = \begin{cases} 1/5, & 0 \le E \le 5 \, mm/day \\ 0, & otherwise \end{cases}$

The probability that E lies in between 2 and 4 mm/day in a day in the watershed is (in decimal)_____ [CE 2014]

Answer: 0.4

$$P(2 \le E \le 4) = \int_{2}^{4} f(E)dE = \frac{1}{5}\int_{2}^{4} dE = \frac{1}{5}(4-2) = 0.4$$

Example 22. Let X be a random variable with probability density function

$$f(x) = \begin{cases} 0.2, & |x| \le 1 \\ 0.1, & 1 < x \le 4 \\ 0, & otherwise \end{cases}$$. The probability $P(0.5 < X < 5)$ is ___ [EE 2014]

Answer: 0.4

$$P(0.5 < X < 5) = \int_{0.5}^{5} f(x)dx = \int_{0.5}^{1} f(x)dx + \int_{1}^{4} f(x)dx = \int_{0.5}^{1} 0.2\,dx + \int_{1}^{4} 0.1\,dx$$

$$= 0.2(1 - 0.5) + 0.1(4 - 1) = 0.4$$

Example 23. Consider the following probability mass function (p.m.f.)

of a random variable $X : p(x,q) = \begin{cases} q & if \ X = 0 \\ 1-q & if \ X = 1 \\ 0, & otherwise \end{cases}$

If $q = 0.4$, the variance of X is _____.   [CE 2015]

Answer: 0.24

$$E(X) = 0.q + 1.(1-q) = 1-q = 0.6$$

$$E(X^2) = 0^2.q + 1^2.(1-q) = 1-q = 0.6$$

$$var(X) = E(X^2) - \{E(X)\}^2 = 0.6 - (0.6)^2 = 0.24$$

Example 24. A continuous random variable X has a probability density function $f(x) = e^{-x}, 0 < x < \infty$. . Then $P(X > 1)$ is

(A)   0.368   (B)   0.5   (C)   0.632   (D)   1.0   [EE 2013]

Answer (A)

$$P(X > 1) = \int_{1}^{\infty} f(x)dx = \int_{1}^{\infty} e^{-x}dx = -\left[e^{-x}\right]_{1}^{\infty} = -(0 - e^{-1}) = \frac{1}{e} = 0.368$$

Example 25. Given that x is a random variable in the range $[0, \infty]$ with a probability density function $\dfrac{e^{-x/2}}{k}$, the value of the constant k is

_____

Answer 2

$$\int_0^\infty \frac{e^{-x/2}}{k}\,dx = 1 \Rightarrow \frac{1}{k}\left[-2e^{-x/2}\right]_0^\infty = 1 \Rightarrow -\frac{2}{k}.(0-1) = 1 \Rightarrow \frac{2}{k} = 1 \Rightarrow k = 2$$

Example 26. For a random variable X, $\mathrm{var}(X) = 1$, then $\mathrm{var}(2X + 3) =$

(A) 1 (B) 4 (C) 16 (D) 1/2

Answer: (B)

$\mathrm{var}(2X + 3) = 2^2.\mathrm{var}(X) = 4.1 = 4$

Example 27. A random variable X has the density function

$$f(x) = \begin{cases} x, & 0 < x < 1 \\ \dfrac{1}{2}, & 1 < x < 2 \end{cases}.$$ The mean of X is

(A) $\dfrac{7}{6}$ (B) $\dfrac{11}{6}$ (C) $\dfrac{11}{12}$ (D) $\dfrac{13}{12}$

Answer: (D)

Mean of $X = E(X) = \int_{-\infty}^{\infty} xf(x)dx = \int_0^1 x^2\,dx + \int_1^2 \frac{1}{2}xdx = \frac{1}{3} + \frac{1}{2}.\frac{1}{2}(4-1) = \frac{13}{12}$

Example 28. The annual precipitation data of a city is normally distributed with mean and standard deviation 1000 mm and 200 mm respectively. The probability that the annual precipitation is more than 1200 mm is

(A) < 50%   (B) 50%   (C) 75%   (D) 100%   [CS/IT 2012]

Answer: (A)

$$P(X > 1200) = P\left(\frac{X - 1000}{200} > \frac{1200 - 1000}{200}\right) = P(Z > 1) = P(1 < z < \infty)$$

$< 0.5 \Rightarrow < 50\%$ [As $P(0 < z < \infty) = 0.5$]

Example 29. The probability density function of a continuous distribution is given by $f(x) = \dfrac{3}{4}x(2-x), 0 < x < 2.$ The mean of the distribution is

(A)   1   (B)   1.5   (C)   2   (D)   2.5

Answer: (A)

Mean of X $= E(x) = \displaystyle\int_{-\infty}^{\infty} xf(x)dx = \dfrac{3}{4}\int_{0}^{2} x.x(2-x)dx = 1$

Example 30. Lifetime of an electric bulb is a random variable with density $f(x) = kx^2$, where x is measured in years. If the minimum and maximum lifetimes of bulb are 1 and 2 years respectively, then the value of k is _____.   [EE 2014]

Answer: $\dfrac{3}{7}$

Lifetime of an electric bulb is a random variable X with p.d.f.

$f(x) = kx^2, 1 \le x \le 2$

so, $\displaystyle\int_{1}^{2} f(x)dx = 1 \Rightarrow k\int_{1}^{2} x^2 dx = 1 \Rightarrow \dfrac{k}{3}\left[2^3 - 1^3\right] = 1 \Rightarrow k = \dfrac{3}{7}$

Example 31. For the probability density $P(x) = 0.5e^{-0.5x}$, the integral

$\displaystyle\int_{0}^{\infty} P(x)dx = $ _____.   [BT 2017]

Answer: 1

$\displaystyle\int_{0}^{\infty} P(x)dx = -\left[e^{-0.5x}\right]_{0}^{\infty} = -(0-1) = 1$

Example 32. If a random variable X has mean m and variance σ^2,

then $E\left(\dfrac{x-m}{\sigma}\right)^2 = $

(A) $\dfrac{1}{\sigma^2}$ (B) $\dfrac{1}{\sigma}$ (C)   1   (D)   0

Answer: (C)

$$E\left(\frac{x-m}{\sigma}\right)^2 = E\left[\frac{1}{\sigma^2}(x-m)^2\right] = \left[\frac{1}{\sigma^2}E(x-m)^2\right] \text{(as } \sigma \text{ is constant)}$$

$$= \frac{1}{\sigma^2}.\sigma^2 = 1$$

Example 33. The mean and variance of a Binomial (n, p) distribution are 20 and 16 respectively. Then (n, p) =

(A) (200, 1/10) (B) (100, 1/5) (C) (50, 2/5) (D) (150, 2/15)

Answer (B)

Mean = np = 20 ------ (i) and variance = npq = 16 ------------(ii)

Dividing (ii) by (i), q = 4/5 i.e. p = 1 – q = 1/5

From (i), n = 20/p = 100 ⟹ (n, p) = (100, 1/5)

Example 34. The probability of a defective bolt is $\frac{1}{10}$. For the distribution of defective bolts in a total of 400, the (mean, variance) =

(A)   (20, 9)   (B)   (80, 16) (C) (50, 25) (D)   (40, 36)

Answer (D)

$$n = 400, p = \frac{1}{10}, q = 1 - \frac{1}{10} = \frac{9}{10} \quad \text{Mean } = np = 40,$$

variance = $npq = 36$ So, (mean, variance) = (40, 36)

Example 35. The second moment of a Poisson- distributed random variable is 2. The mean of the random variable is _____.   [EC 2016]

Answer: 0.9 – 1.1

For a Poisson- distributed random variable, second moment = variance = mean = 2

Example 36. Consider Poisson distribution for the tossing of a biased coin. The mean for this distribution is μ. The standard deviation for this distribution is given by

(A) $\sqrt{\mu}$ (B) μ^2 (C) μ (D) $1/\mu$ [ME 2016]

Answer: (A)

For Poisson distribution, variance = mean = μ s. d. $= \sqrt{\mu}$

Example 37. The number of parameters in the uni variate exponential and Gaussian distributions, respectively are

(A)   2 and 2   (B)   1 and 2   (C)   2 and 1   (D)   1 and 1   [CE 2017]

Answer: (B)

Example 38. $f(x) = \dfrac{5}{\sqrt{\pi}} e^{-25x^2}, -\infty < x < \infty$ is the p.d.f. of a normal

distribution, the variance of the distribution is

(A)   25   (B)   50(C)   1/25   (D)   1/50

Answer (D)

The p.d.f. of a normal distribution is $f(x) = \dfrac{1}{\sigma\sqrt{2\pi}} e^{-\frac{(x-\mu)^2}{2\sigma^2}}$

Comparing with the given p.d.f.

$$\dfrac{1}{\sigma\sqrt{2\pi}} = \dfrac{5}{\sqrt{\pi}} \Rightarrow \sigma = \dfrac{1}{5\sqrt{2}} \qquad \text{Variance} = \sigma^2 = \dfrac{1}{50}$$

Example 39. Let X be a Gaussian random variable with mean 0 and variance σ^2. Let $Y = \max(X,0)$ where $\max(a,b)$ is the maximum of a and b. Then the median of Y is _____.   [CS/IT 2017]

Answer: 0

Half of the values of Y are in the left of the mean $X = 0$ and the remaining half of the values of Y is to the right of the the mean $X = 0$.

So, the median of Y is 0.

Example 40. If X be a normal variate with mean 12 gms and standard deviation 4 gms, then the value of the standard normal variate corresponding to X =20 gms is

(A)   2   (B)   3   (C) 2 gms   (D) 6 gms

Answer (A)

$$z = \dfrac{x - \mu}{\sigma} = \dfrac{20 - 12}{4} = 2 \ \ \text{[z is always unit less]}$$

Example 41. If X be a normal variate with zero mean and unit variance, then the expectation of X^2 is

(A)   0   (B)   1   (C)   2   (D)   1/2

Answer (B)

Mean= $E(X) = 0$

$Variance = E\left[X^2\right] - \{E[X]\}^2 \Rightarrow 1 = E\left[X^2\right] - 0 \Rightarrow E\left[X^2\right] = 1$

Example 42. If $3z - x = 5$ where z is a standard normal variate then for the normal variate x, (mean, variance) =

(A) (5, 9)   (B) (3, 25)   (C)   (-5, 9)   (D) (-3, 25)

Answer: (C)

As z is a standard normal variate, $\bar{z} = 0$ & $var(z) = 1$

$x = 3z - 5 \Rightarrow \bar{x} = 3\bar{z} - 5 = 3.0 - 5 = -5$ & $var(x) = 3^2 . var(z) = 9.1 = 9$

So, (mean, variance) = (- 5, 9)

Example 43. The area (in percentage) under standard normal distribution of random variable Z within limits from -3 to $+3$ is ____. [ME 2016]

Answer: 99.6 to 99.8

$\mu = 0, \sigma = 1 \Rightarrow (\mu - 3\sigma, \mu + 3\sigma) = (-3, 3)$ Area = 99.73%

Example 44. For a certain normal distribution, the fourth central moment is 48. The variance of the distribution is

(A)   2   (B)   4   (C)   8   (D)   16

Answer (B)

For a normal distribution, the central moments (μ_{2n}) satisfy the relation

$\mu_{2n} = 1.3.5. --- (2n-1)\sigma^{2n}$ Taking $n = 2$,

$\mu_4 = 1.3.\sigma^4 \Rightarrow 48 = 3\sigma^4 \Rightarrow \sigma^2 = 4$ So, variance = 4

Question of two marks:

Example 45. The probability density function of a continuous distribution is given by $f(x) = \dfrac{3}{4}x(2 - x), 0 < x < 2$. The variance of the distribution is

(A)   0.1   (B)   0.2   (C)   0.5   (D)   1

Answer: (B)

Variance $(X) = E(X^2) - \{E(X)\}^2$ $- - - - - (i)$

$$E(X^2) = \int_{-\infty}^{\infty} x^2 f(x)dx = \frac{3}{4}\int_0^2 x^2.x(2-x)dx = \frac{3}{4}\int_0^2 (2x^3 - x^4)dx$$

$$= \frac{3}{4}\left[\frac{x^4}{2} - \frac{x^5}{5}\right]_0^2 = \frac{3}{4}\left[8 - \frac{32}{5}\right] = \frac{6}{5}$$

$$E(X) = \int_{-\infty}^{\infty} xf(x)dx = \frac{3}{4}\int_0^2 x.x(2-x)dx = \frac{3}{4}\int_0^2 (2x^2 - x^3)dx$$

$$= \frac{3}{4}\left[\frac{2}{3}x^3 - \frac{x^4}{4}\right]_0^2 = \frac{3}{4}\left[\frac{16}{3} - 4\right] = 1$$

From (i), variance $(X) = \dfrac{6}{5} - 1 = 0.2$

Example 46. A continuous random variable X has the p.d.f.

$f(x) = \dfrac{1}{2} - ax, (0 \le x \le 4)$, where a is a constant. Then $a =$

(A)   1/2 (B)   1/4   (C)   1/6   (D)   1/8

Answer: (D)

As $f(x)$ is the p. d. f. of a random variable x,

(i) $f(x) \ge 0$ and (ii) $\displaystyle\int_{-\infty}^{\infty} f(x)dx = 1$

Using (ii), $\displaystyle\int_{-\infty}^{\infty} f(x)dx = \int_{-\infty}^{0} f(x)dx + \int_0^4 f(x)dx + \int_4^{\infty} f(x)dx$

$$= 0 + \int_0^4 \left(\frac{1}{2} - ax\right)dx + 0 = \left[\frac{1}{2}x - \frac{ax^2}{2}\right]_0^4 = 2 - 8a \Rightarrow 2 - 8a = 1 \Rightarrow a = \frac{1}{8}$$

Example 47. A continuous random variable X has the p.d.f.

$f(x) = \dfrac{1}{2} - ax, (0 \le x \le 4)$, where a is a constant. The probability that X

lies between 2 and 3 is

(A)   3/4 (B)   3/8   (C)   3/16   (D)   3/32

Answer: (C)

284

From the previous example, $a = \dfrac{1}{8}$

$$P(2 < X < 3) = \int_{2}^{3}\left(\frac{1}{2} - \frac{1}{8}x\right)dx = \left[\frac{1}{2}x - \frac{1}{16}x^{2}\right]_{2}^{3} = \frac{1}{2}(3-2) - \frac{1}{16}(9-4) = \frac{3}{16}$$

Example 48. A fair die with faces {1, 2, 3, 4, 5, 6} is thrown repeatedly till '3' is observed for the first time. Let X denote the number of times the die is thrown. The expected value of X is ____. [EC 2015]

Answer: 6

Probability o f throwing '3' (S) is $\dfrac{1}{6}$ and not throwing '3' (F) is $\dfrac{5}{6}$

Events:	S	FS	FFS	--
X:	1	2	3	---
Prob.	$\dfrac{1}{6}$	$\dfrac{5}{6}\times\dfrac{1}{6}$	$\dfrac{5}{6}\times\dfrac{5}{6}\times\dfrac{1}{6}$	___

$$E(X) = 1\times\frac{1}{6} + 2\times\left(\frac{5}{6}\times\frac{1}{6}\right) + 3\left(\frac{5}{6}\times\frac{5}{6}\times\frac{1}{6}\right) + - - \qquad \text{Taking } \frac{5}{6} = x$$

$$E(X) = \frac{1}{6}\left(1 + 2x + 3x^{2} + - -\right) = \frac{1}{6}(1-x)^{-2} = \frac{1}{6}\times\left(1-\frac{5}{6}\right)^{-2} = 6$$

Example 49. The length of life time of a Tyre manufactured by a company follows a continuous distribution given by the density function

$$f(x) = \begin{cases} \dfrac{k}{x^{3}}, & 1000 \le x \le 1500 \\ 0, & elsewhere \end{cases}$$

The probability that a randomly selected tire would function for at least 1200 hours is

(A) $\dfrac{3}{20}$ (B) $\dfrac{7}{20}$ (C) $\dfrac{9}{20}$ (D) $\dfrac{1}{20}$

Answer: (C)

As $f(x)$ is the p.d.f. of x,

$$\int_{-\infty}^{\infty} f(x)dx = 1 \Rightarrow \int_{1000}^{1500} \frac{k}{x^3} dx = 1 \Rightarrow k = 36 \times 10^5$$

$$P(X \geq 1200) = \int_{1200}^{1500} \frac{k}{x^3} dx = -\frac{1}{2}k \left(\frac{1}{x^2}\right)_{1200}^{1500} = -\frac{1}{2} \times 36 \times 10^5 \left[\frac{1}{1500^2} - \frac{1}{1200^2}\right]$$

$$= -18 \times 10^5 \times \frac{1}{10^4} \left(\frac{1}{225} - \frac{1}{144}\right) = 180 \times \frac{81}{225 \times 144} = \frac{9}{20}$$

Example 50. For any discrete random variable X, the probability mass

function $P(X = j) = p_j, p_j \geq 0, j \in \{0, N\}$ and $\sum_{j=0}^{N} p_j = 1$, define the

polynomial function $g_X(z) = \sum_{j=0}^{N} p_j z^j$. For a certain discrete random

variable Y, there exists a scalar $\beta \in [0,1]$ such that $g_Y(z) = (1 - \beta + \beta z)^N$.
The expectation of Y is
(A) $N\beta(1 - \beta)$ (B) $N\beta$
(C) $N(1 - \beta)$ (D) Not expressible in terms of N and β only [CS/IT 2017]
Answer: (B)
$g_Y(z) = [(1 - \beta) + \beta z]^N$. The expansion of RHS of $g_Y(z)$ would lead to a
Binomial distribution with $n = N$ & $p = \beta$. So, $E(Y) = np = N\beta$

Example 51. Let the random variable X represents the number of times a
fair coin needs to be tossed till two consecutive heads appear for the first
time. The expectation of X is _____ [EC 2015]
Answer: 1.5

The outcomes will be HH, $P(X = 2) = \dfrac{1}{2} \times \dfrac{1}{2} = \dfrac{1}{4}$

THH, $P(X = 3) = \dfrac{1}{2} \times \dfrac{1}{2} \times \dfrac{1}{2} = \dfrac{1}{8}$

TTHH, $P(X = 4) = \dfrac{1}{2} \times \dfrac{1}{2} \times \dfrac{1}{2} \times \dfrac{1}{2} = \dfrac{1}{16}$ ----------

Required expectation $= 2 \times \dfrac{1}{4} + 3 \times \dfrac{1}{8} + 4 \times \dfrac{1}{16} + - - -$

$$= \frac{1}{2}\left[2 \times \frac{1}{2} + 3 \times \frac{1}{2^2} + 4 \times \frac{1}{2^3} + --\right] = \frac{1}{2}\left[\left(1 + 2 \times \frac{1}{2} + 3 \times \frac{1}{2^2} + --\right) - 1\right]$$

$$= \frac{1}{2}\left[\left(1 - \frac{1}{2}\right)^{-2} - 1\right] = \frac{1}{2}\left[\left(\frac{1}{2}\right)^{-2} - 1\right] = \frac{1}{2}\left[2^2 - 1\right] = 1.5$$

Example 52. Let X be a random variable which is uniformly chosen from the set of positive odd numbers less than 100. The expectation $E(X)$ is

_____. [EC 2014]

Answer: 50

The positive odd integers less than 100 are 1, 3, 5, ----, 99.
Total number of observations = 50 and the probability of occurrence of any

of the observations = $p = \frac{1}{n} = \frac{1}{50}$

$$E(X) = \sum px = \frac{1}{50}.1 + \frac{1}{50}.3 + ---- + \frac{1}{50}.99$$

$$= \frac{1}{50}(1 + 3 + 5 + ---- + 99) = \frac{1}{50} \times \frac{50}{2}(1 + 99) = 50$$

$$\left[Sum \ of \ an \ A.P. = \frac{n}{2}(a + l)\right]$$

Example 53. For a discrete random variable $X, ran\{X\} = 0,1,2,3$ and the cumulative probability $F(X)$ is shown below:

X	0	1	2	3
$F(X)$	0.5	0.6	0.8	1.0

The mean value of X is _____. [BT 2015]

Answer: 1.1
The corresponding pdf $f(X)$ table is given by

X	0	1	2	3	Total
$f(X)$	0.5 = 0.5	0.6-0.5= 0.1	0.8 – 0.6= 0.2	1.0 – 0.8 =0.2	1

Mean value of $X = E(X) = 0 \times 0.5 + 1 \times 0.1 + 2 \times 0.2 + 3 \times 0.2 = 1.1$

Example 54. If it rains a taxi driver can earn Rs. 100 per day. If it is fair, he can lose Rs. 10 per day. If the probability of rain is 0.4, his mathematical expectation is

(A)   Rs. 40 (B)   Rs. 38   (C) Rs. 36 (D)   Rs. 34

Answer (D)

Probability of rain = 0.4, Probability of fair weather = 1 − 0.4 = 0.6

If the gain of the taxi driver be denoted by x and the respective probability by p, then the Probability distribution of x:

X (Rs.)	100	-10
P	0.4	0.6

Loss of Rs. 10 = Gain of Rs. (- 10)

Expectation of the taxi driver = 0.4 × 100 + 0.6 × (-10) = Rs. 34

Example 55. A machine produces 0, 1 or 2 defective pieces in a day with associated probability of 1/6, 2/3 and 1/6, respectively. The mean value and the variance of the number of defective pieces produced by the machine in a day, respectively, are

(A) $1 \& \dfrac{1}{3}$ (B) $\dfrac{1}{3} \& 1$ (C) $1 \& \dfrac{4}{3}$ (D) $\dfrac{1}{3} \& \dfrac{4}{3}$ [ME 2014]

Answer: (A)

Let x denotes the number of defective pieces produced in a day.

The probability distribution of x:

x	0	1	2	Total
p(x)	1/6	2/3	1/6	1

Mean of $x = \mu = \sum x.p(x) = 0.\dfrac{1}{6} + 1.\dfrac{2}{3} + 2.\dfrac{1}{6} = 1$

Variance of $x = = \sum x^2.p(x) - \mu^2 = 0^2.\dfrac{1}{6} + 1^2.\dfrac{2}{3} + 4.\dfrac{1}{6} - 1^2 = \dfrac{1}{3}$

Example 56. In rolling of two fair dice, the outcome of an experiment is considered to be the sum of the numbers appearing on the dice. The probability is highest for the outcome of _____   [CH 2014]

Answer: 7

Let X is a random variable denoting the sum of the numbers appearing on the dice and $P(X)$ is the corresponding probability of occurrence. The probability distribution of X is

x	2	3	4	5	6	7	8	9	10	11	12
$P(X)$	$\dfrac{1}{36}$	$\dfrac{2}{36}$	$\dfrac{3}{36}$	$\dfrac{4}{36}$	$\dfrac{5}{36}$	$\dfrac{6}{36}$	$\dfrac{5}{36}$	$\dfrac{4}{36}$	$\dfrac{3}{36}$	$\dfrac{2}{36}$	$\dfrac{1}{36}$

So, the probability is highest i.e. $\dfrac{6}{36}$ for the outcome "7".

Example 57. The function $f(x)$ defined by $f(x) = \begin{cases} x, & 0 \le x < 1 \\ k - x, & 1 \le x < 2 \\ 0, & elsewhere \end{cases}$

is a p.d.f. of a random variable X for a suitable value of k.
The probability that the random variable X lies between 1/2 & 3/2 is
(A)   2/3   (B)   3/4   (C)   5/6   (D)   3/5
Answer: (B)
The continuous random variable X has the p.d.f. $f(x)$. So,

$$\int_{-\infty}^{\infty} f(x)dx = 1 \Rightarrow \int_0^1 xdx + \int_1^2 (k-x)dx = 1 \Rightarrow \frac{1}{2} + k - \frac{1}{2}(4-1) = 1 \Rightarrow k=2.$$

$$P\left(\frac{1}{2} < X < \frac{3}{2}\right) = \int_{1/2}^{3/2} f(x)dx = \int_{1/2}^1 xdx + \int_1^{3/2} (2-x)dx = \frac{3}{4}.$$

Example 58. The function $f(x) = \dfrac{x^n}{n!}e^{-x}, x > 0, n$ being a positive integer, is a p.d.f. of X. The mean of X is
(A)   n − 1   (B)   n   (C)   n + 1   (D)   1
Answer: (C)

289

Mean of $X = E(X) = \int_0^\infty xf(x)dx = \frac{1}{n!}\int_0^\infty x^{n+1}e^{-x}dx = \frac{1}{n!}\int_0^\infty x^{(n+2)-1}e^{-x}dx$

$= \frac{\Gamma(n+2)}{n!} = \frac{(n+1)!}{n!} = n+1$

Example 59.. The function $f(x) = \frac{x^n}{n!}e^{-x}, x > 0, n$ being a positive integer,

is a p.d.f. of X. The variance of X is

(A) $(n-1)^2$ (B) n^2 (C) $(n+1)$ (D)   1

Answer: (C)

$\text{var}(X) = E(X^2) - \{E(X)\}^2 ----(i)$

$E(X^2) = \int_0^\infty x^2 f(x)dx = \frac{1}{n!}\int_0^\infty x^{n+2}e^{-x}dx = \frac{1}{n!}\int_0^\infty x^{(n+3)-1}e^{-x}dx = \frac{\Gamma(n+3)}{n!}$

$\frac{(n+2)!}{n!} = (n+2)(n+1)$

From (i), $\text{var}(X) = (n+2)(n+1) - (n+1)^2 = (n+1)(n+2-n-1) = n+1$

Example 60. The diameters of sand particles in a sample range from 50 to 150 microns. The number of particles of diameter x in the sample is

proportional to $\frac{1}{50+x}$. The average diameter (in microns), up to one

decimal place is _____.   [CH 2015]

Answer: 94.3

The probability density function of x is $f(x) = k.\frac{1}{50+x}$

$\int_{50}^{150} k.\left(\frac{1}{50+x}\right)dx = 1 \Rightarrow k.[\ln(50+x)]_{50}^{150} = 1 \Rightarrow k.\ln 2 = 1 \Rightarrow k = \frac{1}{\ln 2}$

Average diameter = mean of x

$$= \int_{50}^{150} xf(x)dx = k\int_{50}^{150} \frac{x}{50+x}dx = k\int_{50}^{150}\left(1-\frac{50}{50+x}\right)dx$$

$$= \frac{1}{\ln 2}[x-50\ln(50+x)]_{50}^{150} = \frac{1}{\ln 2}(100-50\ln 2) = \frac{100}{\ln 2}-50 = 94.3$$

Example 61. A random variable X has probability density function

$f(x)$ as given below: $f(x) = \begin{cases} a+bx, 0 < x < 1 \\ 0, \quad\quad otherwise \end{cases}$

If the expected value $E[X] = \dfrac{2}{3}$, then $\Pr[X < 0.5]$ is _____.[EE 2015]

Answer: 0.25

As $f(x)$ is pdf, $\displaystyle\int_0^1 f(x)dx = 1 \Rightarrow \int_0^1 (a+bx)dx = 1$

or, $a + \dfrac{b}{2} = 1 \Rightarrow 2a + b = 2 --(i)$

$$E[X] = \int_0^1 xf(x)dx = \int_0^1 (ax+bx^2)dx = \frac{a}{2}+\frac{b}{3} = \frac{2}{3} \text{ (given)}$$

or, $3a + 2b = 4 --- (ii)$

Solving (i) & $(ii), a = 0, b = 2$

$$\Pr[X < 0.5] = \int_0^{0.5}(a+bx)dx = \int_0^{0.5} 2x\,dx = [x^2]_0^{0.5} = 0.25$$

Example 62. The probability density function of a random variable,

x is $f(x) = \begin{cases} \dfrac{x}{4}(4-x^2), 0 \le x \le 2 \\ 0, \quad\quad otherwise \end{cases}$

The mean μ_x of the random variable is _____.   [CE 2015]

Answer: $\dfrac{16}{15}$

$$\mu_x = E(X) = \int_0^2 xf(x)dx = \frac{1}{4}\int_0^2 \left(4x^2 - x^4\right)dx = \frac{1}{4}\left[\frac{4}{3}x^3 - \frac{x^5}{5}\right]_0^2$$

$$= \frac{1}{4}\left[\frac{4}{3}.8 - \frac{32}{5}\right] = \frac{1}{4}.32\left[\frac{1}{3} - \frac{1}{5}\right] = 8.\frac{2}{15} = \frac{16}{15}$$

Example 63. A zero mean random signal is uniformly distributed between limits $-a$ and $+a$ and its mean square value is equal to its variance. Then the r.m.s value of the signal is

(A) $\dfrac{a}{\sqrt{3}}$ (B) $\dfrac{a}{\sqrt{2}}$ (C) $a\sqrt{2}$ (D) $a\sqrt{3}$ [EE 2011]

Answer (A)

The p.d.f. of x is $f(x) = \dfrac{1}{b-a} = \dfrac{1}{a-(-a)} = \dfrac{1}{2a}$ [Uniform distribution]

Variance $= \dfrac{(b-a)^2}{12} = \dfrac{\{a-(-a)\}^2}{12} = \dfrac{4a^2}{12} = \dfrac{a^2}{3} =$ Mean square value

Root mean square value (r.m.s.) $= \sqrt{\dfrac{a^2}{3}} = \dfrac{a}{\sqrt{3}}$

Example 64. Demand during lead time with associated probabilities is shown below:

Demand	50	70	75	80	85
Probability	0.15	0.14	0.21	0.20	0.30

Expected demand during lead time is _____.   [ME 2014]

Answer: 74 -75

Expected demand = 50 × 0.15 + 70 × 0.14 + 75 × 0.21 + 80 × 0.20 + 85 × 0.30

= 74.55

Example 65. Consider a continuous random variable with probability density

function $f(t) = \begin{cases} 1+t, & -1 \le t \le 0 \\ 1-t, & 0 \le t \le 1 \end{cases}$

The standard deviation of the random variable is

(A) $\dfrac{1}{\sqrt{3}}$ (B) $\dfrac{1}{\sqrt{6}}$ (C) $\dfrac{1}{3}$ (D) $\dfrac{1}{6}$ [ME 2006]

Answer (B)

Variance $= E(t^2) - \{E(t))\}^2 \ ---\ (i)$

Mean $= E(t) = \int\limits_{-\infty}^{\infty} t.f(t)dt = \int\limits_{-1}^{0} t.(1+t)dt + \int\limits_{0}^{1} t.(1-t)dt$

$= \int\limits_{-1}^{0}\left(t + t^2\right)dt + \int\limits_{0}^{1}(t - t^2)dt = 0$ and

$E(t^2) = \int\limits_{-\infty}^{\infty} t^2.f(t)dt = \int\limits_{-1}^{0} t^2.(1+t)dt + \int\limits_{0}^{1} t^2.(1-t)dt$

$= \int\limits_{-1}^{0}(t^2 + t^3)dt + \int\limits_{0}^{1}(t^2 - t^3)dt = \dfrac{1}{6}$

From (i), variance $= \dfrac{1}{6}$ and standard deviation $= \dfrac{1}{\sqrt{6}}$

Example 66. For a random variable X, $E(X) = 10$ and $\mathrm{var}(X) = 25$. If $y = mx - n, m > 0, n > 0$ have expectation 0 and variance 1, then $(m,n) =$

(A) (1/5, 2) (B) (2/5, 1) (C) (1, 2/5) (D) (2, 1/5)

Answer: (A)

$y = mx - n \Rightarrow E(y) = mE(X) - n \Rightarrow 0 = m.10 - n \Rightarrow n = 10m$

$\mathrm{var}(y) = m^2\,\mathrm{var}(X) - 0 \Rightarrow 1 = m^2.25 \Rightarrow m = \dfrac{1}{5}$ as $m > 0$ and

$n = 10 \times \dfrac{1}{5} = 2 \Rightarrow (m,n) = (1/5, 2)$

Example 67. A continuous random variable X has the p.d.f.

$f(x) = \dfrac{1}{2} - \dfrac{1}{8}x, (0 \le x \le 4)$, then $P(|X - 2| < 0.5) =$

(A) 1/2 (B) 1/4 (C) 1/6 (D) 1/8

Answer (B)

$|X - 2| < 0.5 \Rightarrow -0.5 < X - 2 < 0.5 \Rightarrow 2 - 0.5 < X < 2 + 0.5$

$\Rightarrow 1.5 < X < 2.5$

Now, $P(| X - 2 |< .5 = P(1.5 < X < 2.5) = \int_{1.5}^{2.5}\left(\frac{1}{2} - \frac{1}{8}x\right)dx = \frac{1}{4}$

Example 68. A continuous random variable X has the p.d.f.

$f(x) = \frac{1}{2} - \frac{1}{8}x, (0 \le x \le 4)$, then $P(X \le 1) =$

(A)   7/8   (B)   7/16   (C)   5/8   (D)   5/16

Answer: (B)

The cumulative distribution function (c. d. f.) is given by

$F(x) = \int_{0}^{x} f(t)dt = \int_{0}^{x}\left(\frac{1}{2} - \frac{1}{8}t\right)dt = \frac{x(8-x)}{16}$

$P(X \le 1) = F(1) = \frac{7}{16}$

Example 69. For a Binomial (6, p) distribution, $P(X = 2) = 9P(X = 4)$

Then $P(X = 0)$ is

(A)   729/4096 (B)   243/1024 (C) 81/256 (D) 27/128

Answer: (A)

$P(X = 2) = 9P(X = 4) \Rightarrow {}^6C_2 p^2 q^{6-2} = 9.{}^6C_4 p^4 q^{6-4}$

$\Rightarrow \frac{q^2}{p^2} = 9$ [As ${}^6C_2 = {}^6C_4$] $\Rightarrow q = 3p$ [As $p, q > 0$]

Now $p + q = 1 \Rightarrow$ p $= \frac{1}{4}, q = 1 - p = \frac{3}{4}$ and $n = 6$

$P(X = 0) = {}^6C_0 p^0 q^6 = \left(\frac{3}{4}\right)^6 = \frac{729}{4096}$

Example 70. The probability of obtaining at least two "SIX" in throwing a fair dice four times is

(A)   425/432 (B)   19/144 (C)   13/144 (D)   125/432   [ME 2015]

Answer: (B)

Probability in throwing a six is $\frac{1}{6} = p$ and a non-six is $\frac{5}{6} = q, n = 4$

Favourable cases: (two 6, 2 non-six) or (three 6, 1 non-six) or (four 6, 0 non-six)

Required probability

$$= {}^4C_2 \left\{ \left(\frac{1}{6}\right)^2 \times \left(\frac{5}{6}\right)^2 \right\} + {}^4C_3 \left\{ \left(\frac{1}{6}\right)^3 \times \frac{5}{6} \right\} + {}^4C_4 \left(\frac{1}{6}\right)^4$$

$$= \frac{6 \times 25 + 4 \times 5 + 1 \times 1}{36 \times 36} = \frac{171}{36 \times 36} = \frac{19}{144}$$

Example 71. A random variable follows Binomial Distribution with mean 4 and variance 2. The probability of assuming non-zero value of the variate is

(A) 1/64   (B) 63/64   (C) 1/256   (D)   255/256

Answer: (D)

Mean = $np = 4$ ------ (i)   and   variance = $npq = 2$ ------------(ii)

Dividing (ii) by (i), $q = 1/2$ i.e. $p = 1 - q = =1/2$ From (i), $n = 4 / p = 8$

$$P(X = 0) = {}^8C_0 p^0 q^{8-0} = q^8 = \left(\frac{1}{2}\right)^8 = \frac{1}{256}$$

Probability of assuming non-zero of the variate = $1 - P(X = 0)$

$$= 1 - \frac{1}{256} = \frac{255}{256}$$

Example 72. A fair coin is tossed independently four times. The probability of the event "the number of time heads shown up is more than the number of times tails shown up" is

(A)   1/16 (B)   1/8   (C)   1/4 (D) 5/16   [EC 2010]

Answer: (D)

Out of the 4 tosses, the outcome of (4 tosses will be 'H', no toss will be 'T') or (3 tosses will be 'H', 1 toss will be 'T').

Probability $= {}^4C_4 \left(\frac{1}{2}\right)^4 + {}^4C_3 \left(\frac{1}{2}\right)^3 \left(\frac{1}{2}\right)^1 = \frac{1}{16} + \frac{4}{16} = \frac{5}{16}$

Example 73. *A* and *B* play a game in which their chances of winning are in the ratio 3:2. Out of five games played *A*'s chance of winning three games is

(A)   0.246 (B)   0.346   (C)   0.446 (D)   0.546

Answer: (B)

Let X denotes the number of games won by *A*.

$$P(A) = \frac{3}{5}, P(B) = \frac{2}{5}, n = 5$$

$$P(X = 5) = {}^5c_3 \left(\frac{3}{5}\right)^3 \left(\frac{2}{5}\right)^{5-3} = 10 \times \frac{27}{125} \times \frac{4}{25} = 0.346$$

***Example* 74.** Consider an unbiased cubic dice with opposite faces coloured identically and each face coloured red, blue or green such that each colour appears only two times on the dice. If the dice is thrown thrice, the probability of obtaining red colour on the top face of the dice at least twice is [ME 2014]
Answer: 0.25 − 0.27
Let X be the event of obtaining red colour on the top face of the dice and

obtaining red color is 'success' then $n = 3, p = \dfrac{1}{3}, q = \dfrac{2}{3}$

Probability $= P(X \geq 2) = 1 - P(X < 2) = 1 - [P(X = 0) + P(X = 1)]$

$$= 1 - \left[{}^3C_0 \left(\frac{1}{3}\right)^0 \left(\frac{2}{3}\right)^3 + {}^3C_1 \left(\frac{1}{3}\right)\left(\frac{2}{3}\right)^2 \right] = 1 - \left[\frac{8}{27} + \frac{4}{9}\right] = \frac{7}{27} = 0.2593$$

***Example* 75.** An unbiased coin is tossed is tossed infinite number of times. The probability that the fourth head appears in the tenth toss is
(A) 0.067   (B) 0.073   (C) 0.082   (D) 0,091   [EC 2014]
Answer (C)
In the first 9 tosses, 3 heads and at the 10th toss, head.

Probability of getting a head = Probability of getting a tail $= \dfrac{1}{2}$

Required Probability $= {}^9C_3 \left(\dfrac{1}{2}\right)^6 \left(\dfrac{1}{2}\right)^3 \times \dfrac{1}{2} = 7 \times 12 \times \dfrac{1}{512} \times \dfrac{1}{2} = .082$

***Example* 76.** The probability of hitting a target by a bullet is 1/2.Two direct hits are required to destroy the target. If 4 bullets are thrown independently, the probability of destroying the target is
(A)   3/8   (B)   5/8   (C)   11/16   (D)   5/16
Answer: (C)
The random variable X denotes the number of bullets hitting the target.

X follows Binomial Distribution with $n = 4$ and $p = \dfrac{1}{2}$ and $q = \dfrac{1}{2}$.

Required probability

$$= P(X \geq 2) = 1 - P(X < 2) = 1 - \left[P(X = 0) + P(X = 1)\right]$$

$$= 1 - \left[{}^4C_0\left(\frac{1}{2}\right)^0\left(\frac{1}{2}\right)^4 + {}^4C_1\left(\frac{1}{2}\right)^1\left(\frac{1}{2}\right)^3\right] = 1 - \frac{5}{16} = \frac{11}{16}$$

Example 77. The number of accidents occurring in a plant in a month follows Poisson distribution with mean 5.2. The probability of occurrence of less than 2 accidents in the plant during a randomly selected month is
(A)   0,029   (B)   0.034   (C)   0.039   (D)   0.044   [ME 2014]
Answer: (B)

Let X denotes the number of accidents in a month.

$$P(X < 2) = P(X = 0) + P(X = 1) = e^{-5.2}\left[\frac{(5.2)^0}{0!} + \frac{(5.2)^1}{1!}\right]$$

$$= 0.0055 \times 6.2 = 0.034$$

Example 78. Let X be a Poisson variate with $P(1) = P(2)$, the variance of X is

(A)   1   (B)   2   (C)   3   (D)   4

Answer (B)

$$P(X = x) = e^{-\mu}.\frac{\mu^x}{x!} \qquad P(1) = P(2) \Rightarrow e^{-\mu}.\mu = e^{-\mu}\frac{\mu^2}{2!} \Rightarrow \mu = 2$$

Variance $= \mu = 2$

Example 79. If X is a Poisson variable such that

$P(X = 2) = 9P(X = 4) + 90P(X = 6)$, then the variance of X is

(A)   0   (B)   1   (C)   1/2   (D)   1/3

Answer: (B)

$$P(X = 2) = 9P(X = 4) + 90P(X = 6) \Rightarrow f(2) = 9f(4) + 90f(6)$$

$$\Rightarrow \frac{e^{-m}.m^2}{2!} = 9.\frac{e^{-m}m^4}{4!} + 90.\frac{e^{-m}m^6}{6!} \Rightarrow$$

$$\frac{1}{2} = \frac{9m^2}{24} + 90.\frac{m^4}{720} \Rightarrow m^4 + 3m^2 - 4 = 0 \Rightarrow m = 1 \text{ [As m > 0]}$$

Mean = variance = m = 1

Example 80. An observer counts 240 veh/hr at a specific highway location. Assume that the vehicle arrival at the location is Poisson distributed, the probability of having one vehicle arriving over a 30- second time interval is_____

Answer: 0.27

Mean of the distribution

$$= \mu = 240\,veh/hr = \frac{240}{60} = 4\,veh/min = 2\,veh/30\sec$$

$$P(X = 1) = \frac{e^{-\mu}.\mu^{1}}{1!} = e^{-2} \times 2 = 0.27$$

Example 81. Suppose p is number of cars per minute passing through a certain road junction between 5 PM and 6PM, and p has a Poisson distribution with mean 3. What is the probability of observing fewer than 3 cars during any given minute in this interval?

(A) $\dfrac{8}{2e^{3}}$ (B) $\dfrac{9}{2e^{3}}$ (C) $\dfrac{17}{2e^{3}}$ (D) $\dfrac{26}{2e^{3}}$ [CS/IT 2013]

Answer (C)

Here, mean of the distribution $= \lambda = 3$

By Poisson distribution, $P(X = x) = e^{-\lambda}.\dfrac{\lambda^{x}}{x!}, x = 0,1,2,----$

Required probability $= P(x < 3)$

$$= P(x = 0) + P(x = 1) + P(x = 2) = e^{-3}\left[\frac{3^{0}}{0!} + \frac{3^{1}}{1!} + \frac{3^{2}}{2!}\right] = \frac{17}{2e^{3}}$$

Example 82. If a random variable X has a Poisson distribution with mean 5, then $E[(X + 2)^{2}]$ equals _____.   [CS/IT 2017]

Answer: 54

For a poisson distribution, mean = variance = 5, $E(X) = 5,$

Variance $= E(X^{2}) - \{E(X)\}^{2} \Rightarrow 5 = E(X^{2}) - 25 \Rightarrow E(X^{2}) = 30$

$E[(X + 2)^{2}] = E(X^{2}) + 4E(X) + 4 = 30 + 20 + 4 = 54$

Example 83. Assume that in a traffic junction, the cycle of the traffic signal lights is 2 minutes of green (vehicle does not stop) and 3 minutes of red (vehicle stops). Consider that the arrival time of the vehicles time at the junction is uniformly distributed over a 5 minutes cycle. The expected waiting time (in minutes) for the vehicle at the junction is _ .   [EE 2017]

Answer: 0.9

The pdf of the waiting time (x) is $f(x) = \dfrac{1}{5}, 0 < x < 3$

$$\text{Prob.} = \int_0^3 xf(x)dx = \frac{1}{5}\left[\frac{x^2}{2}\right]_0^3 = 0.9$$

Example 84. Let X be a nominal variable with mean 1 and variance 4. The probability $P(X < 0)$ is

(A) 0.5   (B) Greater than zero and less than 0.5
(C) Greater than 0.5 and less than 1   (D) 1.0   [ME 2013]

Answer: (B)

Here $\mu = 1, \sigma = 2$ If z is a standard normal variable, $P(z < 0) = P(z > 0) = 0.5$

$$P(x < 0) = P\left(\frac{x-\mu}{\sigma} < \frac{0-\mu}{\sigma}\right) = P(z < -0.5) = P(z > 0.5)$$

$= P(z > 0) - P(0 < z < 0.5) = 0.5 - (\text{a positive quantity} < 0.5)$ which is greater than zero and less than 0.5.

Example 85. If $\{x\}$ is a continuous, real valued random variable defined over the interval $(-\infty, \infty)$ and its occurrence is defined by the density function

given as: $f(x) = \dfrac{1}{\sqrt{2\pi} * b} e^{-\frac{1}{2}\left(\frac{x-a}{b}\right)^2}$ where $'a'$ and $'b'$ are the statistical

attributes of the random variable $\{x\}$. The value of the integral

$$\int_{-\infty}^{a} \frac{1}{\sqrt{2\pi} * b} e^{-\frac{1}{2}\left(\frac{x-a}{b}\right)^2} dx \text{ is}$$

(A)   1   (B)   0.5   (C) π (D) 2π [CE 2014]

Answer: (B)

$f(x)$ is the pdf of the continuous random variable x,

299

so $\int_{-\infty}^{\infty} \frac{1}{\sqrt{2\pi} * b} e^{-\frac{1}{2}\left(\frac{x-a}{b}\right)^2} dx = 1$

The mean of x is a so the curve representing $f(x)$ is symmetrical about $x = a$.

Now, $\int_{-\infty}^{a} \frac{1}{\sqrt{2\pi} * b} e^{-\frac{1}{2}\left(\frac{x-a}{b}\right)^2} dx + \int_{a}^{\infty} \frac{1}{\sqrt{2\pi} * b} e^{-\frac{1}{2}\left(\frac{x-a}{b}\right)^2} dx = 1$

$\Rightarrow 2\int_{-\infty}^{a} \frac{1}{\sqrt{2\pi} * b} e^{-\frac{1}{2}\left(\frac{x-a}{b}\right)^2} dx = 1 \Rightarrow \int_{-\infty}^{a} \frac{1}{\sqrt{2\pi} * b} e^{-\frac{1}{2}\left(\frac{x-a}{b}\right)^2} dx = 0.5$

Example 86. A nationalized bank has found that the daily balance available in its savings accounts follows a normal distribution with a mean of Rs. 500 and a standard deviation of Rs. 50. The percentage of savings account holders, who maintain an average daily balance more than Rs. 500 is __[ME 2014]

Answer: 50%

If X represents daily balance in Rs, then the mean of $X = \mu = 500$ and s.d. of $X = \sigma = 50$. $z = \dfrac{X - \mu}{\sigma} = \dfrac{X - 500}{50}$

Now, $P(X > 500) = P\left(\dfrac{X - 500}{50} > \dfrac{500 - 500}{50}\right) = P(z > 0) = 0.5$

The required percentage = 50%

Example 87. Let U and V be two independent zero mean Gaussian random variables of variances $\dfrac{1}{4}$ & $\dfrac{1}{9}$ respectively. The probability $P(3V \geq 2U)$ is

(A) $\dfrac{4}{9}$ (B) $\dfrac{1}{2}$ (C) $\dfrac{2}{3}$ (D) $\dfrac{5}{9}$ [EC 2013]

Answer: (B)

$P(3V \geq 2U) = P(3V - 2U \geq 0) = P(W \geq 0), W = 3V - 2U$

$U \rightarrow N\left(0, \dfrac{1}{4}\right)$ & $V \rightarrow N\left(0, \dfrac{1}{9}\right) \Rightarrow W \rightarrow N(0, \sigma)$

[As U & V are independent random variables]

So, W has mean 0 and s.d. σ

$$P(W \geq 0) = P\left(\frac{W - \mu}{\sigma} \geq \frac{0 - \mu}{\sigma}\right) = P\left(Z \geq \frac{0 - 0}{\sigma}\right) = P(Z \geq 0) = 0.5$$

Example 88. Consider the following two normal distributions:

$$f_1(x) = \exp\left(-\pi x^2\right) \& \ f_2(x) = \frac{1}{2\pi}\exp\left\{-\frac{1}{4\pi}\left(x^2 + 2x + 1\right)\right\} \text{ If } \mu \text{ and } \sigma$$

denote the mean and standard deviation, respectively, then

(A) $\mu_1 < \mu_2$ & $\sigma_1^2 < \sigma_2^2$ (B) $\mu_1 < \mu_2$ & $\sigma_1^2 > \sigma_2^2$

(C) $\mu_1 > \mu_2$ & $\sigma_1^2 < \sigma_2^2$ (D) $\mu_1 > \mu_2$ & $\sigma_1^2 > \sigma_2^2$ [CH 2014]

Answer: (C)

The p.d.f. of a normal distribution is $f(x) = \frac{1}{\sigma\sqrt{2\pi}}\exp\left\{-\frac{(x-\mu)^2}{2\sigma^2}\right\}$

Comparing with $f(x)$ with $f_1(x) = \exp\left(-\pi x^2\right)$

$$\mu_1 = 0, \frac{1}{\sigma_1\sqrt{2\pi}} = 1 \Rightarrow \sigma_1 = \frac{1}{\sqrt{2\pi}}, \ \sigma_1^2 = \frac{1}{2\pi}$$

Comparing $f(x)$ with

$$f_2(x) = \frac{1}{2\pi}\exp\left\{-\frac{1}{4\pi}\left(x^2 + 2x + 1\right)\right\} = \frac{1}{2\pi}\exp\left\{-\frac{1}{4\pi}(x+1)^2\right\}$$

$$\mu_2 = -1, \frac{1}{\sigma_2\sqrt{2\pi}} = \frac{1}{2\pi} \Rightarrow \sigma_2 = \sqrt{2\pi}, \ \sigma_2^2 = 2\pi$$

So, $\mu_1 > \mu_2$ & $\sigma_1^2 < \sigma_2^2$

Example 89. Let X be a random variable following normal distribution with mean $+ 1$ and variance 4. Let Y be another normal variable with mean $- 1$ and variance unknown. If $P(X \leq -1) = P(Y \geq 2)$, the standard deviation of Y is

(A) 3 (B) 2 (C) $\sqrt{2}$ (D) 1 [CS/IT 2008]

301

Answer (A)

$$P(X \le -1) = P(Y \ge 2) \Rightarrow P\left(\frac{X-1}{2} \le \frac{-1-1}{2}\right) = P\left(\frac{Y+1}{\sigma} \ge \frac{2+1}{\sigma}\right)$$

$$P(z \le -1) = P\left(z \ge \frac{2+1}{\sigma}\right) \Rightarrow P(z \ge 1) = P\left(z \ge \frac{3}{\sigma}\right)$$

[As the standard normal curve is symmetrical about the mean $(z = 0)$, so $P(z \le -1) = P(z \ge 1)$]

$$\Rightarrow 1 = \frac{3}{\sigma} \Rightarrow \sigma = 3$$

Example 90.The mean I.Q. of a group of students is 90 with a standard deviation of 20. Assuming that I.Q. is normally distributed, the percentage of students with I.Q. over 100 is _____. $\left[\phi(0.5) = 0.6915\right]$

Answer: 30.8% -30.9%

If X is the normal variable denoting the I. Q. of the students,

$$P(X > 100) = P\left(\frac{X-90}{20} > \frac{100-90}{20}\right) = P(Z > 0.5) = 1 - P(Z \le 0.5)$$

$$= 1 - 0.6915 = 0.3085 \approx 30.85\% \quad \left[\phi(0.5) = P(z \le 0.5)\right]$$

Example 91.The joint probability density function of the random variable

$$X, Y \text{ is } f(x, y) = \begin{cases} k(3x + y), 1 \le x \le 3, 0 \le y \le 2 \\ 0, \qquad\qquad elsewhere \end{cases}. \text{ Then } k =$$

(A) $\dfrac{1}{7}$ (B) $\dfrac{1}{14}$ (C) $\dfrac{1}{21}$ (D) $\dfrac{1}{28}$

Answer (D)

$$\int_{x=1}^{3} \int_{y=0}^{2} k(3x + y)dxdy = 1 \Rightarrow k \int_{x=1}^{3} \left[3xy + \frac{y^2}{2}\right]_{y=0}^{2} dx = 1$$

$$\Rightarrow k \int_{1}^{3} (6x + 2)dx = 1 \Rightarrow 28k = 1 \Rightarrow k = \frac{1}{28}$$

Example 92. Two random variables X & Y are distributed according to

$$f_{X,Y}(x,y) = \begin{cases} (x+y), & 0 \le x \le 1, 0 \le y \le 1 \\ 0, & otherwise \end{cases}$$

The probability $P(X+Y \le 1)$ is _____. [EC 2016]

Answer: $1/3$

$$P(X+Y<1) = \int_{x=0}^{1}\int_{y=0}^{1-x}(x+y)dydx = \int_{x=0}^{1}\left[xy + \frac{y^2}{2} \right]_{y=0}^{1-x}dx$$

$$= \int_{x=0}^{1}\left[x(1-x) + \frac{(1-x)^2}{2} \right]dx = \int_{0}^{1}\left[\frac{1}{2} - \frac{x^2}{2} \right]dx = \left[\frac{x}{2} - \frac{x^3}{6} \right]_{0}^{1} = \frac{1}{3}$$

EXERCISE

1. A discrete random variable X has the following probability mass functions:

Values of X	0	1	2	3	4	5	6
P(X)	0	k	2k	3k	4k	5k	6k

Then $P(X \ge 4) =$

(A)   3/5   (B)   3/7   (C)   5/7   (D)   4/7

2. If $f(x) = \begin{cases} kx(1-x), & 0 < x < 1 \\ 0, elsewhere \end{cases}$ is a probability density function then $k =$

(A) 6   (B)   4   (C)   2   (D)   1

3. A random variable X has the probability density function

$$f(x) = \begin{cases} \frac{1}{4}, & -2 < x < 2 \\ 0, elsewhere \end{cases}$$ then $P(|X|>1)=$

(A) 1/4   (B)   1/2   (C)   1/3 (D)   1/6

$[P(|x|>1) = 1 - P(|x|<1) = 1 - P(-1<x<1) = 1 - \frac{1}{4}\int_{-1}^{1}dx = 1 - \frac{1}{4}.2 = \frac{1}{2}]$

4. Four unbiased coins are thrown. The mathematical expectation of the sum points on the dice is

(A) 8 (B)   10   (C)   12   (D)   14

[Mathematical expectation of the sum points on n dice is $\dfrac{7n}{2}$]

5. The probability density function of a random variable X is

$f(x) = 6(x-1)(2-x), 1 \le x \le 2$, then $P\left(\dfrac{5}{4} \le X \le \dfrac{3}{2}\right)$ is

(A) 11/16   (B)   11/32   (C)   5/8   (D) 3/8

6. A fair coin is tossed 10 times. What is the probability that the first 2 tosses will be head?

(A) $\left(\dfrac{1}{2}\right)^2$ (B) $^{10}C_2\left(\dfrac{1}{2}\right)^2$ (C) $\left(\dfrac{1}{2}\right)^{10}$ (D) $^{10}C_2\left(\dfrac{1}{2}\right)^{10}$ [EC 2009]

7. A box contains 25 parts of which 10 are defective. Two parts are drawn in a random manner from the box. The probability of both the parts being good is

(A) $\dfrac{7}{20}$ (B) $\dfrac{42}{125}$ (C) $\dfrac{25}{29}$ (D) $\dfrac{5}{9}$ [ME 2014]

8. The joint probability density function of the random variable X, Y is

$f(x,y) = \begin{cases} \dfrac{1}{28}(3x+y), 1 \le x \le 3, 0 \le y \le 2 \\ 0, \qquad elsewhere \end{cases}$. Then $P(X+Y<2) =$

(A) $\dfrac{13}{168}$ (B) $\dfrac{11}{164}$ (C) $\dfrac{9}{160}$ (D) $\dfrac{7}{142}$

9. The average number of misprints per page of a book is 2. The probability that a particular page is free from misprints is _____.

10. For a certain normal distribution, the first moment about 10 is 40. The mean of the distribution is

(A)   30   (B)   40   (C)   50   (D)   20

Answers:

1. (C) 2. (A) 3. (B) 4. (D) 5. (B) 6. (B) 7. (A) 8. (A) 9.(0.135 -0.136) 10. (C)
Mean = A + first moment about A = 10 + 40 = 50 [Here, A = 10]

Question of one mark:

Example 1. Consider $p(s) = s^3 + a_2 s^2 + a_1 s + a_0$ with real coefficients. It is known that its derivative $p'(s)$ has no real roots. The number of real roots of $p(s) = 0$ is

(A)   0   (B)   1   (C)   2   (D)   3 [EC 2018]

Answer: (B)

Between any two real roots of the equation $p(s) = 0$ there must be at least one real root of $p'(s) = 0$. As $p'(s) = 0$ has no real root, so the number of real roots of $p(s) = 0$ is either 0 or 1. But every equation of degree 3 has at least one root, so the number of real roots of $p(s) = 0$ is 1.

Example 2. Match the correct pairs:

Numerical Integration Scheme	Order of Fitting Polynomial
P. Simpson's 3/8 Rule	1. First
Q. Trapezoidal Rule	2. Second
R. Simpson's 1/3 Rule	3. Third

(A) P-2, Q-1, R-3   (B) P-3, Q-2, R-1

(C) P-1, Q-2, R-3   (D) P-3, Q-1, R-2   [ME 2013]

Answer: (D)

Example 3. If $f(0) = 12, f(3) = 6, f(4) = 8,$ then $f(x)$ will be

(A) $x^2 - 3x + 12$ (B) $x^2 - 5x$ (C) $x^3 - x^2 - 5x$ (D) $x^2 - 5x + 12$

Answer: (D)

Since three points are given so degree of the polynomial = 2

Let $f(x) = ax^2 + bx + c --- (i)$

$f(0) = 12 \Rightarrow c = 12, f(3) = 6 \Rightarrow 9a + 3b = -6 --- (ii)$

and $f(4) = 8 \Rightarrow 16a + 4b = -4 --- (iii)$

Solving (ii) & (iii), $a = 1, b = -5$ From (i), $f(x) = x^2 - 5x + 12$

Example 4. Which of the following is NOT TRUE?

(A) $\Delta = E - 1$ (B) $\Delta . \nabla = \Delta - \nabla$ (C) $\Delta / \nabla = \Delta + \nabla$ (D) $\nabla = 1 - E^{-1}$

Answer: (C)

Example 5. If $h = \pi$, then $\Delta(x + \cos x) =$

(A) $\pi + 2\cos x$ (B) $x - \sin x$ (C) $\pi - 2\cos x$ (D) $1 - \sin x$

Answer: (C)

$\Delta(x + \cos x) = \Delta x + \Delta \cos x = [(x + \pi) - x] + [\cos(x + \pi) - \cos x]$
$= \pi + [-\cos x - \cos x] = \pi - 2\cos x$

Example 6. $(\Delta - \nabla)(x^2) =$

(A) h^2 (B) $-2h^2$ (C) $2h^2$ (D) $-h^2$

Answer: (C)

$(\Delta - \nabla)(x^2) = [(x + h)^2 - x^2] - [x^2 - (x - h)^2] = 2h^2$

Example 7. In trapezoidal rule of evaluating the approximate value

of $\int_a^b f(x)dx$, the area given by the integral is approximated by the sum of

areas of a

(A) rectangle (B) sector (C) square (D) trapezium

Answer: (D)

Example 8. In trapezoidal rule of evaluating the approximate value

of $\int_a^b f(x)dx$, there exists no error if $f(x)$ is a

(A) parabolic function (B) linear function
(C) exponential function (D) logarithmic function

Answer: (B)

Example 9. In Simpson's 1/3rd rule of evaluating the approximate value

of $\int_a^b f(x)dx$, $f(x)$ is approximated by

(A) line segment (B) parabola (C) circular sector (D) part of ellipse

Answer: (B)

Example 10. The degrees of the approximating polynomial corresponding to trapezoidal and Simpson's 1/3rd rules, respectively are

(A) 2, 3 (B) 1, 2 (C) 1, 6 (D) 2, 6

Answer: (B)

Example 11. The error in the Weddle's Rule is

(A) h^4 (B) h^5 (C) h^6 (D) h^7

Answer: (D)

Example 12. The value of $\int\limits_{-2}^{2} f(x)dx$, where $f(-2) = 0, f(-1) = 3,$

$f(0) = 8, f(1) = 4$ & $f(2) = 1$ using Simpson's 1/3rd rule is

(A) 46 (B) 23 (C) 15 (D) 16.25

Answer: (C)

$$h = 1, \ I_S = \frac{1}{3}\left[(0+1)+4(3+4)+2(8)\right] = 15$$

Example 13. If the total error committed in Simpson's 1/3rd rule in

evaluating the integral $\int\limits_{a}^{b} f(x)dx$ be $-\dfrac{b-a}{m}h^4 f^{iv}(\xi), a < \xi < b$, then the

value of m is

(A) 120 (B) 160 (C) 180 (D) 60

Answer: (C)

Example 14. The order of convergence in the Newton-Raphson method is

(A) 3 (B) 2 (C) 1 (D) 4

Answer: (B)

Example 15. A root of the equation $x^4 - 3x + 1 = 0$ needs to be found using the Newton- Raphson method. If the initial guess x_0 is taken as 0, then the new estimate x_1 after the first iteration is

(A) 1/3 (B) - 1/3 (C) 3 (D) - 3 [CH 2010]

Answer: (A)

$$f(x) = x^4 - 3x + 1, f'(x) = 4x^3 - 3, f(0) = 1, f'(0) = -3$$

$$x_1 = x_0 - \frac{f(x_0)}{f'(x_0)} = 0 - \frac{f(0)}{f'(0)} = 0 - \frac{1}{-3} = \frac{1}{3}$$

***Example* 16.** Newton- Raphson method is used to compute a root of the equation $x^2 - 13 = 0$ with 3.5 as the initial value. The approximation after one iteration is

(A) 3.575　　(B) 3.676　　(C) 3.667　　(D) 3.607　[CS 2010]

Answer: (D)

$$f(x) = x^2 - 13 \Rightarrow f'(x) = 2x, \ f(3.5) = (3.5)^2 - 13 = -0.75, f'(3.5) = 7$$

$$x_{n+1} = x_n - \frac{f(x_n)}{f'(x_n)} \Rightarrow x_1 = 3.5 - \frac{-0.75}{7} = 3.5 + 0.107 = 3.607$$

***Example* 17.** The acceleration $f(t)$ (in cm/sec^2) of a particle in time t sec is measured as

$$f(0) = 0, f(1) = 10, f(2) = 15, f(3) = 28, \ f(4) = 35 \ \& \ f(5) = 54, \text{then}$$

its velocity in cm/s after 5 sec is _____.

Answer: 115

Acceleration is integrated to find velocity.

$$h = 1, \ I_T = \frac{1}{2}[(0+54) + 2(10+15+28+35)] = 115 \ cm/s$$

***Example* 18.** The definite integral $\int_1^3 \frac{1}{x} dx$ is evaluated using Trapezoidal rule with a step size of 1. The correct answer is _____. [ME 2014]

Answer: 1.1 to 1.2

$$a = 1, b = 3, n = 2, h = 1, n = (3-1)/1 = 3 \ \& \ f(x) = \frac{1}{x}$$

x	$f(x) = y = 1/x$
1	$y_0 = 1$
2	$y_1 = 0.5$
3	$y_2 = 0.333$

$$\int_1^3 \frac{1}{x} dx = \frac{h}{2}[(y_0 + y_2) + 2y_1] = \frac{1}{2}[1 + 0.333 + 2 \times 0.5] = 1.1667$$

Example 19. The integral $\int_1^3 \frac{1}{x} dx$ when evaluated by using Simpson's 1/3rd

rule on two equal sub-intervals each of length 1, equals
(A) 1.000 (B) 1.098 (C) 1.111 (D) 1.120   [ME 2011]
Answer: (C)

$$a = 1, b = 3, n = 2, h = (b-a)/n = 1 \ \& \ f(x) = \frac{1}{x}$$

x	$f(x) = y = 1/x$
1	$y_0 = 1$
2	$y_1 = 0.5$
3	$y_2 = 0.333$

$$\int_1^3 \frac{1}{x} dx = \frac{h}{3}[(y_0 + y_2) + 4y_1] = \frac{1}{3}[1 + 0.333 + 4 \times 0.5] = 1.111$$

Example 20. In order to evaluate the integral $\int_0^1 e^x dx$ with Simpson's 1/3rd

rule, values of the function e^x are used at $x = 0.0, 0.5 \& 1.0$. The absolute
value of the error of numerical integration is
(A) 0.000171   (B) 0.000440   (C) 0.000579   (D) 0.002718   [PI 2018]
Answer: (C)

x	0.0	0.5	1.0
$y = f(x) = e^x$	1	1.6487	2.7183

$$h = 0.5, y = f(x) = e^x \Rightarrow y^{iv}(x) = e^x \Rightarrow y^{iv}(\xi) = e^{0.5} = 1.6487$$
$$|E| = \frac{h^5}{180} y^{iv}(\xi), 0 < \xi < 1. \ \ \text{So,} \ |E| = \frac{(0.5)^4}{180} \times 1.6487 = 0.00057$$

Example 21. The velocity (v) of a tractor, which starts from rest, is given
at fixed interval of time (t) as follows:

309

t (min)	0	2	4	6	8	10	12	14	16	18	20
m min^{-1}	0	0.8	1.5	2.1	2.4	2.7	1.7	0.9	0.4	0.2	0

Using Simpson's 1/3rd rule, the distance covered by the tractor in 20 minutes will be _____ m. [AE 2018]

Answer: 25.86

$$d = \frac{2}{3}[(0+0)+4(0.8+2.1+2.7+0.9+0.2)+2(1.5+2.4+1.7+0.4)]$$

$$= \frac{2}{3}[0+4\times6.7+2\times6.0] = 25.86\,m$$

Example 22. The velocity v (in kilometer/minute) of a motorbike which starts from rest , is given at fixed interval of time (in minutes) as follows:

t	2	4	6	8	10	12	14	16	18	20
v	10	18	25	29	32	20	11	5	2	0

Using Simpson's 1/3rd rule, the distance covered by the bike in 20 minutes will be _____ km. [CS/IT 2015]

Answer: 308 - 310

At time 0, velocity = 0

$$d = \frac{2}{3}[(0+0)+4(10+25+32+11+2)+2(18+29+20+5)]$$

$$= \frac{2}{3}(320+144) = 309\,km$$

Example 23. Torque exerted on a flywheel over a cycle is listed in the table. Flywheel energy (in J per unit cycle) using Simpson's rule is

Angle(Degree)	0	60^0	120^0	180^0	240^0	300^0	360^0
Torque(N-m)	0	1066	-323	0	323	- 355	0

(A)   542   (B)   993   (C)   1444   (D)   1986

Answer: (B)

$$h = 60^0 = 60 \times \frac{\pi}{180} = \left(\frac{22}{21}\right)^c = 1.047^c$$

$$y_0 = 0, y_1 = 1066, y_2 = -323, y_3 = 0, y_4 = 323, y_5 = -355, y_6 = 0$$

$$E = \frac{h}{3}\left[(y_0 + y_6) + 4(y_1 + y_3 + y_5) + 2(y_2 + y_4)\right]$$

$$= \frac{1.047}{3}\left[(0 + 0) + 4(1066 + 0 - 355) + 2(-323 + 323)\right]$$

$$= \frac{1.047}{3} \times 2844 = 993(N - m)$$

Example 24. Function f is known at the following points:.

x	0	0.3	0.6	0.9	1.2	1.5	1.8	2.1	2.4
$f(x)$	0	0.09	0.36	0.81	1.44	2.25	3.24	4.41	5.76

2.7	3.0
7.29	9.00

The value of $\int_0^3 f(x)dx$ computed using trapezoidal rule is

(A) 8.983   (B)   9.033   (C) 9.017   (D) 9.045   [CS/IT 2013]

Answer: (D)

Here, $h = \dfrac{b-a}{n} = \dfrac{3-0}{10} = 0.3$

$$I_T = \frac{0.3}{2}\left[(0+9) + 2(0.09 + 0.36 + 0.81 + 1.44 + 2.25\right.$$

$$\left. + 3.24 + 4.41 + 5.76 + 7.29)\right] = 0.15(60.3) = 9.045$$

Example 25. A curve is drawn to pass through the points given in the following table:

x	1	1.5	2	2.5	3	3.5	4
$f(x)$	2	2.4	2.7	2.8	3	2.6	2.1

By Simpson's $1/3^{rd}$ rule, the area bounded by the curve, the x-axis and the lines $x=1$ and $x=4$ (in sq. units) is

(A) 7.2   (B) 7.4   (C) 7.6   (D) 7.8

Answer: (D)

$h = 0.5$

Area $I = \int\limits_{1}^{4} f(x)dx$

$$= \frac{0.5}{3}\left[(2+2.1)+4(2.4+2.8+2.6)+2(2.7+3)\right] = 7.8 \text{ sq. units}$$

Example 26. The area bounded by the curve described in the following table and the $x-$ axis from 1.47 t0 1.52 using Simpson's $3/8^{th}$ rule

x	1.47	1.48	1.49	1.50	1.51	1.52
$f(x)$	3.86	3.90	3.96	4.02	4.06	4.12

is_____.

Answer: 0.1942

$h = 0.01,$ Area $I = \int\limits_{1.47}^{1.52} f(x)dx$

$$= \frac{3}{8}(0.01)\left[(3.86+4.12)+2(4.02)+3(3.90+3.96+4.06)\right] = 1.942$$

Example 27. P(0, 3), Q(0.5, 4) and R(1, 5) are three points on the curve defined by $f(x)$. Numerical integration is carried out using both Trapezoidal rule and Simpson's rule within limits $x=0$ & $x=1$ for the curve. The difference between the two results will be

(A) 0   (B) 0.25   (C) 0.5   (D) 1   [ME 2017]

Answer: (A)

Here, $y_0 = 3, y_1 = 4, y_2 = 5, h = 0.5$

x	0	0.5	1
y	3	4	5

$$I_T = \frac{h}{2}[(y_0 + y_2) + 2y_1] = \frac{1}{4}(8+8) = 4$$

$$I_s = \frac{h}{3}[(y_0 + y_2) + 4y_1] = \frac{1}{6}(8+16) = 4 \quad \text{So, } I_T - I_S = 0$$

Example 28. With respect to numerical evaluation of the definite integral

$K = \int_a^b x^2 dx$, where a and b are given, which of the following

statements is/are TRUE?

(I) The value of K obtained by using trapezoidal rule is always greater than or equal to the exact value of the definite integral.

(II) The value of K obtained by using Simpson's rule is always equal to the exact value of the definite integral.

(A) I only (B) II only (C) both I and II (D) neither I nor II   [CS/IT 2014]

Answer: (C)

Here, $f(x) = x^3 \Rightarrow f''(x) = 2, f'''(x) = 0$

Let $I, I_T \& I_S$ denote respectively the exact value, the value obtained by using trapezoidal rule and by using Simpson's rule.

$$I - I_T = -\frac{nh^3}{12} f''(x) \le 0 \Rightarrow I_T \ge I \quad \text{which is (i).}$$

$$I - I_S = -\frac{h^5}{90} f^{iv}(x) = 0 \Rightarrow I_T = I \text{ which is (ii).}$$

Example 29. Gauss-Seidel method is used to solve the following equations (as per the given order):

$x_1 + 2x_2 + 3x_3 = 5, \ 2x_1 + 3x_2 + x_3 = 1, \ 3x_1 + 2x_2 + x_3 = 3$

Assuming initial guess $x_1 = x_2 = x_3 = 0$, the value of x_3 after the first

iteration is _____.   [ME 2016]

Answer: - 6

$x_1 = 5 - 2x_2 - 3x_3 = 5 - 2.0 - 3.0 = 5$ [Taking initial guess $x_2 = x_3 = 0$]

313

$$x_2 = \frac{1}{3}(1-2x_1-x_3) = \frac{1}{3}(1-2\times5-0) = -3 \text{ [Taking } x_1 = 5, x_3 = 0]$$

$$x_3 = 3-3x_1-2x_2 = 3-3\times5-2\times(-3) = -6 \text{ [Taking } x_1 = 5, x_2 = -3]$$

Example 30. The bisection method is applied to compute a zero of the function $f(x) = x^4 - x^3 - x^2 - 4$ in the interval [1, 9]. The method converges to a solution after ____ iterations.

(A) 1 (B) 3 (C) 5 (D) 7 [CS/IT 2012]

Answer: (B)

$$f(x) = x^4 - x^3 - x^2 - 4 \qquad a = 1, b = 9 \Rightarrow c_1 = \frac{a+b}{2} = 5$$

$f(1) < 0, f(5) > 0 \Rightarrow$ a root lies between 1 and 5

$$c_2 = \frac{1+5}{2} = 3$$

Again $f(1) < 0, f(3) > 0 \Rightarrow$ a root lies between 1 and 5

$$c_3 = \frac{1+3}{2} = 2 \text{ and } f(c_3) = f(2) = 0$$

The method converges to a solution for c_3 after 3rd iteration.

Example 31. The formula used in Newton-Raphson method to evaluate the positive root of the equation $x^2 = a$ is $x_{n+1} =$

(A) $\frac{1}{2}\left(x_n + \frac{a}{x_n}\right)$ (B) $\frac{1}{2}\left(x_n - \frac{a}{x_n}\right)$ (C) $\frac{1}{2}\left(x_n + \frac{a}{x_{n-1}}\right)$ (D) $\frac{1}{2}\left(x_n - \frac{a}{x_{n-1}}\right)$

Answer: (A)

$$f(x) = x^2 - a \Rightarrow f'(x) = 2x$$

$$x_{n+1} = x_n - \frac{f(x_n)}{f'(x_n)} = x_n - \frac{x_n^2 - a}{2x_n} = \frac{1}{2}\left(x_n + \frac{a}{x_n}\right)$$

Example 32. When the Newton-Raphson method is applied to solve the equation $f(x) = x^3 + 2x - 1 = 0$, the solution at the end of the first iteration with the initial guess value as $x_0 = 1.2$ is

(A) - 0.82 (B) 0.49 (C) 0.705 (D) 1.69 [EE 2013]

314

Answer: (C)

$$f(x) = x^3 + 2x - 1 \Rightarrow f'(x) = 3x^2 + 2$$

$x_0 = 1.2,$ By Newton Raphson method,

$$x_{n+1} = x_n - \frac{f(x_n)}{f'(x_n)} = x_n - \frac{x_n^3 + 2x_n - 1}{3x_n^2 + 2} = \frac{2x_n^3 + 1}{3x_n^2 + 2}$$

$$n = 0, \ x_1 = \frac{2x_0^3 + 1}{3x_0^2 + 2} = \frac{2(1.2)^3 + 1}{3(1.2)^2 + 2} = \frac{4.456}{6.32} = 0.705$$

Example 33. The Newton-Raphson method is used to solve the equation $f(x) = x^3 - 5x^2 + 6x - 8 = 0$. Taking the initial guess as $x = 5$, the solution obtained at the end of the first iteration is____. [EC 2015]

Answer: 4.2903

$$f(x) = x^3 - 5x^2 + 6x - 8 \Rightarrow f'(x) = 3x^2 - 10x + 6 \text{ and}$$

$$x_0 = 5 \ f(5) = 125 - 125 + 30 - 8 = 22, \ f'(5) = 3 \times 25 - 10 \times 5 + 6 = 31$$

$$x_1 = x_0 - \frac{f(x_0)}{f'(x_0)} = 5 - \frac{f(5)}{f'(5)} = 5 - \frac{22}{31} = 5 - 0.7097 = 4.2903$$

Example 34. The real root of the equation $5x - 2\cos x - 1 = 0$ (up to two decimal accuracy) is _____. [ME 2014]

Answer: 0.53 to 0.56

$$f(x) = 5x - 2\cos x - 1 \Rightarrow f'(x) = 5 + 2\sin x$$

$f(0) = -3 < 0, f(1) = 2.9 > 0$ A root lies between 0 & 1. Initial guess

$x_0 = 0.$ By Newton Raphson method,

$$x_{n+1} = x_n - \frac{f(x_n)}{f'(x_n)} = x_n - \frac{5x_n - 2\cos x_n - 1}{5 + 2\sin x_n} = \frac{2x_n \sin x_n + 2\cos x_n + 1}{5 + 2\sin x_n}$$

$$n = 0, \ x_1 = \frac{3}{5} == 0.6 \quad n = 1, x_2 = 0.5424, n = 2, x_3 = 0.5492$$

The real root (up to two decimal accuracy) = 0.54

Example 35. Consider the following differential equation $dy / dx = x + \ln y; y = 2$ at $x = 0$. The solution of this equation at $x = 0.4$ using Euler method with a step size of h= 0.2 is [CH 2014]

Answer: 2.3 to 2.4

315

$f(x, y) = x + \ln y, x_0 = 0, y_0 = 2, h = 0.2$

Formula: $y_{n+1} = y(x_{n+1}) = y(x_n + h) = y_n + h f(x_n, y_n)$

$n = 0, y_1 = y(x_1) = y(x_0 + h) = y_0 + hf(x_0, y_0)$

$\Rightarrow y(0.2) = 2 + 0.2 \times (0 + \ln 2) = 2 + 0.2 \times 0.6931 = 2.13862$

$n = 1, y_2 = y(x_2) = y(x_1 + h) = y_1 + hf(x_1, y_1)$

$\Rightarrow y(0.2 + 0.2) = 2.13862 + 0.2 \times (0.2 + \ln 2)$

$\Rightarrow y(0.4) = 2.13862 + 0.2 \times (0.2 + \ln 2.1386) = 2.3306$

EXERCISE

Question of one mark:

1. If $h = \pi$, then $\Delta(x + \cos x) =$

(A) $\pi + 2\cos x$ (B) $x - \sin x$ (C) $\pi - 2\cos x$ (D) $1 - \sin x$

2. In Gaussian elimination method, the given system of equations represented by $AX = B$ is converted to other system $UX = Y$, where U is a/an

(A) diagonal matrix   (B) null matrix

(C) identity matrix   (D) upper triangular matrix

Question of two marks:

3. The evaluated value of the integral $\int_0^1 \frac{1}{1+x} dx$ using Simpson's 1/3rd rule

with three points is

(A) 0.6875   (B) 0.6925   (C) 0.6950   (D) 0.6945

4. The area bounded by the curve described in the following table and the x – axis from 1.47 t0 1.52 using Simpson's 1/3rd rule

x	1.47	1.48	1.49	1.50	1.51	1.52
$f(x)$	3.86	3.90	3.96	4.02	4.06	4.12

Is_____.

5. Using Newton-Raphson method, the real root of

$3x - \cos x - 1 = 0$, correct up to three significant figures with $x_0 = 0$

is _____.

Answers:

1. (C)   2. (D)   3. (D)   4. 0.18567   5. 0.607

Chapter – 11

VERBAL ABILITY

- English Grammar
- Sentence Completion
- Verbal Analogies
- Word Groups
- Instructions
- Critical Reasoning & Verbal Deduction

Question of one mark:

Example 1. Out of the following four sentences, select the most suitable

sentence with respect to grammar and usage:

(A) I will not leave the place until the minister does not meet me.

(B) I will not leave the place until the minister doesn't meet me.

(C) I will not leave the place until the minister meet me.

(D) I will not leave the place until the minister meets me. [CS/IT 2016]

Answer: (D)

Example 2. Choose the grammatically CORRECT sentence:
(A) Two and two add four.
(B) Two and two become four.
(C) Two and two are four.
(D) Two and two make four.   [EC 2013]
 Answer: (D)
Example 3. Choose the option with words which are not synonyms.
(A) aversion, dislike   (B) luminous, radiant
(C) plunder, loot   (D) yielding, resistant   [EE 2017]
Answer: (D)

Example 4. Saturn is_____ to be seen on a clear night with the naked eye.
(A) enough bright   (B) bright enough
(C) as enough bright   (D)   bright as enough [EE 2017]
Answer: (B)

Example 5. After Rajendra Chola returned from his voyage to Indonesia, he ____to visit the temple in Thanjavur.
(A) was wishing (B) is wishing (C) wished (D) had wished   [EE 2017]
Answer: (C)

Example 6. Choose the word that is opposite in meaning to the word "coherent".
(A) sticky (B) well-connected (C) rambling (D) friendly
Answer: (C)

Example 7. Which one of the following options is the closest in meaning to the word **quarrel**?
(A) make out (B) call out (C) dig out (D) fall out   [EC 2013]
Answer: (D)

Example 8. A rewording of something written or spoken is a make out
(A) paraphrase (B) paradox (C) paradigm (D) paraffin   [CS/IT 2016]
Answer: (A)

Example 9. Which of the following is CORRECT with respect to grammar and usage?
Mount Everest is _____ .
(A)   the highest peak in the world
(B)   highest peak in the world
(C)   one of highest peak in the world
(D)   one of the highest peak in the world   [EC 2016]
Answer: (A)

Example 10. "Her _____ should not be confused with miserliness; she is ever willing to assist those in need."
The word that best fills the blank in the above sentence is
(A) cleanliness (B) punctuality (C) frugality (D) greatness [ME 2018]
Answer: (C)

Example 11. "Although it does contain some pioneering ideas, one would hardly characterize the ideas as _____."
The word that best fills the blank in the above sentence is
(A) innovative   (B) simple   (C) dull   (D) boring   [CE 2018]
Answer: (A)

Example 12. "A ____ investigation can sometimes yield new facts, but typically organized are more successful."
The word that best fills the blank in the above sentence is
(A) meandering (B) timely (C) consistent (D) systematic [CS/IT 2018]
Answer: (A)

Example 13. "A common misconception among writers is that sentence structure mirrors thought; the more _____ the structure, the more complicated the ideas."
The word that best fills the blank in the above sentence is
(A) detailed (B) simple (C) clear (D) convoluted   [EE 2018]
Answer: (D)

Example 14. A _____ investigation can sometimes yield new facts, but typically organized once are more successful.
(A) Meandering (B) Timely (C) Consistent (D) Systematic   [CS/IT 2018]
Answer: (A)
Meandering (Adjective) means proceeding in an undirected way.

Example 15.. The policeman asked the victim of a theft, "what did you _____".
(A) loose   (B) lose   (C) loss   (D) louse   [EC 2016]
Answer: (B)

Example 16. "From where are they bringing their books?
_____ bringing _____ books from ____." The words that best fill the blanks in the above sentence are
(A)   Their, they're, there   (B)   They're, their, there
(C)   There, their they're   (D)   They're, there, there [CS/IT 2018]
Answer: (B)

They're bringing their books from there.

Example 17. The unruly crowd demanded that the accused be _____ without trial.
(A) hanged   (B) hanging   (C) hankering   (D) hung   [ME 2016]
Answer: (A)

Example 18. The ways in which this game can be played _____ potentially infinite.
(A)   is   (B)   is being   (C)   are (D)   are being   [ME 2017]
Answer: (C)

Example 19. The volume of a sphere of diameter 1 unit is _____ than the volume of a cube of side 1 unit.

(A)   least   (B)   less   (C)   lesser (D)   low [ME 2016]
Answer: (B)

Example 20. A rewording of something written or spoken is a _____.

(A) paraphrase (B) paradox (C) paradigm (D) paraffin [CS/IT 2016]
Answer (A)

Example 21. Choose the correct verb to fill in the blank below:
Let us _____.
(A) Introvert (B) alternate (C) atheist (D) altruist [EC 2015]
Answer: (B)

Example 22. Choose the most appropriate word from the options given below to complete the following sentence?
If the athlete had wanted to come first in the race, he_____ several hours every day.
(A)   Should practice   (B)   Should have practised
(C)   Practised   (D)   Should be practicing [EC 2015]
Answer: (B)

Example 23. Choose the most suitable one word substitute for the following expression
Connotation of a road or way
(A) Pertinacious (B) Viaticum (C) Clandestine (D) Ravenous [EC 2015]
Answer: (A)
No word (option) is close to the meaning of the given group of words. But 'Pertinacious' has some relevance to the given group of words.

Example 24. Complete the sentence: Dare _____ mistakes.
(A) commit (B) to commit (C) committed (D) committing [EC 2013]
Answer: (B)

Example 25. As the two speakers became increasingly agitated, the debate became
(A) lukewarm (B) poetic (C) forgiving (D) heated   [ME 2017]
Answer: (D)

Example 26. Choose the most appropriate phrase from the options given below to complete the following sentence:
The aircraft _____ take off as soon as its flight plan was filed.
(A)   is allowed to   (B)   will be allowed to
(C)   was allowed to   (D)   has been allowed to [EC 2014]
Answer: (C)

Example 27. He was one of my best _____ and I felt his loss_____.
(A)   friend, keenly   (B)   friends, keen

(C)   friend, keener (D)   friends, keenly   [ME 2017]
Answer: (D)

Example 28. Research in the work place reveals that people work for many reasons _____.
(A)   money beside   (B)   money besides
(C)   beside money   (D)   besides money   [EE 2017]
Answer: (D)

Example 29. Choose the most appropriate word from the options given below to complete the following sentence:
Many ancient cultures attributed disease to supernatural causes.
However, modern science has largely helped _____ such notions.
(A) impel   (B) dispel   (C)   propel (D)   repel   [EC 2014]
Answer: (B)

Example 30. Choose the most appropriate word from the options given below to complete the following sentence.
Communication and interpersonal skills are _____ important in their own ways.
(A)   each   (B)   both   (C)   all   (D)   either [ME 2014]
Answer: (B)

Example 31. Choose the appropriate answer to complete the following sentence:
To those of us who had always thought him timid, his _____ came as a surprise.
(A) intrepidity (B) inevitability (C) inability (D) inertness
Answer: (A)

Example 32. Choose the most appropriate word from the options given below to complete the following sentence.
Communication and interpersonal skills are_____ important in their own ways.
(A)   each   (B)   both   (C)   all   (D)   either
Answer: (B)

Example 33. Choose the appropriate answer to complete the following sentence:
Medicine is to illness as law is to _____
(A) discipline (B) anarchy (C) treason (D) etiquette
Answer: (B)

Example 34. The unruly crowd demanded that the accused be _____ without trial.
(A) hanged   (B) hanging   (C) hankering   (D) hung   [ME 2016]
Answer: (A)

Example **35.** Choose the most appropriate word from the options given below to complete the following sentence.
He could not understand the judges awarding her the first prize, because he thought that her performance was quite _____.
(A) superb (B) medium (C) mediocre (D) exhilarating [EE 2014]
Answer: (C)

Example **36.** Complete the sentence:
Universalism is to particularism as diffuseness is to _____.
(A) specificity (B) neutrality (C) generality (D) adaptation [ME 2013]
Answer: (A)

Example **37.** Which one of the following options is the closest in meaning to the word given below?

Nadir

(A) Highest (B) Lowest (C) Medium (D) Integration [ME 2013]
Answer: (B)

Example **38.** While receiving the award, the scientist said, "I feel vindicated." Which of the following is closest in meaning to the word 'vindicated'?
(A) punished (B) substantiated (C) appreciated (D) chastened [EC 2014]
Answer: (B)

Example **39.** Choose the most appropriate pair of words from the options given below to complete the following sentence. She could not _____ the thought of _____ the election to her bitter rival.
(A) bear, loosing (B) bare, loosing (C) bear, losing (D) bare, losing
Answer: (C)

Example **40.** Which of the options given below best completes the following sentence?
She will feel much better if she _____.
(A)   will get some rest   (B) gets some rest
(C)   will be getting some rest (D) is getting some rest   [ME 2014]
Answer: (B)

Example **41.** Were you a bird, you _____ in the sky.
(A) would fly (B) shall fly (C) should fly (D) shall have flown   [ME 2013]
Answer: (A)

Example **42.** In a press meet on the recent scam, the minister said, "The buck stops here". What did the minister convey by the statement?

(A)   He wants all the money
(B)   He will return the money
(C)   He will assume final responsibility

322

(D) He will resist all inquiries   [EE 2014]

Answer: (C)

Example 43. If she _____ how to calibrate the instrument, she _____ done the experiment.

(A) knows, will have   (B) knew, had

(C) had known, could have (D) should have known, would have

Answer: (C)

Example 44. If you choose plan P, you will have to ____ plan Q as these two are mutually _____.

(A) forgo, exclusive   (B) forget, inclusive

(C) accept, exhaustive   (D) adopt, intrusive   [ME 2017]

Answer: (A)

Example 45. "Going by the ____ that many hands make light work, the school ____ involved all the students in the class."

The words that best fills the blank in the above sentence are

(A) principle, principal (B) principal, principle

(C) principle, principle (D) principal, principal   [ME 2018]

Answer: (A)

Example 46. "His face _____ with joy when the solution of the puzzle was _____ to him."

The words that best fill the blanks in the above sentence are

(A)   shone, shown   (B)   shone, shone

(C)   shown, shone   (D)   shown, shown   [CE 2018]

Answer: (A)

Example 47. "Since you have gone off the _____, the ____ sand is likely to damage the car."

The words that best fills the blank in the above sentence are

(A)   course, coarse   (B)   course, course

(C)   coarse, course (D)   coarse, coarse   [EE 2018]

Answer: (A)

Example 48. "By giving him the last ___ of the cake, you will ensure lasting _____ in our house today."

The words that best fill the blanks in the above sentence are

(A)   peas, piece   (B)   piece, peace

(C)   peace, piece   (D)   peace, peas   [EC 2018]

Answer: (B)

Example 49. "Even though there is a vast scope for its ___, tourism has remained a/an ____ area".

The words that best fill the blanks in the above sentence are

(A) improvement, neglected (B) rejection, approved
(C) flame, glum (D) interest, disinterested[EC 2018]
Answer: (A)

Example 50. Despite the new medicine's _____ in treating diabetes, it is not _____ widely.
(A) effectiveness --- prescribed (B) availability --- used
(C) prescription --- available (D) acceptance ____ prescribed [EC 2016]
Answer: (A)

Example 51. "From where are they bringing their books?" ____ bringing ____ books from ____.
The words that best fill the blanks in the above sentence are
(A) Their, they're, there (B) They're, their, there
(C) There, their, there (D) They're, there, their [CS/IT 2018]
Answer: (B)

Example 52. Choose the statement(s) where the underlined word is used correctly:
(i) A <u>prone</u> is a dried plum. (ii) He was lying <u>prone</u> on the floor.
(iii) People who eat a lot of fat are <u>prone</u> to heart disease.
(A) (i) and (iii) only (B) (iii) only
(C) (i) and (ii) only (D) (ii) and (iii) only [ME 2016]
Answer: (D)

Example 53. Fact: If it rains, then the field is wet. Read the following statements:
(i) It rains (ii) The field is not wet
(iii) The field is wet (iv) It did not rain
Which one of the options given below is NOT logically possible, based on the given fact?
(A) If (iii), then (iv). (B) If (i), then (iii).
(C) If (i), then (ii). (D) If (ii), then (iv). [ME 2016]
Answer: (B)

Example 54. Rahul, Murali, Srinivas and Arul are seated around a square table. Rahul is sitting to the left of Murali. Srinivas is sitting to the right of Arul. Which of the following pairs are seated opposite each other?
(A) Rahul and Murali (B) Srinivas and Arul
(C) Srinivas and Murali (D) Srinivas and Rahul [EE 2017]
Answer: (C)

Example 55. "The hold of the nationalist imagination on our colonial past is such that anything inadequately or improperly nationalist is just not

324

history."
Which of the following statements best reflects the author's opinion?
(A) Nationalists are highly imaginative.
(B) History is viewed through the filter of nationalism.
(C) Our colonial past never happens.
(D)Nationalism has to be both adequately and properly imagined.[EE 17]
Answer: (B)

Example **56. Statement:** You can always give me a ring whenever you need.
Which one of the following is the best inference from the above statement?
(A) Because I have a nice caller tune.
(B) Because I have a better telephone facility.
(C) Because a friend in need is a friend indeed.
(D) Because you need not pay towards the telephone bills when you give me a ring.   [EC 2013]
Answer: (C)

Example **57. Statement:** There were different streams of freedom movements in colonial India carried out by the moderates, liberals, radicals, socialists, and so on.
Which one of the following is the best inference from the above statement?
(A) The emergence of nationalism in colonial India led to our Independence.
(B) Nationalism in India emerged in the context of colonialism.
(C) Nationalism in India is homogeneous.
(D) Nationalism in India is heterogeneous.   [EC 2013]
Answer: (D)

Example **58.** While trying to collect an envelope from under the table,
          I   II
Mr. X fell down and was losing consciousness.
      III   IV
Which one of the above underlined parts of the sentence is NOT appropriate?
(A)   I   (B)   II   (C)   III   (D)   IV
Answer: (D)
Example **59.** Read the statements:
All women are entrepreneurs. Some women are doctors.

325

Which of the following conclusions can be logically inferred from the above statements?
(A) All women are doctors   (B) All doctors are entrepreneurs
(C) All entrepreneurs are women   (D) Some entrepreneurs are doctors
[EC 2014]
Answer: (D)

Example 60. Choose the statement(s) where the underlined word is used correctly:
(i)   A _prone_ is a dried plum.   (ii)   He was lying _prone_ on the floor.
(iii) People who eat a lot of fat are _prone_ to heart disease.
(A)   (i) and (iii) only   (B)   (iii) only
(C)   (i) and (ii) only   (D)   (ii) and (iii) only   [ME 2016]
Answer: (D)

Example 61. Fact: If it rains, then the field is wet. Read the following statements:
(i)   It rains   (ii)   The field is not wet
(iii)   The field is wet   (iv)   It did not rain
Which one of the options given below is NOT logically possible, based on the given fact?
(A)   If (iii), then (iv). (B)   If (i), then (iii).
(C)   If (i), then (ii).   (D)   If (ii), then (iv).   [ME 2016]
Answer: (B)

Example 62. The students _____ the teacher on teachers' day for twenty years of dedicated teaching.
(A) facilitated (B) felicitated (C) fantasized (D) facilitated [EC 2016]
Answer: (B)

Example 63. After India's cricket world cup victory in 1985, Shrotria who was playing both tennis and cricket till then, decided to concentrate only on cricket. And the _rest is history_. What does the underlined phrase mean in this context?
(A) history will rest in peace   (B) rest is recorded in history books
(C) rest is well known   (D) rest is archaic [EC 2016]
Answer: (C)

Example 64. The chain snatchers _took to their heels_ when the police party arrived.
(A)   took shelter in a thick jungle   (B)   open indiscriminate fire
(C)   took to fight
(D)   unconditionally surrendered   [CS/IT 2015]
Answer: (C)

Example 65. Choose the grammatically **INCORRECT** sentence:
(A) He is of Asian origin. (B) They belonged to Africa. [ME 2013]

(C) She is an European.   (D) They migrated from India to Australia.

Answer: (C)

Example 66. Out of the following four sentences, select the most suitable sentence with respect to grammar and usage.

(A) I will not leave the place until the minister does not meet me.

(B) I will not leave the place until the minister doesn't meet me.

(C) I will not leave the place until the minister meet me.

(D) I will not leave the place until the minister meets me. [CS/IT 2016]
Answer: (D)

Example 67. Archimedes said, "Give me a lever long enough and a fulcrum on which to place it, and I will move the world."
The sentence above is an example of a _____ statement.

(A) figurative (B) collateral   (C) literal (D) figurine [CS/IT 2016]
Answer: (A)

Example 68. All hill-stations have a lake. Ooty has two lakes. Which of the statement(s) below is/are logically valid and can be inferred from the above sentences?
(i) Ooty is not a hill-station
(ii) No hill-station can have more than one lake.
(A)   (i) only   (B)   (ii) only
(C)   both (i) and (ii)   (D)   neither (i) nor (ii)   [CS/IT 2016]
Answer: (D)
Example 69. If 'relftaga' means carefree, 'otaga' means careful and 'fertaga' means careless, which of the following could mean 'aftercare'?

(A) zentaga   (B) tagafer (C) tagazen   (D) relffer [CS/IT 2016]
Answer: (C)

Example 70. The man who is now Municipal Commissioner worked as
_____.

(A) the security guard at a university
(B) a security guard at the university
(C) a security guard at university

(D) the security guard at the university   [CS/IT 2016]
Answer: (B)

Example 71. Nobody knows how the Indian cricket team is going to <u>cope with</u> the difficult and seamer-friendly wickets in Australia.
Choose the option which is closest in meaning to the underlined phrase in the above sentence.
(A) put up with   (B) put in with
(C) put down to   (D) put up against   [CS/IT 2016]
Answer: (A)

Example 72. Which of the following options is the closest in meaning to the phrase underlined in the sentence below?
It is fascinating to see life forms **cope with** varied environmental conditions.
(A) adopt to (B) adapt to (C) adept in (D) accept with [EE 2014]
Answer: (B)

Example 73. The chairman requested the aggrieved shareholders to _____him.
(A) bear with (B) bore with (C) bear with (D) bare   [EE 2016]
Answer (C)

Example 74. Didn't you buy _____ when you went shopping?
(A) any paper (B) much paper (C) no paper (D) any paper [CS 2015]
Answer: (A)

Example 75. What is the adverb for the given word below?
Misogynous

(A)   Misogynousness   (B)   Misogynity

(C)   Misogynous   (D)   Misogynous   [EC 2015]
Answer: (C)

Example 76. Choose the appropriate word–phrase out of the four options given below, to complete the following sentence
Dhoni, as well as the other team members of Indian team _____ present on the occasion
(A)   Were   (B)   Was (C)   Has   (D)   Have [EC 2015]
Answer: (A)

Example 77. Choose the word most similar in meaning to the given word:
Awkward
(A) Inept (B) Graceful (C) Suitable (D) Dreadful   [EC 2015]
Answer: (A)

Example **78**. Choose the word most similar in meaning to the given word:
Educe
(A) Exert (B) Educate (C) Extract (D) Extend   [EC 2015]
Answer: (C)

Example **79**. Choose the most appropriate word from the options given below to complete the following sentence.
The principal presented the chief guest with a _____, as token of appreciation.
(A) momento (B) memento (C) momentum (D) moment [EC 2015]
Answer: (B)

Example **80.** Choose the appropriate word/phrase, out of the four options given below, to complete the following sentence:

 Frogs _____.
(A) Croak (B) Roar (C) Hiss (D) Patter   [EC 2015]
Answer: (A)

Example **81.** Identify the correct spelling out of the given options:
(A)   Managable   (B)   Manageable
(C)   Manageble   (D)   Managible   [EE 2016]
Answer: (B)

Example **82.** If she _____ how to calibrate the instrument, she _____ done the experiment.
(A) knows, will have   (B) knew, had
(C) had known, could have
(D) should have known, would have   [CS/IT 2014]

Answer: (C)

 Example **83.** Choose the word that is opposite in meaning to the word "coherent".
(A) sticky (B) well-connected (C) rambling (D) friendly [CS/IT 2014]

Answer: (C)

Example **84.** Didn't you buy _____ when you went shopping?
(A) any paper (B) much paper (C) no paper (D) a few paper [CS/IT 2015]
Answer: (A)

Example **85.** Which of the following options is the closest in meaning to the sentence below?
She enjoyed herself immensely at the party.
(A)   She had a terrible time at the party

(B)   She had a horrible time at the party
(C)   She had a terrific time at the party
(D)   She had a terrifying time at the party   [CS/IT 2015]
Answer (C)

Example 86. Which one of the following combinations is incorrect?
(A) Acquiescence - Submission (B) Wheedle – Round about
(C) Flippancy - Lightness   (D) Profligate – Extravagant [CS 15]
Answer: (B)

Example 87. Based on the following statements, select the following options to solve the given question.
If two floors in a certain building are 9 feet apart, how many steps are there in a set of stairs that extends from the first floor to the second floor of the building?
(I)   Each step is 3/4 foot high   (II) Each step is 1 foot wide
(A) Statement I alone is sufficient, but Statement II alone is not sufficient
(B) Statement II alone is not sufficient, but Statement I alone is sufficient
(C) Both statements together are sufficient but neither statement alone is sufficient
(D) Statements I and II together are not sufficient   [CS/IT 2015]
Answer: (A)
Width of the stairs is not required in the calculation, only height of each step is sufficient.

Example 88. If I were you, I _____ that laptop. It's much too expensive.
(A)   won't buy   (B)   shan't buy
(C)   wouldn't buy   (D)   would buy   [CE 2016]
Answer (C)

Example 89. He turned a deaf ear to my request. What does the underlined phrasal verb mean?
(A) ignored (B) appreciated (C) twisted (D) returned   [CE 2016]
Answer: (A)

Example 90. Choose the most appropriate set of words from the options given below to complete the following sentence.
_____, _____ is a will, _____ is a way.
(A)   Wear, there, their   (B)   Were, their, there
(C)   Where, there, there   (D)   Where, their, their [CE 2016]
Answer: (C)

Example **91.** Find the odd one in the following group of words:
mock, deride, praise, jeer
(A) mock (B) deride (C) praise (D) jeer   [CS/IT 2016]
Answer: (C)

Question of two marks:

Example **92.** R2D2 is a robot. R2D2 can repair aeroplanes. No other robot
can repair aeroplanes.
Which of the following can be logically inferred from the above
statements?
(A) R2D2 is a robot which can only repair aeroplanes.
(B) R2D2is the only robot which can repair aeroplanes.
(C) R2D2 is a robot which can repair only aeroplanes.
(D) Only R2D2 is a robot.   [EE 2016]
Answer (B)

Example **93.** Pick the odd one from the following options:
(A) CADBE (B) JHKIL (C) XVYWZ (D) ONPMQ   [CS/IT 2016]
Answer: (D)
The positions of the letters in the alphabet are written in the second row:

C A D B E   J H K I L   X V Y W Z

3 1 4 2 5   10 8 11 9 12   24 22 25 23 26

O N P M Q

15 14 16 13 17
Second digit = First digit -2, third digit = second digit + 3
fourth digit = third digit – 2, fifth digit = fourth digit + 2 which is not true
for (D).

Example **94.** The Palghat Gap (or Palakkad Gap), a region about 30 km
wide in the southern part of the Western Ghats in India, is lower than the
hilly terrain to its north and south. The exact reasons for the formation of
this gap are not clear. It results in the neighbouring regions of Tamil Nadu
getting more rainfall from the South West monsoon and the neighbouring
regions of Kerala having higher summer temperatures.
What can be inferred from this passage?
(A) The Palghat gap is caused by high rainfall and high temperatures in
southern Tamil Nadu and Kerala

(B) The regions in Tamil Nadu and Kerala that are near the Palghat Gap are low-lying
(C) The low terrain of the Palghat Gap has a significant impact on weather patterns in neighbouring parts of Tamil Nadu and Kerala
(D) Higher summer temperatures result in higher rainfall near the Palghat Gap area [EE 2014]
Answer: (C)

Example 95. Geneticists say that they are very close to confirming the genetic roots of psychiatric illnesses such as depression and schizophrenia, and consequently, that doctors will be able to eradicate these diseases through early identification and gene therapy.
On which of the following assumptions does the statement above rely?
(A) Strategies are now available for eliminating psychiatric illnesses
(B) Certain psychiatric illnesses have a genetic basis
(C) All human diseases can be traced back to genes and how they are expressed
(D) In the future, genetics will become the only relevant field for identifying psychiatric illnesses [EE 2014]
Answer: (B)

Example 96. After several defeats in wars, Robert Bruce went in exile and wanted to commit suicide. Just before committing suicide, he came across a spider attempting tirelessly to have its net. Time and again, the spider failed but that did not deter it to refrain from making attempts. Such attempts by the spider made Bruce curious. Thus, Bruce started observing the near-impossible goal of the spider to have the net. Ultimately, the spider succeeded in having its net despite several failures. Such act of the spider encouraged Bruce not to commit suicide. And then, Bruce went back again and won many a battle, and the rest is history. Which of the following assertions is best supported by the above information?
(A)   Failure is the pillar of success.
(B)   Honesty is the best policy.
(C)   Life begins and ends with adventures.
(D)   No adversity justifies giving up hope. [ME 2013]
Answer: (D)

Example 97. The Cricket Board has long recognized John's potential as a leader of the team. However his on- field temper has always been a matter of concern for them since his junior days. While this aggression has filled stadia with die-hard fans, it has taken a toll on his own batting.

Until recently, it appeared that he found it difficult to convert his aggression into big scores. Over the past three seasons, though that picture of John has been replaced by a cerebral, calculative and successful batsman-captain. After many years, it appears that the team has finally found a complete captain.

Which of the following statements can be logically inferred from the above paragraph?

(i) Even as a junior cricketer John was considered a good captain.
(ii) Finding a complete captain is a challenge.
(iii) Fans and the Cricket Board have differing views on what they want in a captain.
(iv) Over the past three seasons John has accumulated big scores.

(A)   (i), (ii) & (iii) only   (B)   (iii) & (iv) only
(C)   (ii) & (iv) only   (D)   (i), (ii), (iii) & (iv) [EC 2018]
Answer: (C)

Example 98. Read the following paragraph:

"The ordinary form of mercury thermometer is used for temperature ranging from -40^0 F to 500^0 F. For measuring temperature below -40^0 F, thermometers filled with alcohol are used. These are, however, not satisfactory for use in high temperatures. When a mercury thermometer is used for temperature above 500^0 F, the space above the mercury is filled with some inert gas, usually nitrogen or carbon dioxide, placed in the thermometer under pressure. As the mercury rises, the gas pressures is increased, so that it is possible to use these thermometers for temperatures as high as 1000^0 F."

With what, besides mercury, would a thermometer be filled if it was designed to be used for measuring temperature of about 500^0 F?

(A) Pyrometer (B) Inert gas (C) Iron and brass (D) Gas
Answer: (B)

Example 99. The cost of manufacturing tractors in Korea is twenty percent less than the cost of manufacturing tractors in Germany. Even after transportation fees and import taxes are added, it is still cheaper to import tractors from Korea to Germany than to produce tractors in Germany.

Which of the following assertions is best supported by the above information?

(A) Labour costs in Korea are twenty percent below those in Germany.
(B) Importing tractors into Germany will eliminate twenty percent of the manufacturing jobs in Germany.

(C) The costs of transporting a tractor from Korea to Germany is more than twenty percent.

(D) The import taxes on a tractor imported from Korea to Germany is less than twenty percent of the cost of manufacturing the tractor in Germany.

Answer: (D)

Example 100. "We lived in a culture that denied any merit to literary works, considering them important only when they were hand maidens to something seemingly more urgent- namely ideology. This was a country where all gestures, even the most private, were interpreted in political terms.

The author's belief that ideology is not as important as literature is revealed by the word

(A) 'culture' (B) 'seemingly' (C) 'urgent' (D) 'political'   [EE 2017]

Answer: (B)

Example 101. There were three boxes. One contains apples, another contains oranges and the last one contains both apples and oranges. All three are known to be incorrectly levelled. If you are permitted to open just one box and then pull out and inspect only one fruit, which box would you open to determine the contents of all the three boxes?

(A) The box levelled 'Apples'   (B) The box levelled 'Apples and Oranges'

(C) The box levelled 'Oranges' (D) Cannot be determined   [EE 2017]

Answer: (B)

We rename the boxes as:

Box I $\rightarrow$ wrongly levelled as 'Apples'

Box II $\rightarrow$ wrongly levelled as 'Oranges'

Box III $\rightarrow$ wrongly levelled as 'Apples and Oranges'

If the Box III levelled 'Apples and Oranges' is opened then it will rightly contain either 'only Apple' or ' only orange'. If one fruit is inspected, then it will be rightly a box of 'Apple' or 'orange'. Suppose it is a box of 'Apples'.

Box III $\rightarrow$ 'Apples' Box II $\rightarrow$ 'Apples and Oranges' Box I $\rightarrow$ 'Oranges'

Example 102. "If you are looking for a history of India, or for an account of the rise or fall of the British Raj or for the reason of the cleaving of the subcontinent into two mutually antagonistic parts and the effects this mutilation will have in the respective sections, and the effects ultimately on Asia, you will not find it in these ages: for though I have spent a life time in the country. I lived too near the seat of events, and was too intimately associated with the actors, to get the perspective needed for the impartial recording of these matters".

Which of the following is closest in meaning to 'cleaving'?

(A) deteriorating (B) arguing (C) departing (D) splitting [ME 2017]
Answer: (D)

Example 103. Two very famous sports men Marks and Steve happened to be brothers and played for the country K. Mark teased James, an opponent from country E, "There is no way you are good enough to play for your country". James replied "may be not but at least I am the best player in my own family."
Which one of the following can be inferred from this conversation?
(A) Mark was known to play better than James
(B) Steve was known to play better than Mark
(C) James and Steve were good friends
(D) James played better than Steve   [ME 2017]
Answer: (B)

Example 104. A coastal region with unparalleled beauty is home to many species of animals. It is dotted with coral leafs and unspoilt white sandy beaches. It has remained inaccessible to tourists due to poor connectivity and lack of accommodation. A company has spotted the opportunity and is planning to develop a luxury resort with helicopter service to the nearest city airport. Environmentalists are upset that this would lead to the region becoming crowded and polluted like any other major beach resorts.
Which one of the following statements can be logically inferred from the information given in the above paragraph?
(A) The culture and tradition of the local people will be influenced by the tourists.
(B) The region will become crowded and polluted due to tourism.
(C) The coral reefs are on the decline and could soon vanish.
(D) Helicopter connectivity would lead to an increase in tourists coming to the region.   [EC 2018]
Answer: (D)

Example 105. In a world filled with uncertainty, he was glad to have many friends. He had always assisted them in times of need and he was confident that they would reciprocate. Howevver, the events of the last week proved him wrong.
Which of the following inference(s) is/are logically valid and can be inferred from the above passage?
(i) His friends were always asking him to help
(ii) He felt that when in need of help, his friends would let him down.
(iii) He was sure that his friends would help him when in need.

(iv) His friends did not help him last week.

(A) (i) and (ii) (B) (iii) and (iv) (C) (iii) only (D) (iv) only   [EC 2016]

Answer: (B)

Example 106. Leela is older than her cousin Pavithra. Pavithra's brother Shiva is older than Leela. When Pavithra and Shiva are visiting Leela, all three like to play chess. Pavithra wins more often than Leela does. Which one of the following statements must be TRUE based on the above?

(A) When Shiva plays chess with Leela and Pavithra, he often loses.

(B) Leela is the oldest of the three.

(C) Shiva is a better chess player than Pavithra.

(D) Pavithra is the youngest of the three   [EC 2016]

Answer: (D)

Older   to   Younger

Shiva $\rightarrow$ Leela $\rightarrow$ Pavithra

Example 107. A smart city integrates all modes of transport, uses clean energy and promotes sustainable use of resources. It also uses technology to ensure safety and security of the city, something which critics argue, will lead to a surveillance state. Which of the following can be logically inferred from the above paragraph?

(i) All smart cities encourage the formation of surveillance states.

(ii) Surveillance is an integral part of a smart city.

(iii) Sustainability and surveillance go hand in hand in a smart city.

(iv) There is a perception that smart cities promote surveillance.

(A) (i) and (iv) only   (B)   (ii) and (iii) only

(C) (iv) only   (D)   (i) only   [ME 2016]

Answer: (C)

Example 108. Social science disciplines were in existence in an amorphous form until the colonial period when they were institutionalized. In varying degrees, they were intended to further the colonial interest. In the time of globalization and the economic rise of postcolonial countries like India, conventional ways of knowledge production have become obsolete.

Which of the following can be logically inferred from the above statements?

(i)   Social science disciplines have become obsolete

(ii) Social science disciplines had a pre-colonial origin

(iii) Social science disciplines always promote colonialism

336

(iv) Social science must maintain disciplinary boundaries

(A)   (ii) only   (B)   (i) and (ii) only

(B)   (ii) and (iv) only   (D)   (iii) and (iv) only   [EC 2016]

Answer: (A)

Example 109. Indian currency notes show the denomination indicated in at least seventeen languages. If this is not an indication of the nation's diversity, nothing else is. Which of the following can be logically inferred from the above sentences?
(A) India is a country of exactly seventeen languages.
(B) Linguistic pluralism is the only indicator of a nation's diversity.
(C) Indian currency notes have sufficient space for all the Indian languages.
(D) Linguistic pluralism is strong evidence of India's diversity.[CS/IT 2016]
Answer: (D)

Example 110. Consider the following statements relating to the level of poker play of four players P, Q, R and S.
I.   P always beats Q   II. R always beats S
III. S loses to P only sometimes   IV. R always loses to Q
Which of the following can be logically inferred from the above statements?
(i) P is likely to beat all the three other players
(ii) S is the absolute worst player in the set
(A)   (i) only   (B)   (ii) only
(C)   (i) and (ii)   (D)   neither (i) nor (ii)   [CS/IT 2016]

Answer: (D)

Examples 111. Computers were invented for performing only high-end useful computations. However, it is no understatement that they have taken over our world today. The internet, for example, is ubiquitous. Many believe that the internet itself is an unintended consequence of the original invention with the advent of mobile computing on our phones, a whole new dimension is now enabled. One is left wondering if all these developments are good or more importantly, required. Which of the statement(s) below is/are logically valid and can be inferred from the above paragraph?
(i) The author believes that computers are not good for us
(ii) Mobile computers and the internet are both intended inventions
(A)   (i) only   (B)   (ii) only
(C)   both (i) and (ii)   (D) neither (i) nor (ii)   [CS/IT 2016]
Answer: (D)

Example 112. A poll of students appearing for masters in engineering indicated that 60% of the students believed that mechanical engineering is a profession unsuitable for women. A research study on women with master or higher degrees in mechanical engineering found that 99% of such women were successful in their professions. Which of the following can be logically inferred from the above paragraph?
(A) Many students have misconceptions regarding various engineering disciplines
(B) Men with advanced degrees in mechanical engineering believe women are well suited to be mechanical engineering.
(C) Mechanical engineering is a profession well suited for women with masters or higher degrees in mechanical engineering.
(D) The number of women pursuing high degrees in mechanical engineering is small. [EE 2016]
Answer: (C)

Example 113. Sourya committee had proposed the establishment of Sourya Institutes of Technology (SITs) in line with Indian Institutes of Technology (IITs) to cater to the technological and industrial needs of a developing country. Which of the following can be logically inferred from the above sentence based on the proposal,

(i) In the initial years, SIT students will get degrees from IIT.
(ii) SITs will have a distinct national objective
(iii) SIT like institutions can only be established in consultation with IIT.
(iv) SITs will serve technological needs of a developing country.

(A)   (iii) and (iv) only   (B)   (i) and (iv) only
(C)   (ii) and (iv) only   (D)   (ii) and (iii) only   [EE 2016]
Answer: (C)

Example 114. Today, we consider Ashoka as a great ruler because of the copious evidence he left behind in the form of stone carved edicts. Historians tend to correlate greatness of a king at his time with the availability of evidence today. Which of the following can be logically inferred from the above sentences?
(A) Emperors who do not leave significant sculpted evidence are completely forgotten.
(B) Ashoka produced stone carved edicts to ensure that later historians will respect him.
(C) Statues of kings are a reminder of their greatness.
(D) A king's greatness, as we know him today, is interpreted by historians

Answer: (D)

Example 115. Fact 1: Humans are mammals.

Fact 2: Some humans are engineers.

Fact 3: Engineers build houses.

If the above statements are facts, which of the following can be logically inferred?

I. All mammals build houses.
II. Engineers are mammals.
III. Some humans are not engineers.
(A) II only. (B) III only. (C) I, II and III. (D) I only. | [CE 2016]
Answer: (B)

Example 116. Given below are two statements followed by two conclusions. Assuming these statements to be true, decide which one logically follows.

Statements: I. All film stars are playback singers.

II. All film directors are film stars

Conclusions: I. All film directors are playback singers.
II. Some film stars are film directors.

Which of the following must be true?
(A)   Only conclusion I follows
(B)   Only conclusion II follows
(C)   Neither conclusion I nor II follows
(D)   Both conclusions I and II follow   [EC 2015]
Answer: (D)

Example 117. Lamenting the gradual side-lining of the arts ill school curricula, a group of prominent artists wrote to the Chief Minister last year, asking him to allocate more funds to support arts education in schools. However, no such increase has been announced in this year's Budget. The artists expressed their deep anguish at their request not being approved, but many of them remain optimistic about funding in the future.

Which of the statement(s) below is/are logically valid and can be inferred from the above statements?

(i) The artists expected funding for the arts to increase this year
(ii) The Chief Minister was receptive to the idea of increasing funding for the arts

(iii) The Chief Minister is a prominent artist
(iv) Schools are giving less importance to arts education nowadays
(A) (iii) and (iv)   (B) (i) and (iv)
(C) (i), (ii) and (iv)   (D) (i) and (iii)   [EC 2015]
Answer: (C)

Example **118.** By the beginning of the 20th century, several hypotheses were being proposed, suggesting a paradigm shift in our understanding of the universe. However, the clinching evidence was provided by experimental measurements of the position of a star which was directly behind our sun. Which of the following inference(s) may be drawn from the above passage?

(i) Our understanding of the universe changes based on the positions of stars
(ii) Paradigm shifts usually occur at the beginning of centuries
(iii) Stars are important objects in the universe
(iv) Experimental evidence was important in confirming this paradigm shift

(A) (i), (ii) and (iv) (B) (iii) only (C) (i) and (iv) (D) (iv) only [CS/IT 2014]
Answer: (D)

Example **119.** Ananth takes 6 hours and Bharath takes 4 hours to read a book. Both started reading copies of the book at the same time. After how many hours is the number of pages to be read by Ananth, twice that to be read by Bharath?
Assume Ananth and Bharath read all the pages with constant pace.
(A)   1   (B)   2   (C)   3   (D)   4   [CE 2016]
Answer: (C)

Example **120.** Humpty Dumpty sits on a wall every day while having lunch. The wall sometimes breaks. A person sitting on the wall falls if the wall breaks. Which one of the statements below is logically valid and can be inferred from the above sentences?
(A) Humpty Dumpty always falls while having lunch
(B) Humpty Dumpty does not fall sometimes while having lunch
(C) Humpty Dumpty never falls during dinner
(D) When Humpty Dumpty does not sit on the wall, the wall does not break   [EC 2015]

Answer: (B)

*Example***121.** The following question presents a sentence, part of which is underlined. Beneath the sentence you find four ways of phrasing the underline part. Following the requirements of the standard written English, select the answer that produces the most effective sentence. Tuberculosis, together with its effects, <u>ranks one of the leading causes of death</u> in India.

(A) ranks as one of the leading causes of death
(B) rank as one of the leading causes of death
(C) has the rank of one of the leading causes of death
(D) are one of the leading causes of death   [EC 2015]

Answer: (A)

Example 122. Read the following paragraph and choose the correct statement.

Climate change has reduced human security and threatened human well being. An ignored reality of human progress is that human security largely depends upon environmental security. But on the contrary, human progress seems contradictory to environmental security. To keep up both at the required level is a challenge to be addressed by one and all. One of the ways to curb the climate change may be suitable scientific innovations, while the other may be the Gandhian perspective on small scale progress with focus on sustainability.

(A) Human progress and security are positively associated with environmental security.
(B) Human progress is contradictory to environmental security.
(C) Human security is contradictory to environmental security.
(D) Human progress depends upon environmental security.   [EC 2015]
Answer: (B)

Example 123. In the following sentence certain parts are underlined and marked P, Q and R. One of the parts may contain certain error or may not be acceptable in standard written communication. Select the part containing an error.

Choose D as your answer if there is no error. The student corrected <u>all the errors</u> that <u>the instructor marked</u> on the <u>answer book</u>.

    P   *Q*   *R*

(A)   P   (B)   Q   (C)   R   (D)   No error   [EC 2015]
Answer: (B)
'Q' is not required.

Example 124. In the following question, the first and the last sentence of the passage are in order and numbered 1 and 6. The rest of the passage is split into 4 parts and numbered as 2,3,4, and 5. These 4 parts are not arranged in proper order. Read the sentences and arrange them in a logical sequence to make a passage and choose the correct sequence from the given options.

1. One Diwali, the family rises early in the morning.

2. The whole family, including the young and the old enjoy doing this,

3. Children let off fireworks later in the night with their friends.

4. At sunset, the lamps are lit and the family performs various rituals

5. Father, mother, and children visit relatives and exchange gifts and sweets.

6. Houses look so pretty with lighted lamps all around.

(A) 2, 5, 3, 4 (B) 5, 2, 4, 3 (C) 3, 5, 4, 2 (D) 4, 5, 2, 3   [EC 2015]

Answer: (B)

***Example* 125.** Ms. X will be in Bagdogra from 01/05/2014 to 20/05/2014 and from 22/05/2014 to 31/05/2014. On the morning of 21/05/2014, she will reach Kochi via Mumbai.
Which one of the statements below is logically valid and can be inferred from the above sentences?
(A) Ms. X will be in Kochi for one day, only in May
(B) Ms. X will be in Kochi for only one day in May
(C) Ms. X will be only in Kochi for one day in May
(D) Only Ms. X will be in Kochi for one day in May. [EC 2015]
Answer: (A)
Second sentence: Ms. X reaches Kochi on 21/05/2014. She will be in Bagdogra again on 22/5/2014.
So, Ms. X will be in Kochi only on 22/5/2014.

***Example* 126.** Ram and Shyam shared a secret and promised to each other that it would remain between them. Ram expressed himself in one of the following ways as given in the choices below.
Identify the correct way as per standard English.
(A) It would remain between you and me.
(B) It would remain between I and you
(C) It would remain between you and I
(D) It would remain with me.   [EC 2015]
Answer: (A)

342

Example **127.** The given statement is followed by some courses of action. Assuming the statement to be true, decide the correct option.
Statement:
There has been a significant drop in the water level in the lakes supplying water to the city.
Course of action:
(I) The water supply authority should impose a partial cut in supply to tackle the situation.
(II) The government should appeal to all the residents through mass media for minimal use of water.
(III) The government should ban water supply in lower areas.

(A) Statements I and II follow.   (B) Statements I and III follow.
(C) Statements II and III follow. (D) All statements follow. [CS/IT 2015]
Answer: (A)

Example **128.** All people in a certain island are either 'Knights' or 'Knaves' and each person knows every other person's identity. Knights NEVER lie and Knaves ALWAYS lie.
P says "Both of us are knights" and Q says "None of us are knaves".
Which one of the following can be logically inferred from the above?
(A) Both P and Q are knights
(B) P is a knight; Q is a knave
(C) Both P and Q are knaves
(D) The identities of P and Q cannot be determined   [ME 2017]
Answer: (D)
The statements of both P and Q are same.

Example **129.** Lights of four colors (red, blue, green, yellow) is hung on a ladder. On every step of the ladder there are two lights. If one of the lights is red, the other light on that step will always be blue. If one of the lights on a step is green, the other light on that step will always be yellow. Which of the following statements is not necessarily correct?
(A) The number of red lights is equal to the number of blue lights
(B) The number of green lights is equal to the number of yellow lights
(C) The sum of the red and green lights is equal to the sum of the yellow and blue lights
(D) The sum of the red and blue lights is equal to the sum of the green and yellow lights
Answer: (D)

The following topics will be presented in the Chapter:
I. Arithmetic   II. Algebra   III. Mensuration
IV. Probability   V. Numerical Reasoning   VI. Data Interpretation

I. Arithmetic

Question of one mark:

Example 1. Given $(9 \text{ inches})^{1/2} = (0.25 \text{ yards})^{1/2}$, which one of the following statements is TRUE?
(A) 3 inches = 0.5 yards   (B)   9 inches = 1.5 yards
(C) 9 inches = 0.25 yards   (D) 81 inches = 0.0625 yards [EC 2016]
Answer: (C)

Example 2. If $(1.001)^{1259} = 3.52$ and $(1.001)^{2062} = 7.85$, then $(1.001)^{3321}$
(A)   2.23   (B)   4.33   (C)   11.37   (D)   27.64 [EC 2012]
Answer: (D)
$$(1.001)^{1259} \times (1.001)^{2062} = (1.001)^{3321} = 3.52 \times 7.85 = 27.64$$

Example 3. Find the smallest number y such that $y \times 162$ is a perfect cube.
(A)   24   (B)   27   (C)   32   (D)   36   [EE 2017]
Answer: (D)
$162 = 81 \times 2 = 27 \times (3 \times 2) = 27 \times 6 \Rightarrow y = 36$

Example 4. What would be the smallest number which when divided by either by 20 or by 42 or by 76 leaves a remainder of 7 in each case?
(A)   3047 (B)   6047   (C)   7987   (D)   63847 [CS/IT 2018]
Answer: (C)
The lcm of 22, 34 & 56 is 7980 which is the smallest number exactly divisible by 22, 34 & 56.
For 7 as the remainder in each case, the number = 7980 + 7 = 7987

Example 5. If the number 715 * 423 is divisible by 3 (* denotes the missing digit in the thousandths place), then the smallest whole number in the place of * is
(A)   0   (B)   2   (C)   5   (D)   6 [EC 2018]

344

Answer: (B)

$7 + 1 + 5 + * + 4 + 2 + 3 = 22 + *$ which should be divisible by 3. So, $* = 2$

Example 6. The sum of the digits of a two digit number is 12. If the new number formed by reversing the digits is greater than the original number by 54, find the original number.

(A)   39   (B)   57   (C)   66   (D) 93   [CE 2016]

Answer: (A)

If the options are checked, the sum of the digits in each number is 12

As 93 = 39 + 54, so the second condition is valid for option (A) only.

Example 7. Two numbers are in the ratio 11:13. After 12 is subtracted from each, the remainders are in the ratio 7:9. The numbers are

(A)   33, 39   (B)   55, 65 (C)   44, 52 (D)   22, 26

Answer: (A)

(33-12)/(39-12) = 21:27 = 7:9

Example 8. A test has twenty questions worth 100 marks in total. There are two types of questions. Multiple choice questions are worth 3 marks each and essay questions are worth 11 marks each. How many multiple type of questions does the examination have?

(A)   12   (B)   15   (C)   18   (D)   19   [EE 2017]

Answer (B)

If the no. of MCQ (3 marks question) = x,

then the no. of Essay type (11 marks question) = $20 - x$,

$x \times 3 + (20 - x) \times 11 = 100 \Rightarrow 8x = 120 \Rightarrow x = 15$

Example 9. A train covers a distance of 80 km at a speed of 40 km/hr for the first 60 km and the remaining distance at the speed of 20 km/hr. What is the average speed of the train in km/hr?

(A)   32   (B)   30   (C)   40   (D)   48

Answer (A)

Total time in covering a distance of 80 kms = 60/40 + 20/20 = 5/2 hrs

The average speed = 80 × (2/5) = 32 km/hr

Example 10. A train travels from A to B @ 20 km/hr and from B to A @ 30 km/hr. The average speed for the whole journey is

(A) 25 km/hr   (B) 24 km/hr   (C) 26 km/hr   (D) 28 km/hr

Answer (B)

Average speed = $\dfrac{2 \times 20 \times 30}{20 + 30} = 24$ km/hr

Example 11. A train 280 meters long is moving at 60 km/hr. The time taken by the train to cross a bridge of 220 meters length is

(A) 40 seconds (B) 30 seconds (C) 45 seconds (D) 60 seconds

Answer (B)

In crossing the bridge 220 meters long, 280 meters long train will have to cover a distance of (280 + 220) = 500 m at 60 km/hr = 60 x 5/18 m/s.

Time = Distance / Speed = (500 x 18) / (60 x 5) = 30 seconds.

Example 12. A train that is 280 metres long, travelling at a uniform speed, crosses a platform in 60 seconds and passes a man standing on the platform in 20 seconds. What is the length of the platform in metres?

[EC 2014]

Answer: 560 m

Let the length of the platform $= x$

$280 = v \times 20 \Rightarrow v = 14 \quad 280 + x = v \times 60 = 14 \times 60 = 840 \Rightarrow x = 560\,m$

Example 13. A car travels 8 km in the first quarter of an hour, 6 km in the second quarter and 16 km in the third quarter. The average speed of the car in km per hour over the entire journey is

(A)   30   (B)   36   (C)   40   (D)   24   [EC 2013]

Answer (C)

Total distance = 8 + 6 + 16 = 30 km & total time = 3/4 hrs

Average speed = (total distance)/(total time) = $30 \times \dfrac{4}{3} = 40$ km per hour

Example 14. The speeds of two trains are in the ratio 3:4. They are moving on the opposite directions on parallel tracks. If each takes 3 seconds to cross a telegraph post, then the time taken by the trains to cross each other completely is

(A) 2.5 sec   (B) 3 sec   (C) 3.5 sec   (D) 4 sec

Answer (B)

The speeds are $3x$, $4x$ Length of the trains $= 9x$, $12x$

Time to cross each other =

$$\frac{sum\ of\ lengths\ of\ the\ trains}{Sum\ of\ speeds} = \frac{9x + 12x}{3x + 4x} = 3\sec$$

Example 15. A motor car does a journey in 10 hrs, the first half at 42 km/hr and the second half at 48 km/hr. The length of the journey is

(A) 448 km   (B) 224 km   (C) 336 km   (D) 336 km

Answer (A)

Average speed $= \dfrac{2 \times 42 \times 48}{42 + 48}$ km/hr. Length of the journey = average

speed $\times$ time $= \dfrac{2 \times 42 \times 48}{42 + 48} \times 10 = 448$ km.

Example 16. A tourist covers half of his journey by train at 60 km/h, half of the remainder by bus at 30 km/h and the rest by cycle at 10 km/h. The average speed of the tourist in km/h during his entire journey is

(A)   36   (B)   30   (C)   24   (D)   18   [ME 2013]

Answer (C)

Total distance = d

Total time taken $= \dfrac{d/2}{60} + \dfrac{d/4}{30} + \dfrac{d/4}{10} = d\left(\dfrac{1}{120} + \dfrac{1}{120} + \dfrac{1}{40}\right) = \dfrac{d}{24}$

Average speed $= \dfrac{d}{d/24} = 24$ km/h

Example 17. A person travels three equal distances at a speed of x km/hr, y km/hr and z km/hr respectively. The average speed during the whole journey (in km/hr) is

(A) $\dfrac{xyz}{xy + yz + zx}$ (B) $\dfrac{xy + yz + zx}{xyz}$ (C) $\dfrac{3xyz}{xy + yz + zx}$ (D) $2xyz$

Answer (C)

Each equal distance = s

average speed $= \dfrac{total\ dist}{total\ time} = \dfrac{3s}{s/x + s/y + s/z} = \dfrac{3xyz}{xy + yz + zx}$

Example 18. Walking $\dfrac{3}{4}$ of his usual speed, a person is 10 min late to his

office. His usual time to cover the distance is

(A) 20 min   (B) 30 min   (C) 40 min   (D) 50 min

Answer (B)

The usual time = x min. As time is inversely proportional to speed, so

time taken at $\dfrac{3}{4}$ of the usual speed = $\dfrac{4x}{3}$ min. $\dfrac{4x}{3} - x = 10 \Rightarrow x = 30$ min

Example 19. Two men A and B start apart from a place P walking at 3 km and 3.5 km an hour respectively. If they walk in the same direction, what time will they take to be 16 km apart?

(A) 16 hrs   (B) 24 hrs   (C) 32 hrs   (D) 40 hrs

Answer (C)

As they walk in the same direction, they will be apart by (3.5 – 3) = 0.5 km in 1 hour. Required time = 16/0.5 = 32 hrs.

Example 20. The upstream speed of a boat is 36 km/hr and the speed of the boat in still water is 45 km/ hr. What is the downstream speed of the boat?

(A) 8 km/hr   (B) 28 km /hr   (C) 54 km/hr   (D) 58 km/hr

Answer (C)

Speed of the stream = speed of the boat in still water - upstream speed
= 45 - 36 = 9 km/hr

Downstream speed = Speed of the boat in still water + speed of stream =
45 + 9 = 54 km/hr

Example 21. A man can row 7 km/hr in still water. In a stream flowing at 3 km/hr, he takes 7 hrs to row to a place and back. How far is the place?

(A) 20 km   (B) 16 km   (C) 12 km   (D) 24 km

Answer (A)

Downstream speed = (7 + 3) =10 km/hr, Upstream speed = (7 – 3) = 4

km/hr   Distance = x, then $\dfrac{x}{10} + \dfrac{x}{4} = 7 \Rightarrow \dfrac{7x}{20} = 7 \Rightarrow x = 20$ km

Example 22. Two pipes P and Q fill a cistern separately in 24 hrs and 32 hours. Both pipes are opened together. When the first pipe must be turned off so that the cistern may be just filled in 16 hrs?

(A) 8 hrs   (B) 9 hrs   (C) 10 hrs   (D) 12 hrs

Answer: (D)

The first pipe was closed after x hrs.

First x hrs' supply + second's 16 hrs' supply = 1

or, $\dfrac{x}{24} + \dfrac{16}{32} = 1 \Rightarrow x = 12$ hrs

Example 23. S, M, E and F are working in shifts in a team to finish a project. M works with twice the efficiency of others but for half as many days as E worked. S and M have 6 hour shifts in a day, whereas E and F have 12 hours shifts. What is the ratio of contribution of M to contribution of E in the project?

(A) 1:1   (B)   1:2   (C)   1:4   (D)   2:1 [EC 2016]

Answer (B)

Efficiency: $M \rightarrow 2x$ S, E, F $\rightarrow x$ Days: $M \rightarrow y/2$ E $\rightarrow y$

Time:

$S \rightarrow 6, M \rightarrow 6, E \rightarrow 12, F \rightarrow 12$

$$\dfrac{contribution\ of\ M}{contribution\ of\ E} = \dfrac{2x \times y/2 \times 6}{x \times y \times 12} = 1:2$$

Example 24. In a survey, 3/16 of the people surveyed told that they preferred to use public transport while commuting daily to office. 5/8 of the people surveyed told that they preferred to use their own vehicles. The remaining 75 respondents said that they had no clear preference. How many people preferred to use public transport?

(A)   75   (B)   100   (C)   125   (D)   133

Answer (A)

Remaining part $= 1 - \left(\dfrac{3}{16} + \dfrac{5}{8}\right) = \dfrac{3}{16}$ So, $\dfrac{3}{16} \rightarrow 75$

People preferred to use public transport = 3/16 part,   No. = 75

Example 25. The current erection cost of a structure is *Rs.* 13,200. If the labour wages per day increase by 1/5 of the current wages and the working hours decrease by 1/24 of the current period, then the new cost of erection in *Rs.* is

(A) 16,500   (B) 15,180 (C) 11,000 (D) 10,120   [ME 2013]

Answer (B)

The new cost of erection in *Rs* $= 13200 \times \left(1 + \dfrac{1}{5}\right) \times \left(1 - \dfrac{1}{24}\right) = 15180$

Example 26. Mini buys oranges at the rate of Rs. 10 per orange. How many oranges must she sell for Rs.100 so that she makes a profit of 25% ?

(A)   5   (B)   6   (C)   7   (D)   8

Answer (D)

If Mini has to make a profit of 25%, her selling price per orange = 10 + 2.5 = Rs. 12.5

If the selling price of an orange is 12.5, then for Rs. 100 she will sell 100/12.5 = 8 oranges.

Example 27. If the cost price of 10 oranges be equal to the selling price of 8 oranges, then the percentage of profit in selling oranges is

(A)   25%   (B)   20%   (C)   22.5%   (D)   27.5%

Answer (A)

If the investment for 10 oranges be Rs. 80, then cost price of 1 orange = Rs. 8 and the selling price of 1 orange = Rs. 10.

Percentage of profit in selling oranges

$$= \left(\frac{S.P. - C.P.}{C.P.} \times 100 \right)\% = \left(\frac{10 - 8}{8} \times 100 \right)\% = 25\,\%$$

Example 28. The salary of a worker is first decreased by 20% and then increased by 10%. The net percentage decrease of his salary is

(A)   10   (B)   12   (C)   8   (D)   15

Answer (B)

Net effect = % increase - % decrease − (% increase × % decrease)/100

$$= = 10 - 20 - \frac{10 \times 20}{100} = -10 - 2 = -12\% \quad \text{Net decrease} = 12\%$$

Example 29. The average of marks obtained by 120 candidates in a certain examination is 35. If the average marks of passed candidates is 39 and that of the failed candidates is 15, then the number of candidates passing the examination is

(A) 80   (B) 90   (C) 100   (D) 85

Answer (C)

Let the number of passed candidates = m, number of failed candidates = n = $120 - m$, $\bar{x}_1 = 39, \bar{x}_2 = 15$ & $\bar{x} = 35$ So,

$$\bar{x} = \frac{m\bar{x}_1 + n\bar{x}_2}{m+n} \Rightarrow 35 = \frac{m \times 39 + (120 - m) \times 15}{120}$$
$$\Rightarrow 4200 = 24m + 1800 \Rightarrow m = 100$$

Example 30. The average run of a batsman in 10 innings is 21.5. If he wants to raise his run average to 24, then the number of runs he has to make in the next innings is

(A) 46　　　　(B) 47　　　　(C) 48　　　　(D) 49

Answer (D)

Total runs in 10 innings = 10 × 21.5 = 215

Total runs in 11 innings = 11 × 24 = 264

Runs in the 11th innings = 264 − 215 = 49

Example 31. The average of 10 numbers is 37. Two of the numbers 32 & 45 are misprinted as 23 and 54 and as a result the new average will be

(A)　　37　　(B)　　36　　(C)　　38　　(D)　　36.5

Answer (A)

The new sum of the 10 numbers = 10 × 37 + (23 + 54) − (32 + 45) = 370

The new average =370/10 = 37

Example 32. Seven machines take 7 minutes to make 7 identical toys. At the same rate, how many minutes would it take for 100 to make 100 toys?

(A)　　1　　(B)　　7　　(C)　　　100　　(D)　　700　[ME 2018]

Answer (B)

More time more toys & more time less no. of machines

time → x, toy → t, machines → m,

$$\frac{x_1}{x_2} = \frac{t_1}{t_2} \times \frac{m_2}{m_1} \Rightarrow \frac{7}{x_2} = \frac{7}{100} \times \frac{100}{7} \Rightarrow x_2 = 7$$

Example 33. Round trip tickets to a tourist destination are eligible for a discount of 10% on the total fare. In addition, groups of 4 or more get a discount of 5% on the total fare. If one way single person fare is Rs. 100, a group of 5 tourists purchasing round-trip tickets will be charged Rs. __.

[EE 2014]

Answer: 850

Total round-trip fare for 5 tickets without discount = 5 × 200 = Rs. 1000

Discount

i) for round trip (10%)　　　　　　　　= Rs. 100

ii) for a group of more than 4 persons (5%) = Rs. 50

Total discount = Rs. 150

Chargeable fare = total fare – discount = Rs.(1000 – 150) = Rs. 850

Example 34. A person moving through a tuberculosis prone zone has a 50% probability of becoming infected. However, only 30% of infected people develop the disease. What percentage of people moving through a tuberculosis prone zone remains infected but does not show symptoms of disease?

(A)　　15　　(B)　　33　　(C)　　35　　(D)　　37 [EC 2016]

Answer (C)

The percentage of infected people for whom the symptom of disease has

developed $= \left(\dfrac{50 \times 30}{100 \times 100} \right)\% = 15\,\%$

So, the percentage of infected people for whom the symptom of disease has not developed = (50 - 15)% = 35%

Example 35. Two men A and B start apart from a place P walking at 3 km and 3.5 km an hour respectively. If they walk in opposite direction, then the number of kilometers they will be apart at the end of 3 hours is

(A) 1.5　　　　(B) 10.5　　　　(C) 14.5　　　　(D) 19.5

Answer (D)

As they walk in opposite directions, their relative speed is (3 + 3.5) = 6.5

At the end of 3 hours, they will be apart by 6.5 × 3 = 19.5 km

Question of two marks:

Example 36. Given that $a\,\&\,b$ are integers and $a + a^2 b^3$ is odd. Which one of the following statements is correct?

(A) $a\,\&\,b$ are both odd (B) $a\,\&\,b$ are even

(C) a is even and b is odd　　(D)　a is odd and b is even [ME 2018]

Answer: (D)

If a is odd and b is even then a^2 is odd and b^3 is even, so

$a^2 b^3 = odd \times even = even$ is and $a + a^2 b^3 = odd + even = odd$

Example 37. Trucks (10 m long) and cars (5 m long) go on a single lane bridge. There must be a gap of at least 20 m after each truck and a gap of 15 m after each car. Trucks and cars travel at a speed of 36 km/h. If cars

352

and trucks go alternately, what is the maximum number of vehicles that can use the bridge in one hour?

(A)   1440   (B)   1200   (C)   720   (D)   600   [EC 2017]

Answer (A)

In one hour, total distance covered = 36 km = 36000 m

Truck (10 m) + gap (20 m) + car (5 m) + gap (15 m) and then again truck.

Total distance = (10 + 20 + 05 + 15) = 50 m for 2 vehicles.

For one vehicle 25m   Total no. of vehicles in 36000 m =36000/25 = 1440

$T \rightarrow \quad C \rightarrow$

10 20 5 15

Example 38. Train A, traveling at 50 km/hr leaves city P for city Q at 5 pm. Train B leaves city Q for city P at 7 pm and travels at 90 km/hr. When will train B overtake train A?

(A) 8:30 pm   (B) 9:15 pm   (C) 9:30 pm   (D) 9:00 pm

Answer (C)

Between 5 pm and 7 pm, train A will travel a distance of 2 × 50 = 100 kms

The difference of speed between train A and train B = 90 - 50 = 40 km/hr

Now this means that train B gains 40 kms every hour over train A. The moment it gains 100 kms, it will overtake train A.

Time taken to cover 100 kms at a speed of 40 km/hr = (100/40)

= 2 hrs 30 mins from the time train B started for city Q. That is 9:30 pm.

Example 39. A cyclist after travelling a distance of 60 kms. observed that if he would run with a speed of 2 km/hr more, he could arrive there 1 hour before. The speed of the cyclist in km/hr is

(A)   5   (B)   12   (C)   8   (D)   9

Answer (C)

Original time = t, original speed = v, then $v.t = (v + 2)(t - 1)$ or, $v = 2(t - 1)$

Now, 40 = $vt = 2t(t - 1)$ or, $t(t - 1) = 20 = 5. (5 - 1)$ or, $t = 5$

v = distance/time = 40/5 = 8 km/hr

Example 40. A boat covers 24 km in upstream and 36 km downstream in 6 hours and 36 km upstream and 24 km downstream in 6.5 hrs. The speed of the current (in km/hr) is

(A) 1.5   (B) 2   (C) 2.5   (D) 3

Answer: (B)

upstream   downstream   time

353

24	36	6
36	24	6.5

Upstream speed of the boat = $\dfrac{24 \times 24 - 36 \times 36}{24 \times 6 - 36 \times 6.5} = \dfrac{720}{90} = 8$ km/hr

Downstream speed of the boat = $\dfrac{24 \times 24 - 36 \times 36}{24 \times 6.5 - 36 \times 6} = \dfrac{720}{60} = 12$ km/hr

Speed of the current = $\dfrac{1}{2}$ (downstream speed – upstream speed) = 2 km/hr

***Example* 41.** 16 workers can do a piece of work in 16 days. 4 days after they started work, 8 more workers joined them. How many days will they now take to complete the remaining work?

(A) 8 days   (B) 10 days   (C) 15 days   (D) 6 days

Answer (A)

Work done by 16 workers in 4 days = 4 /16 = ¼

Remaining work = 1 - 1/4 = ¾

Total no. of workers now = 16 + 8 = 24

24 workers do 3/4 work in = (16 × 16 × 3)/(24 × 4) = 8 days

***Example* 42.** A can do a piece of work in 24 days, while B can do it in 16 days. With the help of C, they finish the work in 8 days. C alone can do the work in

(A) 48 days   (B) 46 days   (C) 42 days   (D) 40 days

Answer (A)

In 1 day A and B together can do $\left(\dfrac{1}{24} + \dfrac{1}{16} \right) = \dfrac{5}{48}$ part of the work.

In 8 days they can do $\dfrac{5}{48} \times 8 = \dfrac{5}{6}$ part of the work.

Remaining part of the work $= 1 - \dfrac{5}{6} = \dfrac{1}{6}$ part is done by C in 8 days.

C alone can do the work in (8 × 6) = 48 days.

***Example* 43.** Pipe A fills a tank in 24 minutes. Pipe B fills the same tank 7 times faster than pipe A. If both the pipes are kept open when the tank is empty, how long will it take for the tank to overflow?

(A) 2 mins   (B) 2.5 mins   (C) 3 mins   (D) 3.5 min

Answer (C)

Pipe B will fill the tank in 24/7 minutes as it is 7 times faster than pipe A. Together, the two pipes will fill 1/24 + 7/24 = 8/24 = 1/3rd of the tank in a minute. So, it will take 3 minutes for the tank to overflow.

Example 44. Three pipes A, B and C can fill a cistern in 6 hours. After working at it together for 2 hrs, C is closed and A and B fill it in 7 hrs more. How many hours will C alone take to fill the cistern?
(A) 10 hrs   (B) 14 hrs   (C) 18 hrs   (D) 20 hrs
Answer (B)
Part of the cistern filled up in 2 hrs by all the three pipes = 2 × (1/6) = 1/3
Part of cistern remaining to be full = 1 - 1/3 = 2/3
Now 2/3 of the cistern is filled by (A + B) in 7 hrs.
Therefore, in 1 hour (A + B)'s contribution = (2/3) × (1/7)= 2/21 of cistern
C's work for 1 hour = (A + B + C)'s for 1 hour - (A+B)'s work for 1 hour = 1/6 - 2/21 = 1/14   So, C's time for filling the cistern = 14 hrs.

Example 45. A dealer allows 10% discount on the list price of a certain article and yet makes a profit of 25% on each article. Find the cost price of the article when list price is Rs. 50.00.
(A)   30   (B)   36   (C)   40   (D)   42
Answer (B)
Let the cost price of the article = Rs. 100.   At 25% profit S.P. = 125
If list price is Rs. 100 at 10% discount, S.P. = Rs. 90
Therefore, S.P. is Rs. 125, list price = (100 × 125) / 90 = 1250 /9
If list price is Rs. 1250 /9, then C.P. = Rs. 100
If list price is Rs. 50, C.P. = (100 × 50 × 9)/1250 = Rs. 36

Example 46. A man buys two horses for Rs. 13,500. He sells one so as to lose 6% and the other so as to gain 7.5%. On the whole he neither gains nor loses. The cost of the second horse is
(A) Rs. 6,000   (B) Rs. 6,500   (C) Rs. 7000   (D) Rs. 7,500
Answer (A)
Loss on one horse = gain on the other
6% of the cost of the first horse = 7.5% of the cost of the second horse

$$\frac{\cos t \ of \ the \ first \ horse}{\cos t \ of \ the \ \sec ond \ horse} = \frac{7.5}{6} = \frac{5}{4} \qquad \text{Cost of the second horse}$$

$$= Rs. \frac{4}{5+4} \times 13500 \quad = Rs. \frac{4}{9} \times 13500 \quad = Rs. 4 \times 1500 \quad = Rs. 6,000$$

Example 47. In a college, 1/5th of the boys and 1/8th of the girls took part in a seminar. Total number of students in the college is 420 and ratio of number of boys to that of girls is 5:2. What is the ratio of no. of students who took part in seminar to total no. of students ?

(A)　　5:42　(B)　　5:28　(C)　　5:13　(D)　　13:20

Answer (B)

Number of boys = 5/7 × 420 = 300 and number of girls = 2/7 × 420 = 120

Number of boys taking part in seminar = 1/5 × 300 = 60

Number of girls taking part in seminar = 1/8 × 120 = 15

Total no. of students who took part in seminar = 60 + 15 = 75

Reqd. ratio = 75:420 = 5:28

Example 48. Amal, Bimal and Kamini enter into partnership. Amal contributes one-third of the capital while Bimal contributes as much as Amal and Kamini together contribute. If the profit at the end of the year amounts to Rs. 840 what would Kamini receive?

(A)　Rs. 100　(B)　Rs. 120　(C)　Rs. 130　(D)　Rs. 140

Answer (D)

As Amal contributes one-third of the capital his profit = 840 /3 = Rs. 280

Now as Bimal contributes as much as Amal and Kamini

Profit of Bimal = Profit of Amal + Profit of Kamini = Rs. 280 + Profit of Kamini

or, Profit of Bimal - Profit of Kamini = Rs. 280

Profit of Bimal + Profit of Kamini = Rs. 840 - Rs. 280

So, 2 × Profit of of Bimal = Rs. 840　or, Profit of Bimal = Rs. 420

So, Profit of Kamini = 840 - 420 - 280 = Rs. 140

Example 49. A, B and C are in a partnership business. A receives 2/5 of the profit and B & C share the remaining profit equally. A's income is increased by Rs.220 when the profit rises from 8% to 10%. The capital invested by C is

(A)　Rs. 8000　(B)　Rs. 8250　(C)　Rs. 8500　(D)　Rs. 8750

Answer (B)

A's share: (10% - 8%) = 2% = Rs. 220　or, 100% = Rs. 11,000

A's capital = Rs. 11,000

B & C's share:

$$\frac{2}{5} \equiv 11000 \Rightarrow 1 \equiv 11000 \times \frac{5}{2} \Rightarrow \frac{3}{5} \equiv 11000 \times \frac{5}{2} \times \frac{3}{5} = 16500$$

C's capital = Rs. 16500/2 = Rs. 8250

Example 50. The incomes of A & B are in the ratio 3:2 and their expenditures are in the ratio 5:3. If each saves Rs. 2000, then the income of A is

(A) Rs. 8000   (B) Rs. 10000   (C) Rs. 12000   (D) Rs. 14000

Answer (C)

	Income	Expenditure	Savings (Rs.)
A	3x	5y	2000
B	2x	3y	2000

$3x - 5y = 2x - 3y$ or, $x = 2y$ Again, $3x - 5y = 2000$ or, $y = 2000$ & $x = 4000$

The income of A = 3x = Rs. 12000

Example 51. A vessel contains liquids A & B in the ratio 5:3. If 16 liters of the mixture are removed and the same quantity of liquid B is added, the ratio becomes 3:5. The original quantity of the liquid in the vessel is (in liters)

(A)   48   (B)   46   (C)   44   (D)   40

Answer (D)

Let The vessel has 5x and 3x of A & B. Out of 16 liters removed quantity,

$$A = \frac{5}{5+3} \times 16 = 10 \, liters \ \& \ B = 16 - 10 = 6 \, liters$$

After removing and adding, $\dfrac{5x - 10}{3x + 10} = \dfrac{3}{5} \Rightarrow x = 5$

The original quantity of the liquid in the vessel = 8x = 40 liters

Example 52. An aptitude test was conducted separately with three groups of students and the following information was obtained:
(i) the average score of Group I is 83
(ii) the average score of Group II is 76
(iii) the average score of Group III is 85
(iv) the average score of Groups I & II taken together is 79 and

(v) the average score of Groups II & III taken together is 81

The average score for all the three Groups taken together is

(A)   81   (B)   81.5   (C)   82   (D)   83.5

Answer (B)

Let the number of students in the three groups are respectively m, n, p

For Groups I & II

$$79 = \frac{m \times 83 + n \times 76}{m + n} \Rightarrow 83m + 76n = 79m + 79n \Rightarrow 4m = 3n \Rightarrow \frac{m}{n} = \frac{3}{4}$$

For Groups II & III:

$$81 = \frac{n \times 76 + p \times 85}{n + p} \Rightarrow 81n + 81p = 76n + 85p \Rightarrow 5n = 4p \Rightarrow \frac{n}{p} = \frac{4}{5}$$

$$\frac{m}{n} = \frac{3}{4}, \frac{n}{p} = \frac{4}{5} \Rightarrow m : n : p = 3 : 4 : 5$$

The average score for all the three Groups taken together

$$= \frac{m \times 83 + n \times 76 + p \times 85}{m + n + p} = \frac{3 \times 83 + 4 \times 76 + 5 \times 85}{3 + 4 + 5} = \frac{978}{12} = 81.5$$

Example 53. There are 35 students in a hostel. If the number of students increases by 7, the expenses of the mess increase by Rs. 42 per day while the average expenditure per head diminishes by Re 1. The original expenditure of the mess is

(A) Rs. 460   (B) Rs. 440   (C) Rs. 420   (D) Rs. 400

Answer (C)

Original average expenditure = x, total expenditure = $35x$

After joining 7 more students, total expenditure = $35x + 42$,

Average $= \frac{35x + 42}{35 + 7} = x - 1 \Rightarrow 42x - 42 = 35x + 42 \Rightarrow x = 12$

The original expenditure of the mess =Rs. (35×12) = Rs. 420

Example 54. The population of a village is 10000. It increases by 10% during the first year and by 15% during the second year. The population of the town after two years will be

(A) 12650   (B) 12250   (C) 12100   (D) 12750

Answer (A)

If x% and y% be the growth for the first and second years, then the population after two years

$$= \frac{10000 \times (100 + 10) \times (100 + 15)}{100 \times 100} = 110 \times 115 = 12650$$

Example 55. After deducting 15% from a certain sum and then 20% from the remainder, there is Rs. 1156 left. The original sum is
(A) Rs. 1500   (B) Rs. 1650   (C) Rs. 1600   (D) Rs. 1700
Answer (D)

The required sum $= Rs. \dfrac{1156 \times 100 \times 100}{(100 - 15) \times (100 - 20)} = Rs. \dfrac{1156 \times 100 \times 100}{85 \times 80}$

$= Rs. \dfrac{1156 \times 20 \times 5}{17 \times 4} = 68 \times 5 \times 5 = Rs. 1700$

Example 56. Due to fall in the manpower, the production of a factory falls by 20%. To restore the original production, the working hours should be increased by percent
(A)   20   (B)   22.5   (C)   25   (D)   27.5
Answer (C)

Production = Manpower × Working hours
Manpower = 100 , Working hours = 100
Let the working hour increases by $x\%$ then $(100 - 20) \times (100 + x) = 100 \times 100$ or, $100 + x = 125$ or, $x = 25$
So, the working hours is to be increased by 25%

Example 57. A tiger is 50 leaps of its own behind a deer. The tiger takes 5 leaps per minute to the deer's 4. If the tiger and the deer cover 8 metre and 5 metre per leap respectively, what distance in metres will be tiger have to run before it catches the deer?   [EC 2015]

Answer: 800

In one min. tiger covers the distance 5× 8 = 40 m
In one min. deer covers the distance 4× 5 = 20 m
Tiger is 50 leaps = 50 × 8 = 400 m behind the deer.
Let the time taken by the tiger to catch the deer = t min

T____D_____O (meeting point)

400 m

TO = 40t, DO = 20t, TD = TO − DO or, 400 = 40t − 20t or, t = 20
Distance covered by the tiger = TO =TD + TO = 400 + 20 × 20 = 800 m

***Example* 58.**The Gross Domestic Product (GDP) in Rupees grew at 7% during 2012-2013. For international comparison, the GDP is compared in US Dollars (USD) after conversion based on the market exchange rate. During the period 2012-2013 the exchange rate for the USD increased from Rs. 50/ USD to Rs. 60/ USD. India's GDP in USD during the period 2012- 2013

(A) increased by 5 %   (B) decreased by 13%
(C) decreased by 20%   (D) decreased by 11%   [CS/IT 2014]
Answer: (D)

	Original value	Final value
Per (Rs.)	100	107
Per (USD)	100/50	107/60
For USD	100	$\left(\dfrac{107}{60} \times \dfrac{50}{100}\right) \times 100\% = 89.16\%$

Decrease = $(100 - 89)\% = 11\%$ [As the final value < original value]

***Example* 59.**In appreciation of the social improvements completed in a town, a wealthy philanthropist decided to gift Rs. 750 to each senior male citizen in the town and Rs. 1000 to each senior female citizen. Altogether, there were 300 senior citizens eligible for this gift. However, only 8/9th of the eligible men and 2/3rd of the eligible women claimed the gift. How much money (in Rs.) did the philanthropist give away in total?

(A) 1,50,000 (B) 2,00,000 (C) 1,75,000 (D) 1,51,000   [CS/IT2018]
Answer (B)

If the total no. of male citizen $= x,$ then total amount of gift (in

$$\text{Rs.}) = \frac{8}{9} x \times 750 + \frac{2}{3}(300 - x) \times 1000 = \frac{2000}{3} x + 200000 - \frac{2000}{x} = 2,00,000$$

***Example* 60.** Two finance companies, P and Q, declared fixed annual rates of interest on the amounts invested with them. The rates of interest offered by these companies may differ from year to year. Year-wise annual rates of interest offered by these companies are shown by the line graph provided below:

If the amounts invested in the companies, P and Q, in 2006 are in the ratio 8:9, then the amounts received after one year as interests from companies P and Q would be in the ratio:

(A)   2:3 (B)   3:4   (C)   6:7 (D)   4:3   [CE 2016]
Answer (D)

Let the deposits in the companies be $8a \& 9a$

Rate of interest on deposits of P is 6% and deposits of Q is 4%

Interest from $8a$ @ 6% $= 8a \times \dfrac{6}{100} = \dfrac{48}{100}a$

Interest from $9a$ @ 4% $= 9a \times \dfrac{4}{100} = \dfrac{36}{100}a$

Ratio of the interests $48 : 36 = 4 : 3$

Example 61 .In a party, 60% of the invited guests are male and 40% are female. If 80% of the invited guests attended the party and if all the invited female guests attended, what would be the ratio of males to females among the attendees in the party?

(A)   2:3   (B)   1:1   (C)   3:2   (D)   2:1   [CS/IT 2018]

Answer (B)

Total invitees =100, male = 60, female = 40
Total attendees = 80, female attendees = 40
so male attendees = 80 - 40 = 40
Among the attendees, ratio of males to females = 40:40 = 1:1

Example 62. A faulty wall clock is known to gain 15 minutes in every 24 hours. It is synchronized to the correct time at 9 AM on 11th July. What will be correct time to the nearest minute when the clock shows 2 PM on 15th July of the same year?

(A) 12:45 PM (B) 12:58 PM (C) 1:00 PM (D) 2:00 PM [CE 2018]

Answer: (B)

From 9th July 9 AM to 15Th July 2 PM time = $(24 \times 4 + 5) = 101$ hrs.
For 24 hrs. gain = 15 min,   For 101 hrs. gain = $(15/24) \times 101 = 63.125$ min
= 1 hr 3.125 min
On 15th July at 2 PM, right time = 1 hr 3.125 min back back from 2 PM
= 12:58 PM (approx.)

Example 63. Leila aspires to buy a car worth Rs. 10,00,000 after 5 years. What is the minimum amount in Rupees that she should deposit now   in a Bank which offers 10% annual rate of interest if the interest is compounded annually?

(A) 5,00,000 (B) 6,21,000 (C) 6,66,667 (D) 7,50,000 [EC 2018]

Answer (B)

If the deposit be Rs. $P,$ then

$$A = P(1+i)^n$$

$$\Rightarrow 1000000 = P(1+0.1)^5 = P \times 1.6105 \Rightarrow P = \frac{1000000}{1.6105} = 620,925 \cong 6,21,000$$

Example 64. A designer uses marbles of four different colours for his designs. The cost of each marble is the same, irrespective of the colour. The table below shows the percentage of marbles of each colour used in the current design. The cost of each marble increased by 25%. Therefore, the designer decided to reduce equal numbers of marbles of each colour to keep the total cost unchanged. What is the percentage of blue marbles in the new design?

Blue	Black	Red	Yellow
40%	25%	20%	15%

(A)   35.75   (B)   40.25   (C)   43.75   (D)   46.25   [EE 2018]
Answer: (C)

Example 65. The ratio of male to female students in a college for five years is plotted in the following line graph. If the number of female students in 2011 and 2012 is equal, what is the ratio of male students in 2012 to male students in 2011?

(A)   1:1   (B)   2:1   (C)   1.5:1   (D)   2.5:1   [CS/IT 2014]
Answer: (C)

2011: Let the number of female students is 100

        Ratio of male and female students=1, the no. of male = 100
2012: No.of female students =No.of female students in 2011 = 100

 Ratio of male and female students=1.5, the no. of male = 150
(Male in 2012)/(Male in 2011) = 150/100 = 1.5:1

Example 66. Two alloys A and B contain gold and copper in the ratio 2:3 and 3:7 by mass, respectively. Equal masses of A and B are melted to make an alloy C. The ratio of gold to copper in alloy C is
(A)   5:10   (B)   7:13   (C)   6:11   (D)   9:13   [EC 2018]
Answer: (B)

$$\frac{gold}{copper} = \frac{2/5+3/10}{3/5+7/10} = 7:13$$

Example 67. x bullocks and y tractors take 8 days to plough a field. If we halve the number of bullocks and double the number of tractors, it takes 5 days to plough the same field. How many days will it take x bullocks

362

alone to plough the field?

(A)   30   (B)   35   (C)   40   (D)   45   [ME 2017]

Answer (A)

1 bullock only can do the work in c days

x bullocks only can do the work in c/x days. In one day, x/c part

1 bullock only can do the work in d days

y bullocks only can do the work in d/y days. In one day, y/d part

In the first case, $8\left(\dfrac{x}{c} + \dfrac{y}{d}\right) = 1$ (total work) $\Rightarrow \left(\dfrac{x}{c} + \dfrac{y}{d}\right) = \dfrac{1}{8}$ $--(i)$

In the second case,

$5\left(\dfrac{x}{2c} + \dfrac{2y}{d}\right) = 1 \Rightarrow \left(\dfrac{x}{2c} + \dfrac{2y}{d}\right) = \dfrac{1}{5}$ $--(ii)$

$(i) \times 2 - (ii)$, $\dfrac{2x}{c} - \dfrac{x}{2c} = \dfrac{1}{4} - \dfrac{1}{5} \Rightarrow \dfrac{x}{c}\left(2 - \dfrac{1}{2}\right) = \dfrac{1}{20} \Rightarrow \dfrac{x}{c} = \dfrac{2}{3} \times \dfrac{1}{20} = \dfrac{1}{30}$

So, x bullocks only can do the work in $\dfrac{c}{x} = 30$ days.

Example 68. In manufacturing industries, loss is usually taken to be the square of the deviation from a target. If the loss is Rs. 4900 for a deviation of 7 units, what would be the loss in Rupees for a deviation of 4 units from the target?

(A)   400   (B)   1200   (C)   1600   (D)   2800   [CE 2018]

Answer (C)

$l = kd^2 \Rightarrow 4900 = k.7^2 \Rightarrow k = 100$

When $d = 4, l = 100 \times 4^2 = 1600$ (Rs.)

Example 69. P, Q, R and S are working on a project. Q can finish the task in 25 days, working alone for 12 hours a day. R can finish the task in 50 days, working alone for 12 hours per day. Q worked 12 hours a day but took sick leave in the beginning for two days. R worked 18 hours a day on all days. What is the ratio of work done by Q and R after 7 days from the start of the project?

(A)   10:11   (B)   11:10   (C)   20:21   (D)   21:20   [EC 2016]

Answer (C)

In 1 hour, Q can do $\dfrac{1}{25 \times 12}$ part and R can do $\dfrac{1}{50 \times 12}$ part of the work.

Q has worked for 5 days working 12 hours/day.

Part of the work done by Q = $\dfrac{5 \times 12}{25 \times 12} = \dfrac{1}{5}$

R has worked for 7 days working 18 hours/day.

Part of the work done by R = $\dfrac{7 \times 18}{50 \times 12}$

Ratio of work done by Q and R = $\dfrac{1}{5} \times \dfrac{50 \times 12}{7 \times 18} = 20 : 21$

Example 70. Arun, Gulab, Neel and Shweta must choose one shirt from a pile of four shirts coloured red, pink, blue and white respectively. Arun dislikes the colour red and Shweta dislikes the colour white. Gulab and Neel like all the colours. In how many different ways can they choose the shirts so that no one has a colour he or she dislikes?

(A)   21   (B)   18   (C)   16   (D)   14   [EE 2017]

Answer (D)

Four persons may choose four shirts in 4 ! = 24 ways

Excluded cases:

i) Arun $\rightarrow$red and Shewta $\rightarrow$ white and pink and blue to two other persons.

No. of ways = 2! = 2

ii) Arun $\rightarrow$red, pink, white and blue to three other persons.

No. of ways = 3! -2! = 4

[The case Arun $\rightarrow$red and Shewta $\rightarrow$ white is to be excluded]

iii) Shweta $\rightarrow$white, pink, red and blue to three other persons.

No. of ways = 3! -2! = 4

[The case Arun $\rightarrow$red and Shewta $\rightarrow$ white is to be excluded]

Total no. of ways = 24 − (2 + 4 + 4) = 14

II. Algebra:

One mark Question:

Example 71. The number of distinct integral factors of 2014 is _____   [CS/IT 2015]

Answer: 8

The prime factors of 2014 are 2, 19, 53

The number of distinct integral factors of 2014 is (1 + 1)(1 + 1)(1 + 1) = 8

Example 72. The value of $\sqrt{12 + \sqrt{12 + \sqrt{12 + ---}}}$ is

(A)   3.464   (B)   3.932   (C)   4.000   (D)   4.444 [EE 2014]

Answer (C)

$$y = \sqrt{12 + \sqrt{12 + \sqrt{12 + ----}}} \Rightarrow y^2 = 12 + y \Rightarrow y^2 - y - 12 = 0$$
$$y = 4, -3 \quad \text{As } y > 0, y = 4.000$$

Example 73. Find the sum of the expression

$$\frac{1}{\sqrt{1} + \sqrt{2}} + \frac{1}{\sqrt{2} + \sqrt{3}} + \frac{1}{\sqrt{3} + \sqrt{4}} + - - + \frac{1}{\sqrt{80} + \sqrt{81}}$$

(A)   7   (B)   8   (C)   9   (D)   10   [ME 2013]

Answer (B)

The expression

$$= \frac{\sqrt{2} - \sqrt{1}}{2 - 1} + \frac{\sqrt{3} - \sqrt{2}}{3 - 2} + \frac{\sqrt{4} - \sqrt{3}}{4 - 3} + - - - + \frac{\sqrt{81} - \sqrt{80}}{81 - 80}$$
$$= \sqrt{81} - \sqrt{1} = 8$$

Example 74. X is 30 digit number starting with 4 followed 7. Then the number X^3 will have

(A)   90 digits   (B)   91 digits   (C)   92 digits   (D)   93 digits   [CS/IT 2017]

Answer: (A)

$$X = 4777 - - - 77\,(one\,4\;and\;twenty\;nine\;\;7) = 4.777 - - - - - \times 10^{29}$$
$$X^3 = \left(4.7777 - - - \times 10^{29}\right)^3 = (4.777 - -)^3 \times 10^{87}$$

So, X^3 will have (3 + 87) = 90 digits.

Example 75. If $f(x) = 2x^7 + 3x - 5$. which of the following is a factor of $f(x)$?

(A) $\left(x^3 + 8\right)$ (B) $(x - 1)$ (C) $(2x - 5)$ (D) $(x + 1)$ [CS/IT 2016]

Answer (B)

$f(1) = 0 \Rightarrow (x - 1)$ is a factor of $f(x)$.

Example 76. Functions $F(a, b)$ and $G(a, b)$ are defined as follows:

$F(a, b) = (a - b)^2$ and $G(a, b) = |a - b|$, where $|x|$ represents the absolute value of x. What would be the value of $G\big(F(1,3), G(1,3)\big)$?

(A)   2   (B)   4   (C)   6   (D)   36   [EE 2018]

Answer (A)

$$F(1,3) = (1 - 3)^2 = 4,$$
$$G(1,3) = |1 - 3| = 2, \; G\big(F(1,3), G(1,3)\big) = G(4,2) = |4 - 2| = 2$$

Example 77. The set of values of p for which the roots of the equation $3x^2 + 2x + p(p-1) = 0$ are of opposite sign is

(A) $(-\infty, 0)$ (B) $(0, 1)$ (C) $(1, \infty)$ (D) $(0, \infty)$ [EC2013]

Answer (B)

If $p \in (0,1)$, then p is a positive proper fraction, so $p - 1 < 0$ & $p(p-1) < 0$.

Example 78. The three roots of the equation $f(x) = 0$ are $x = \{-2,0,3\}$. What are the three values of x for which $f(x-3) = 0$?

(A) - 5, - 3, 0   (B) - 2, 0, 3   (C) 0, 6, 8   (D) 1, 3, 6   [EE 2018]

Answer (D)

$f(x) = c(x + 2)(x - 0)(x - 3), c$ is a const.

$f(x - 3) = c(x + 2 - 3)(x - 0 - 3)(x - 3 - 3) = c(x - 1)(x - 3)(x - 6)$

$f(x - 3) = 0 \Rightarrow x = 1,3,6$

Example 79. What is the value of $1 + \dfrac{1}{4} + \dfrac{1}{16} + \dfrac{1}{64} + \dfrac{1}{256} + ...$?

(A)   2   (B)   7/4   (C)   3/2   (D)   4/3 [EC 2018]

Answer (D)

It is an infinite G.P. with $a = 1, r = \dfrac{1}{4}$. Sum $= \dfrac{a}{1-r} = \dfrac{4}{3}$

Example 80. $\left(\dfrac{a + a + a + - - + a}{n \; times}\right) = a^2 b$ & $\left(\dfrac{b + b + b + - - - - + b}{m \; times}\right) = ab^2$,

where a, b, n, m are natural numbers.

What is the value of $\left(\dfrac{m + m + m + - - + m}{n \; times}\right)\left(\dfrac{n + n + n + - - - - + n}{m \; times}\right)$?

(A) $2a^2 b^2$ (B) $a^4 b^4$ (C) $ab(a+b)$ (D) $a^2 + b^2$ [CE 2018]

Answer (B)

$na = a^2 b \Rightarrow n = ab$ & $mb = ab^2 \Rightarrow m = ab$

Now, $\left(\dfrac{m + m + m + - - + m}{n \; times}\right)\left(\dfrac{n + n + n + - - - - + n}{m \; times}\right) = (nm)(mn) = n^2 m^2 = (ab)^4$

Example 81. If $x > y > 1$ which of the following must be true?

(i) $\ln x > \ln y$ (ii) $e^x > e^y$ (iii) $y^x > x^y$ (iv) $\cos x > \cos y$

(A) (i) and (ii)   (B) (i) and (iii)

(C) (iii) and (iv)   (D) (ii) and (iv)   [EC 2015]

Answer (A)

For numbers >1, ln and exp. functions gradually increase but the argument is not valid for the functions given in (iii) and (iv).

Example 82. If $(z + 1/z)^2 = 98$, compute $(z^2 + 1/z^2)$. [EE 2014]

Answer: 96

$$\left(z + \frac{1}{z}\right)^2 = 98 \Rightarrow z^2 + \frac{1}{z^2} + 2 = 98 \Rightarrow z^2 + \frac{1}{z^2} = 96$$

Example 83. For what values of k given below is $\dfrac{(k+2)^2}{k-3}$ an integer?

(A) 4, 8, 18   (B) 4, 10, 16   (C) 4, 8, 28   (D) 8, 26, 28 [EE 2018]

Answer (C)

Let $P = \dfrac{(k+2)^2}{k-3}$ Taking $k = 4$, P is an integer. So, option D is wrong.

Taking $k = 8$, P is an integer. So, options B & D are wrong.

Taking $k = 28$, P is an integer. So, option C is correct.

Example 84. If $q^{-a} = \dfrac{1}{r}$, $r^{-b} = \dfrac{1}{s}$, $s^{-c} = \dfrac{1}{q}$, the value of abc is

(A) $(rqs)^{-1}$ (B) 0   (C) 1   (D) $r + q + s$ [EC 2016]

Answer (C)

$$r = q^a, s = r^b, q = s^c \Rightarrow q = s^c = \left(r^b\right)^c = r^{bc} = \left(q^a\right)^{bc} = q^{abc}$$

So, $q^{abc} = q^1 \Rightarrow abc = 1$

Example 85. In a quadratic function, the value of the product of the roots (α, β) is 4. Find the value of $\dfrac{\alpha^n + \beta^n}{\alpha^{-n} + \beta^{-n}}$.

(A) n^4 (B) 4^n (C) 2^{2n-1} (D) 4^{n-1} [CS/IT 2016]

Answer (B)

$$\frac{\alpha^n + \beta^n}{\alpha^{-n} + \beta^{-n}} = \frac{\alpha^n + \beta^n}{\alpha^n + \beta^n} \times (\alpha\beta)^n = 4^n$$

Example 86. If $\log_x(5/7) = -1/3$ then the value of x is

(A) 343/125 (B) 125/343 (C) -25/49 (D) - 49/25 [EC 2015]

Answer: (A)

$\log_x(5/7) = -1/3 \Rightarrow x^{-1/3} = 5/7$

$\Rightarrow x^{1/3} = 7/5 \Rightarrow x = (7/5)^3 = 343/125$

Example 87. For non-negative integers, a, b, c, what would be the value of $a+b+c$ if $\log a + \log b + \log c = 0$?

(A) 3 (B) 1 (C) 0 (D) - 1 [CE 2018]

Answer (A)

As a, b, c are non-negative integers, $\log a, \log b, \log c$ are each non-negative. The sum of three non-negative numbers $\log a, \log b, \log c$ is zero $\Rightarrow$ each is zero. So, $\log a = 0, \log b = 0, \log c = 0 \Rightarrow a = b = c = 1$

$a + b + c = 3$

Example 88. If $|9y - 6| = 3$, then $y^2 - \dfrac{4y}{3}$ is _____.

(A) 0 (B) 1/3 (C) $-1/3$ (D) undefined [EE 2016]

Answer (C)

$|9y - 6| = 3 \Rightarrow 9y - 6 = \pm 3$

If $9y - 6 = 3 \Rightarrow y = 1 \Rightarrow y^2 - \dfrac{4y}{3} = 1 - \dfrac{4}{3} = -\dfrac{1}{3}$

If $9y - 6 = -3 \Rightarrow y = \dfrac{1}{3} \Rightarrow y^2 - \dfrac{4y}{3} = \dfrac{1}{9} - \dfrac{4}{9} = -\dfrac{1}{3}$

Example 89. Given that $f(y) = \dfrac{|y|}{y}$ and q is any non-zero real number, the value of $|f(q) - f(-q)|$ is

(A) 0 (B) -1 (C) 1 (D) 2

Answer: (D)

If $q > 0, |q| = q, f(q) = 1$ and $f(-q) = -1 \Rightarrow |f(q) - f(-q)| = 2$

If $q < 0, |q| = -q, f(q) = -1$ and $f(-q) = 1 \Rightarrow |f(q) - f(-q)| = 2$

Example 90. What is the average of all multiples of 10 from 2 to 198?

(A) 90 (B) 100 (C) 110 (D) 120 [EE 2014]

Answer (B)

The no. of multiples of 10 from 2 to 198 (starting from 10 – 190) = 19

Their sum = $= \dfrac{n}{2}(a+l) = \dfrac{19}{2}(10+190) = 100 \times 19$

Average $=\dfrac{100\times19}{19}=100$

Example 91. What will be the maximum sum of 44, 42, 40, ...?

(A)　502　(B)　504　(C)　506　(D)　500　　　[ME 2013]

Answer (C)

If the n-th term is 0, 44 + (n - 1)(- 2) = 0 or, n = 23

All terms after 23rd term will be negative & 23rd term = 0

Max. sum = sum of the first 22 terms $= \dfrac{22}{2}\{44 + (22 - 1).(-2)\} = 506$

Example 92. $(x\%\,of\,\,y)+(y\%\,of\,\,x)$ is equivalent to

(A)　　2 % of xy　　　(B)　　2 % of $(xy/100)$

(C)　　$xy\%$ of 100　(D)　　100 % of xy　　　[CE 2016]

Answer (A)

$(x\%\ of\ y) = y \times \dfrac{x}{100} = \dfrac{xy}{100}$

$(y\%\ of\ x) = x \times \dfrac{y}{100} = \dfrac{xy}{100}$

$(x\%\ of\ y) + (y\%\ of\ x) = \dfrac{xy}{100} + \dfrac{xy}{100} = \dfrac{2}{100}(xy) = 2\%\ of\ xy$

Example 93. What is the value of x when

$81 \times \left(\dfrac{16}{25}\right)^{x+2} \div \left(\dfrac{3}{5}\right)^{2x+4} = 144$?　　　　[CE 2017]

(A)　　1　(B)　　- 1　　(C)　　- 2　　(D) cannot be determined

Answer: (B)

$\dfrac{2^{4x+8}}{5^{2x+4}}\times\dfrac{5^{2x+4}}{3^{2x+4}}=\dfrac{144}{81}\Rightarrow\dfrac{4^{2x+4}}{3^{2x+4}}=\dfrac{16}{9}\Rightarrow\left(\dfrac{4}{3}\right)^{2x+4}=\left(\dfrac{4}{3}\right)^{2}\Rightarrow2x+4=2\Rightarrow x=-1$

Example 94. Given that $\dfrac{\log P}{y-z} = \dfrac{\log Q}{z-x} = \dfrac{\log R}{x-y} = 10$ for $x \neq y \neq z$,

what is the value of the product PQR ?

(A)　0　(B)　1　(C)　xyz　(D)　10^{xyz}　　　[CE 2018]

Answer (B)

$\log P = 10\,(y - z)\,etc$. Now,

$\log(PQR) = \log P + \log Q + \log R$

$= 10\,(y - z + z - x + x - y) = 0 = \log 1 \Rightarrow PQR = 1$

Example 95. A three-member committee has to be formed from a group of 9 people. How many such distinct committee can be formed?

(A) 27   (B) 72   (C) 81   (D) 84   [CE 2018]

Answer: (D)

No. of distinct committee $=^9c_3 = \dfrac{9!}{3!6!} = 84$

Example 96. The binary operation * is defined as $a * b = ab + (a + b)$, where a and b are any two real numbers. The value of the identity element of this operation, defined as the number X such that $a * x = a$, for any a is

(A) 0   (B) 1   (C) 2   (D) [ME 2016]

Answer (A)

$a * x = a \Rightarrow ax + (a + x) = a \Rightarrow ax + x = 0 \Rightarrow (a + 1)x = 0 \Rightarrow x = 0$

Example 97. For integers a, b and c, what would be the minimum and maximum values respectively of a + b + c if log |a| + log |b| + log |c| = 0?

(A) -3 and 3 (B) -1 and 1 (C) -1 and 3 (D) 1 and 3   [ME 2018]

Answer: (A)

a = b = c = ± 1

Minimum and maximum values respectively of a + b + c are -3, 3

Example 98. A 1.5 m tall person is standing from a distance of 3 m from a lamp post. The light from the lamp at the top of the post casts her shadow. The length of the shadow is twice her height. What is the height of the lamp post in meter?

(A)   1.5   (B)   3   (C)   4.5   (D) 6 [EC 2018]

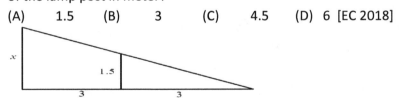

Answer (B)

$\dfrac{x}{1.5} = \dfrac{3 + 3}{3} = 2 \Rightarrow x = 3\,m$

Question of 2 marks:

Example 99. Find the sum to n terms of the series $10 + 84 + 734 + -----$

(A) $\dfrac{9(9^n + 1)}{10}$ (B) $\dfrac{9(9^n - 1)}{8} + 1$ (C) $\dfrac{9(9^n - 1)}{8} + n$ (D) $\dfrac{9(9^n - 1)}{8} + n^2$ [EC2013]

Answer (D)

$10 + 84 + 734 + -----$

$= (9 + 1) + (9^2 + 3) + (9^3 + 5) + - - - = (9 + 9^2 + 9^3 + - - - -) + (1 + 3 + 5 + - -)$

$= \dfrac{9(9^n - 1)}{9 - 1} + \dfrac{n}{2}(2.1 + (n-1).2) = \dfrac{9(9^n - 1)}{8} + n^2$

Example 100. The sum of n terms of the series $4 + 44 + 444 + ----$

(A) $\dfrac{4}{81}\left[10^{n+1} - 9n - 1\right]$ 　　　　(B) $\dfrac{4}{81}\left[10^{n-1} - 9n - 1\right]$

(C) $\dfrac{4}{81}\left[10^{n+1} - 9n - 10\right]$ 　　　　(D) $\dfrac{4}{81}\left[10^n - 9n - 10\right]$

Answer: (C)

$4 + 44 + 444 + -- = \dfrac{4}{9}\left[(10 - 1) + (100 - 1) + (1000 - 1) + - - -\right]$

$= \dfrac{4}{9}\left[(10 + 100 + 1000 + ---) - n\right] = \dfrac{4}{9}\left[\dfrac{10(10^n - 1)}{10 - 1} - n\right]$

$= \dfrac{4}{81}\left[10^{n+1} - 9n - 10\right]$

Example 101. The value of the expression

$\dfrac{1}{1 + \log_u (vw)} + \dfrac{1}{1 + \log_v (wu)} + \dfrac{1}{1 + \log_w (uv)}$ is _____.

(A) 　- 1 　　(B) 　0 　　(C) 　1 　　(D) 3 [CE 2018]

Answer (C)

$\dfrac{1}{1 + \log_u (vw)} + \dfrac{1}{1 + \log_v (wu)} + \dfrac{1}{1 + \log_w (uv)}$

$= \dfrac{1}{\log_u u + \log_u (vw)} + \dfrac{1}{\log_v v + \log_v (wu)} + \dfrac{1}{\log_w w + \log_w (uv)}$

$= \dfrac{1}{\log_u (uvw)} + \dfrac{1}{\log_v (wuv)} + \dfrac{1}{\log_w (uvw)} = \log_{(uvw)} u + - - - - = \log_{(uvw)} (uvw) = 1$

371

Example 102. If $a^2 + b^2 + c^2 = 1$ then $ab + bc + ca$ lies in the interval
(A) [1, 2/3] (B) [−1/2, 1] (C) [−1, 1/2] (D) [2, −4]   [EC 2015]
Answer: (B)

$$ab + bc + ca = \frac{1}{2}\left[(a+b+c)^2 - (a^2 + b^2 + c^2)\right] = \frac{1}{2}\left[(a+b+c)^2 - 1\right] \geq \frac{1}{2}(0-1)$$

$$\Rightarrow ab + bc + ca \geq -\frac{1}{2} \Rightarrow -\frac{1}{2} \leq ab + bc + ca --- (i)$$

$$\frac{1}{2}\left[(a-b)^2 + (b-c)^2 + (c-a)^2\right] \geq 0 \Rightarrow (a^2 + b^2 + c^2) - (ab + bc + ca) \geq 0$$

$$1 - (ab + bc + ca) \geq 0 \Rightarrow 1 \geq ab + bc + ca \Rightarrow ab + bc + ca \leq 1 --(ii) \text{ Co}$$

mbining (i) & (ii), $-\frac{1}{2} \leq ab + bc + ca \leq 1$

Example 103. If $x^2 + x - 1 = 0$, what is the value of $x^4 + \frac{1}{x^4}$?

(A)   1   (B)   5   (C)   7   (D) 9 [CH 2018]
Answer (C)

$$x^2 + x - 1 = 0 \Rightarrow x - \frac{1}{x} = -1 \Rightarrow \left(x - \frac{1}{x}\right)^2 = 1 \Rightarrow x^2 + \frac{1}{x^2} = 3$$

$$\Rightarrow \left(x^2 + \frac{1}{x^2}\right)^2 = 9 \Rightarrow x^4 + \frac{1}{x^4} = 9 - 2 = 7$$

Example 104. The roots of $ax^2 + bx + c = 0$ are real and positive.
a, b and c are real. Then $ax^2 + b\,|\,x\,| + c = 0$ has
(A)no real root (B) 2 real roots (C) 3 real roots (D) 4 real roots [EE 2014]
Answer (D)
The roots of $ax^2 + bx + c = 0$ are real and positive,
discriminant $= b^2 - 4ac \geq 0 - (i)$
The given quadratic equation takes two forms
$ax^2 + bx + c = 0 --(ii)$ and $ax^2 - bx + c = 0 --(ii)$
$[|\,x\,| = x, x > 0 \;\&\; |\,x\,| = -x, x < 0]$
The discriminant of both $(ii)\,\&\,(iii)$ are same. So, $(ii)\,\&\,(iii)$ will have two
roots each, totally 4 roots.
Example 105. If x is real and $|\,x^2 - 2x + 3\,| = 11$ then possible values of
$|-x^3 + x^2 - x\,|$ include
(A)   2, 4   (B)   2, 14   (C)   4, 52   (D)   14, 52   [EE 2014]

372

Answer (D)

$$|x^2 - 2x + 3| = 11 \Rightarrow x^2 - 2x + 3 = \pm 11$$

$$x^2 - 2x + 3 = 11 \Rightarrow x^2 - 2x - 8 = 0 \Rightarrow x = -2, 4$$

$$x^2 - 2x + 3 = -11 \Rightarrow x^2 - 2x + 14 = 0 \Rightarrow Disc. = 4 - 56 < 0 \Rightarrow$$

no real root.

$$x = -2 \Rightarrow |-x^3 + x^2 - x| = |8 + 4 + 2| = 14 \text{ and}$$

$$x = 4 \Rightarrow |-x^3 + x^2 - x| = |-64 + 16 - 4| = |-52| = 52$$

Example 106. Budhan covers a distance of 19 km in 2 hours by cycling one-fourth of the time and walking the rest. The next day he cycles (at the same speed as before) for half the time(at the same speed as before) and walks the rest. The speed in km/h at which Budhan walks is

(A)   3   (B)   4   (C)   5   (D)   6   [EC 2015]

Answer: (D)

Let the speeds in km/hr in cycling and walking are u, v

$$u \times \frac{1}{2} + v \times \frac{3}{2} = 19 \Leftarrow u + 3v = 38 - - - (i) \text{ and}$$

$$u \times 1 + v \times 1 = 26 \Rightarrow u + v = 26 - - - - (ii)$$

$$(i) - (ii) \Rightarrow 2v = 12 \Rightarrow v = 6$$

Example 107. Operators $\bullet$, $*$ and $\rightarrow$ are defined by

$$a \bullet b = \frac{a - b}{a + b}; a * b = \frac{a + b}{a - b} \text{ and } a \rightarrow b = ab$$

Find the value of $(66 \bullet 6) \rightarrow (66 * 6)$

(A)   -2   (B)   -1   (C)   1   (D)   2   [EC 2015]

Answer: (C)

$$(66 \bullet 6) \rightarrow (66 * 6) = \left(\frac{66 - 6}{66 + 6} \right) \rightarrow \left(\frac{66 + 6}{66 - 6} \right) = \frac{60}{72} \times \frac{72}{60} = 1$$

Example 108. The expression $\dfrac{x + y - |x - y|}{2}$ is equal to

(A) the maximum of x and y (B) the minimum of x and y

(C) 1   (D) none of the above   [CS/IT 2017]

Answer (B)

Case i) $x > y \Rightarrow (x - y) > 0 \Rightarrow |x - y| = (x - y)$

$$\frac{x + y - |x - y|}{2} = \frac{x + y - (x - y)}{2} = y \text{ -which is minimum of } x \text{ and } y$$

373

Case ii) $x < y \Rightarrow (x-y) < 0 \Rightarrow |x-y| = -(x-y)$

$$\frac{x+y-|x-y|}{2} = \frac{x+y+(x-y)}{2} = x \text{ -which is minimum of } x \text{ and } y$$

Example 109. The number of roots of $e^{-x} + 0.5x^2 - 2 = 0$ in the range $[-5, 5]$ is

(A)   0   (B)   1   (C)   2   (D)   3   [CS/IT 2017]

Answer (C)

$f(x) = e^{-x} + 0.5x^2 - 2 = 0$

$\Rightarrow f(-5) = e^{-5} + 0.5 \times 25 - 2 > 0, \quad f(0) = 1 - 2 < 0$ and

$f(5) = e^5 + 0.5 \times 25 - 2 > 0$

So, $f(-5)$ & $f(0)$ are of opposite signs, a root lies between -5 and 0.

$f(0)$ & $f(5)$ are of opposite signs, a root lies between 0 and 5.

Two roots are in the range $[-5, 5]$.

Example 110. X is a 30 digit number starting with the digit 4 followed by the digit 7. Then the number X^3 will have

(A) 90 digits (B) 91 digits (C) 92 digits (D) 93 digits [CS/IT,EE 2017]

Answer (A)

The number is of the form 47 A of 3 digits only.

If A = 0, $(470)^3 = 103823000$ which has 9 digits.

If A = 9, $(479)^3 = 109902399$ which has 9 digits.

So, X^3 will have 90 digits.

Example 111. There are 3 Indians and 3 Chinese in a group of 6 people. How many subgroups of this group can we choose so that every group has at least one Indian?

(A)   56   (B)   52   (C)   48   (D)   44   [EC 2017]

Answer (A)

No. of Indians in each subgroup will be 1 or 2 or 3.

It may be selected in $\left({}^3c_1 + {}^3c_2 + {}^3c_3 \right) = (3+3+1) = 7$ ways.

For each selection, the no. of Chinese may be 0 or 1 or 2 or 3

It may be selected in $\left({}^3c_0 + {}^3c_1 + {}^3c_2 + {}^3c_3 \right) = (1+3+3+1) = 8$ ways.

Total no. of subgroups = 7 × 8 = 56

Example 112. An e mail password must contain three characters. The password has to contain one numerical from 0 to 9, one upper case and one lower case character from the English alphabet. How many distinct passwords are possible?

(A) 6,760 (B) 13,520 (C) 40,560 (D) 1,05,456 [EE 2018]

Answer (C)

No. of distinct passwords $= 10 \times 26 \times 26 \times 3! = 40,560$

Example 113. If a, b, c are all positive, $\begin{vmatrix} 1 & \log_a b & \log_a c \\ \log_b a & 1 & \log_b c \\ \log_c a & \log_c b & 1 \end{vmatrix} =$

(A) 1 (B) 0 (C) $(\log a)(\log b)(\log c)$ (D) $\log(abc)$

Answer (B)

$$\Delta = \begin{vmatrix} 1 & \log_a b & \log_a c \\ \log_b a & 1 & \log_b c \\ \log_c a & \log_c b & 1 \end{vmatrix} = \begin{vmatrix} 1 & \log b / \log a & \log c / \log a \\ \log a / \log b & 1 & \log c / \log b \\ \log a / \log c & \log b / \log c & 1 \end{vmatrix}$$

[Multiplying R_1, R_2, R_3 by $(\log a)$, $(\log b)$, $(\log c)$ respectively]

$$\Delta = \frac{1}{\log a \log b \log c} \begin{vmatrix} \log a & \log b & \log c \\ \log a & \log b & \log c \\ \log a & \log b & \log c \end{vmatrix} = 0$$

[As the rows are identical]

Example 114. 500 students are taking one or more course in Chemistry, Physics and Mathematics. Registration records indicate course enrolment as follows: Chemistry (329), Physics (186), Mathematics (295), Chemistry and Physics (83), Chemistry and Mathematics (217), and Physics and Mathematics (63). How many students are taking all three subjects?

(A) 37 (B) 43 (C) 47 (D) 53 [EC 2017]

Answer (D)

If the no. of students taking all subjects $= n(C \cap P \cap M) = x$, then

$$n(C \cup P \cup M) = n(C) + n(P) + n(M) - n(C \cap P) - n(C \cap M)$$
$$- n(P \cap M) + n(C \cap P \cap M)$$
$$\Rightarrow 500 = 329 + 186 + 295 - 83 - 217 - 63 + x \Rightarrow x = 53$$

Example 115. Among 150 faculty members in an Institute, 55 are connected with each other through Face book and 85 are connected through WhatsApp. 30 faculty members do not have Face book or WhatsApp account. The number of faculty members connected only through Face book accounts is

(A)   35   (B)   45   (c)   65   (D)   90 [CS/IT 2016]

Answer (A)

No. of faculty members connected through at least one of the accounts =

$150 - 30 = 120$ $n(F) = 55, n(W) = 85, n(F \cup W) = 120, n(F \cap W) = x$

$n(F \cup W) = n(F) + n(W) - n(F \cap W) \Rightarrow 120 = 55 + 85 - x \Rightarrow x = 20$

Number of faculty members connected only through Face book accounts

$= n(F) - n(F \cap W) = 55 - 20 = 35$

Example 116. Forty students watch films A, B and C over a week. Each student watch only one film or all three. Thirteen students watched film A, sixteen students watched film B and nineteen students watched film C. How many students watched all three films?

(A)   0   (B)   2 (c)   4   (D)   8   [Math 2018]

Answer (C)

No. of students watched all three films = $= x$

No. of students watched film A only $= 13 - x$ etc.

$$40 = (13 - x) + (16 - x) + (19 - x) = x \Rightarrow x = 4$$

[As there is no student watching two films]

Example 117. 1200 men and 500 women can build a bridge in 2 weeks. 900 men and 250 women will take 3 weeks to build the same bridge. How many men will be needed to build the same bridge?

(A)   3000   (B)   3300   (C)   3600   (D)   3900 [EC 2017]

Answer (C)

x men can do the job in one week and y women can do in one week.

Proportion of work in one week by a man is $1/x$ and a woman

is $1/y$..$\dfrac{1200}{x} + \dfrac{500}{y} = \dfrac{1}{2} - (i)$ $\dfrac{900}{x} + \dfrac{250}{y} = \dfrac{1}{3} - (ii)$

$(i) - (ii) \times 2 \Rightarrow \dfrac{600}{x} = \dfrac{1}{6} \Leftarrow x = 3600$

Example 118. In the summer of 2012, in New Delhi, the mean temperature of Monday to Wednesday was 41°C and of Tuesday to Thursday was 43°C. If the temperature on Thursday was 15% higher than that of Monday, then the temperature in °C on Thursday was

(A)   40   (B)   43   (C)   46   (D)   49 [EC 2013]

Answer: (C)

$m + t + w = 3 \times 41 = 123 - -(i)$ &

$$t + w + \frac{115}{100}m = 3 \times 43 = 129 \ - \ -(ii) \quad (ii) - (i), \frac{15\,m}{100} = 6 \Rightarrow m = 40.$$

The temperature in °C on Thursday $= \dfrac{115\,m}{100} = \dfrac{115 \times 40}{100} = 46$

Example 119. The number of 3-digit numbers such that the digit 1 is never to the immediate right of 2 is

(A)　　781　(B)　　791　(C)　　881　(D)　　891　　　[EC 2017]

Answer (C)

Out of 10 digits 3 digits may be selected (with repletion) in

$10^3 = 1000$ ways.

Some of them will start with 0 which are not 3-digit number.

The count of these is $10^2 = 100$

ctual count of 3-digit numbers =1000 – 100 = 900

The count for the numbers for which the digit 1 will be to the right of 2 is

9.So, the required count = 900 - 9 = 881.

Example 120. The sum of eight consecutive odd numbers is 656. The average of four consecutive even numbers is 87. What is the sum of the smallest odd number and second largest even number? [EC 2013]

Answer: 163

Eight consecutive odd numbers are

$x - 6, x - 4, x - 2, x, x + 2, x + 4, x + 6, x + 8$

Their sum $= 8x + 8 = 656 \Rightarrow x = 81,$

Smallest no. of the set $= x - 6 = 75$

Four consecutive even numbers are $y - 2, y, y + 2, y + 4$

Their sum $= 4y + 4 = 87 \times 4 \Rightarrow y = 86,$

Second largest even no. $= y + 2 = 88$, Required sum $= 75 + 88 = 163$

Example 121. Consider a sequence of numbers $a_n = \dfrac{1}{n} - \dfrac{1}{n + 2},$

for each integer $n > 0$. What is the sum of the first 50 terms?

(A)　　$\left(1 + \dfrac{1}{2}\right) - \dfrac{1}{50}$ 　　　　(B)　　$\left(1 + \dfrac{1}{2}\right) + \dfrac{1}{50}$

(C)　　$\left(1 + \dfrac{1}{2}\right) - \left(\dfrac{1}{51} + \dfrac{1}{52}\right)$ 　(D)　　$1 - \left(\dfrac{1}{51} + \dfrac{1}{52}\right)$ 　　　[CE 2018]

Answer (C)

$$\sum_{n=1}^{50} a_n = \left(1 - \frac{1}{3}\right) + \left(\frac{1}{2} - \frac{1}{4}\right) + \left(\frac{1}{3} - \frac{1}{5}\right) + \left(\frac{1}{4} - \frac{1}{6}\right) + - - - -$$

$$+ \left(\frac{1}{48} - \frac{1}{50}\right) + \left(\frac{1}{49} - \frac{1}{51}\right) + \left(\frac{1}{50} - \frac{1}{52}\right) = \left(1 + \frac{1}{2}\right) - \left(\frac{1}{51} + \frac{1}{52}\right)$$

Example 122. $\log\tan 1^0 + \log\tan 2^0 + - - - + \log\tan 89^0$

(A)   1   (B) $1/\sqrt{2}$ (C)   0   (D)   - 1   [EC 2015]

Answer (C)

$\log\tan 1^0 + \log\tan 89^0 = \log\tan 1^0 + \log\tan(90^0 - 1^0)$

$= \log\tan 1^0 + \log\cot 1^0 = \log(\tan 1^0 \times \cot 1^0) = \log 1 = 0$

Each similar pair = 0 and the middle term $= \log\tan 45^0 = \log 1 = 0$

So the sum = 0

Example 123. Consider the equation: $(7526)_8 - (Y)_8 = (4364)_8$, where $(X)_N$ stands for X to the base N. Find Y.

(A)   1634   (B)   1737   (C)   3142   (D)   3162 [CS/IT 2014]

Answer: (C)

$(7526)_8 = 7 \times 8^3 + 5 \times 8^2 + 2 \times 8 + 6 = (3926)_{10}$

$(4364)_8 = 4 \times 8^3 + 3 \times 8^2 + 6 \times 8 + 4 = (2292)_{10}$

$(7526)_8 - (Y)_8 = (4364)_8 \Rightarrow (Y)_8 = (3926 - 2292)_{10}$

Example 124. Operators $*, \bullet$ & $\rightarrow$ are defined by

$a * b = \dfrac{a - b}{a + b}; \ a \bullet b = \dfrac{a = b}{a - b}$ & $a \rightarrow b = ab$. Find the value of

$(66 * 6) \rightarrow (66 \bullet 6)$.

(A)   - 2   (B)   - 1   (C)   1   (D)   2

Answer: (C)

$(66 * 6) \rightarrow (66 \bullet 6) = \left(\dfrac{66 - 6}{66 + 6}\right) \rightarrow \left(\dfrac{66 + 6}{66 - 6}\right) = \dfrac{60}{72} \times \dfrac{72}{60} = 1$

III. Mensuration

Question of one mark:

***Example* 125.** The area of an equilateral triangle is $\sqrt{3}$.
What is the perimeter of the triangle?
(A)   2   (B)   4 (C) 6   (D)   8   [Ch E 2018]
Answer: (C)
If the side of the equilateral triangle is a, then its area

$$= \frac{\sqrt{3}}{4} a^2 = \sqrt{3} \Rightarrow a^2 = 4 \Rightarrow a = 2. \qquad \text{Perimeter} = 3a = 6$$

***Example* 126.** The area of a square is d. What is the area of the circle which has the diagonal of the square as diameter?

(A) πd (B) πd^2 (C) $\dfrac{1}{4} \pi d^2$ (D) $\dfrac{1}{2} \pi d$ [CS/IT 2018]

Answer (D)

If a be the side of a square, then $a^2 = d$.

Diagonal of the square = diameter of the circle = $a\sqrt{2}$,

Area of the circle $= \pi \left(\dfrac{a\sqrt{2}}{2} \right)^2 = \pi \dfrac{a^2}{2} = \dfrac{1}{2} \pi d$

***Example* 127.** The perimeters of a circle, a square and an equilateral triangle are equal. Which one of the following statements is TRUE?
(A) The circle has the largest area
(B) The square has the largest area
(C) The equilateral triangle has the largest area
(D) All the three shapes have the same area
Answer: (A)
$r = $ radius of the circle, $a = $ side of the square,
$b = $ side of the equilateral triangle So,

$$2\pi r = 4a = 3b = k \text{ (say)} \Rightarrow r = \frac{k}{2\pi}, a = \frac{k}{4}, b = \frac{k}{3}$$

Area of circle $= \pi r^2 = k^2 \times 0.0795$ Area of square $= a^2 = k^2 \times 0.0625$

Area of equilateral triangle $= \dfrac{\sqrt{3}}{4} b^2 = k^2 \times 0.0481$

So, the circle has the largest area.

***Example* 128.** A rectangle becomes a square when its length and breadth are reduced by 10 m and 5 m, respectively. During this process, the rectangle loses 650 m^2 of area. What is the area of the original rectangle in square meters?

(A)   1125   (B)   2250   (C)   2924   (D)   4500 [ME 2018]
Answer: (B)
Length = l, breadth = b so, $l - 10 = b - 5 \Rightarrow l = b + 5$
$(b + 5).b - (b - 5)^2 = 650 \Rightarrow b = 45$
Original area $= l.b = (b + 5).b = 2250\, m^2$

Example 129. Tower A is 90 m tall and tower B is 140 m tall. They are 100 m apart. A horizontal sky walk connects the floors at 70 m in both the towers. If a taut rope connects the top of tower A to the bottom of tower B, at what distance (in meters) from tower A will the rope intersect the sky walk?

(A)   22.22 (B)   50   (C)   57.87 (D)   77.78
Answer: (A)

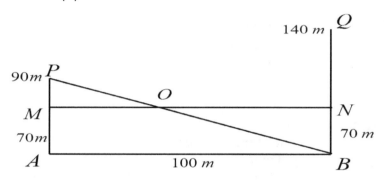

PM = PA − AM = 90 − 70 = 20 m, BN = 140 − 70 = 70 m   Let OM = x
tan POM = tan BON $\Rightarrow$ PM/OM = BN/ON $\Rightarrow$ 20/x = 70/(100− x)
$\Rightarrow x = 22.22$ m

Question of two marks:

Example 130. A square pyramid has a base perimeter x, and the slant height is half of the perimeter. What is the lateral surface area of the pyramid?

(A) x^2 (B) $0.75x^2$ (C) $0.50x^2$ (D) $0.25x^2$ [CE 2016]
Answer (D)
Lateral surface area =

$\dfrac{1}{2} \times (perimeter) \times slant\ height = 0.5 \times x \times \dfrac{x}{2} = 0.25x^2$

Example 131. Two and a quarter hours back, when seen in a mirror, the

380

reflection of a wall clock without number markings seemed to show 1:30.What is the actual current time shown by the clock?

(A) 8:15   (B) 11:15   (C) 12:15   (D) 12:45   [EC 2016]

Answer (D)

Reflection is 1:30(i.e.,on the right half way between the markings of 1 &2). Corresponding actual time (on the left) will be half way between 10 &11 i.e.10:30. This time is 2:30 hours back. Now it will be (10:30+2:15) = 12:45

Example 132. A cube of side 3 units is formed using a set of smaller cubes of side 1 unit. Find the proportion of the number of faces of the smaller cubes visible to those which are NOT visible.

(A) 1 : 4   (B) 1 : 3   (C) 1 : 2   (D) 2 : 3   [EC 2015]

Answer: (C)

In any cube, the no. of faces = 6

The big cube is formed with $3^3 = 27$ small cubes.

The total no. of faces of the big cube (formed with small cubes)

= 27 × 6 = 162

Each visible face of the big cube consists of 9 small cubes.

Total number of visible faces of the small cubes = 9 × 6 = 54

Total number of NOT visible faces of the small cubes = 162 – 54 = 108

Ratio = 54:108 = 1:2

Example 133. A cube is built using 64 cubic blocks of side one unit. After it is built, one cubic block is removed from every corner of the cube. The resulting surface area of the body (in square units) after the removal is

(A)   56   (B)   64   (C)   72   (D)   96 [CS/IT 2016]

Answer (D)

No. of directions of the big cube = 3   No. of cubic blocks $= 124 = 4^3$

So, 4 blocks in each direction are needed to build the big cube.

Area of one side of the big cube = 4 × 4 = 16 sq. units.

There are 6 faces.   Total area = 6 ×16 = 96 sq. units.

If one cubic block is removed from every corner of the cube, they introduce new surfaces equal to exposes surfaces so the area of the bigger cube does not change from 96

Example 134. From a circular sheet of paper of radius 30 cm, a sector of 10% area is removed. If the remaining part is used to make a conical surface, then the ratio of the radius and height of the cone is _____.

Answer: 13.08   [EC 2015]

Radius of the circular sheet = $r = 30\,cm$

90% of the area of the circular sheet = $= 0.9\pi\,r^2 = 0.9\pi \times 900$

If R be the radius of the conical surface, cross sectional area of the cone
$= \pi R \times 30$

So, $0.9\pi \times 900 = \pi R \times 30 \Rightarrow R = 0.9 \times 30 = 27$ cm

Height of the cone
$$= \sqrt{l^2 - r^2} = \sqrt{30^2 - 27^2} = \sqrt{900 - 729} = \sqrt{171} = 13.08$$

Example 135. A wire of length 340 mm is to be cut into two parts. One of the parts is to be made into a square and the other into a rectangle where sides are in the ratio of 1:2. What is the length of the side of the square (in mm) such that the combined area of the square and the rectangle is a MINIMUM?

(A) 30 (B) 40 (C) 120 (D) 180 [EC 2016]

Answer (B)

Side of square $= x$, sides of the rectangle $y, 2y$

Length of the wire $= 4x + 6y = 340 \Rightarrow 2x + 3y = 170 \Rightarrow x = 85 - \dfrac{3y}{2}$

Combined area $= A = x^2 + 2y^2 = \left(85 - \dfrac{3y}{2}\right)^2 + 2y^2$ - to minimize

$\dfrac{dA}{dy} = 2\left(85 - \dfrac{3y}{2}\right)\left(-\dfrac{3}{2}\right) + 4y = -3\left(85 - \dfrac{3y}{2}\right) + 4y = -255 + \dfrac{17y}{2}$

$\dfrac{dA}{dy} = 0 \Rightarrow \dfrac{17y}{2} = 255 \Rightarrow y = \dfrac{255 \times 2}{17} = 30 \Rightarrow x = 85 - \dfrac{3}{2} \times 30 = 40 \, m$

Example 136. A window is made up of a square portion and an equilateral triangle portion above it. The base of the triangular portion coincides with the upper side of the square. If the perimeter of the window is $6\,m$, the area of the window in m^2 is _____.

(A) 1.43 (B) 2.06 (C) 2.68 (D) 2.88 [ME 2016]

Answer (B)

If the side of the square be a, then the side of the equilateral triangle will be a So, Perimeter of the window $= 5a = 6 \Rightarrow a = 6/5$ [The upper side of the square = base of the equilateral triangle and this will not be considered to calculate the perimeter of the window]

Area of the window = area of the square + area of the equilateral triangle

$$= a^2 + \frac{\sqrt{3}}{4}a^2 = \left(1 + \frac{\sqrt{3}}{4}\right)a^2 = 1.433 \times \frac{36}{25} = 2.06 \, m^2$$

Example 137. In a 2 × 4 rectangle grid shown below, each cell is a rectangle. How many rectangles can be observed in the grid?

(A) 21 (B) 27 (C) 30 (D) 36 [CS/IT 2016]

Answer (C)

Small rectangles ABFG, -- = 8,
rectangles AFHC, AFDI, AEJF and similar in the down = 2(3) = 6
rectangles BGID, BGJE and similar in the down = 2(2) = 4
rectangles CHJE and similar in the down = 2
ABLK, BCML, CDNM, DEON = 4   ACMK, ADNK, AEOK = 3
ADNK, AEOK = 2   AEOK = 1   TOTAL = 30

Example 138. A right-angled cone (with base radius 5 cm and height 12 cm), as shown in the figure below, is rolled on the ground keeping the point P fixed until the point Q (at the base of the cone, as shown) touches the ground again.

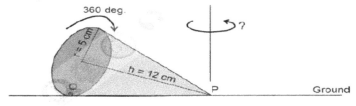

By what angle (in radians) about P does the cone travel?

(A) $\dfrac{5\pi}{12}$ (A) $\dfrac{5\pi}{24}$ (A) $\dfrac{24\pi}{5}$ (A) $\dfrac{10\pi}{13}$ [ME 2017]

Answer (D)

Slant height $= l = \sqrt{r^2 + h^2} = \sqrt{5^2 + 12^2} = 13\,cm$

Angle (in radians) $= \dfrac{2\pi r}{l} = \dfrac{2\pi \times 5}{13} = \dfrac{10\pi}{13}$

Example 139. A contour line joins locations having the same height above the mean sea level. The following is a contour plot of a geographical region. Contour lines are shown at 25 m intervals in this plot.

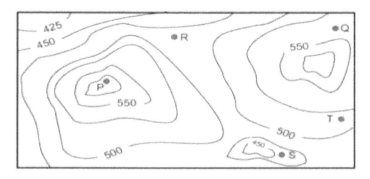

If in a flood, the water level rises to 525 m, which of the villages P, Q, R, S, T get submerged?

(A) P, Q   (B)   P, Q, T   (C)   R, S, T   (D)   Q, R, S   [CS/IT 2017]

Answer: (C)

Height of the villages (in meter):

P→575, Q→525, R→475, S→425, T→500

In flood, water level is 525 m. So, villages R, S, T will be submerged.

Example 140. An air pressure contour line join the locations in the region having the same atmospheric pressure. The following is the air pressure contour of a geographical region. The contour lines are shown as 0.05 bar intervals in the plot.

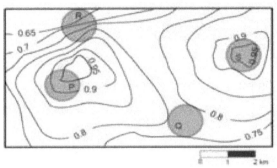

If the probability of thunderstorm is given by how fast air pressure rises or drops over a region. Which of the following region is most likely to have thunderstorm?

(A)   P   (B)   Q   (C)   R   (D)   S   [CS/IT 2017]

Answer: (C)

Through R, the maximum number of lines (4 lines) pass. So, the air pressure changes at 4 places which is the maximum.

IV. Probability:

Question of one mark:

Example 141. A regular die has six sides with numbers 1 to 6 marked on its sides. If a very large number of throws show the following frequencies of occurrence: 1→ 0.167; 2→ 0.167; 3→ 0.152; 4→ 0.166; 5→ 0.168; 6 → 0.180. We call this die

(A) irregular (B) biased (C) Gaussian (D) insufficient
Answer: (B)
For a large number of throws, the frequencies should be equal for an unbiased die.

***Example* 142.** Given Set A = {2, 3, 4, 5} and Set B = {11, 12, 13, 14, 15}, two numbers are randomly selected, one from each set. What is the probability that the sum of the two numbers equals 16?
(A)   0.20   (B)   0.25   (C)   0.30   (D) 0.33 [CS/IT 2015]
Answer: (A)
No. of ways in which one number may be selected from each set = 4×5=20   Sum of the numbers will be 16 for the pairs (2, 14), (3, 13), (4, 12), (5, 11)   No. of ways = 4   Probability = 4/20 = 0.20

***Example* 143.** Out of all the 2-digit integers between 1 and 100, a 2-digit number has to be selected at random. What is the probability that the selected number is not divisible by 7?
(A)   13/90 (B)   12/90   (C)   78/90 (D)   77/90   [ME 2013]
Answer (D)
No. of 2-digit integers between 1 and 100 is 90.
Out of them 13 are divisible by 7, so non-divisible by 7 is (90 – 13) = 77
Required probability = 77/90

***Example* 144.** There are 3 red socks, 4 green socks and 3 blue socks. You choose 2 socks. The probability that they are of the same color is
(A)   1/5   (B)   7/30   (C)   1/4 (D)   4/15 [EE /CS/IT 2017]
Answer (D)

Out of 10 socks,2 may be selected in $^{10}c_2 = 45$ ways $\Rightarrow n = 45$

The favorable cases are:

2 red socks or 2 green socks or 2 blue socks.

No. of favorable events $= ^3c_2 + ^4c_2 + ^3c_2 = 3 + 6 + 3 = 12 \Rightarrow m = 12$

Probability $= \dfrac{m}{n} = \dfrac{12}{45} = \dfrac{4}{15}$

***Example* 145.** The probability that a $k -$ digit number does NOT contain 0, 5 or 9 is
(A) 0.3^k (B) 0.6^k (C) 0.7^k (D) 0.9^k [EE 2017]
Answer (C)
Out of 10 digits (0 – 9) three digits namely 0, 5, 9 will not appear. So, seven digits may appear up to $k -$ times.
Probability $= (7/10)^k = 0.7^k$

Example 146. Ram and Ramesh appeared in an interview for two vacancies in the same department. The probability of Ram's selection is 1/6 and that of Ramesh is 1/8. What is the probability that only one of them will be selected?

(A) 47/48 (B) 1/4 (C) 13/48 (D) 35/48 [EC 2015]

Answer: (B)

Let A be event that "Ram is selected" and B be event that "Ramesh is selected"

The event is: (A selected but B not) or (B selected but A not)

$$P(A) = 1/6 \Rightarrow P(\overline{A}) = 5/6, P(B) = 1/8 \Rightarrow P(\overline{B}) = 7/8$$

Probability =

$$P\left[(A\overline{B}) \ OR \ P(\overline{A}B)\right] = P(A).P(\overline{B}) + P(\overline{A}).P(B)$$

$$= \frac{1}{6} \times \frac{7}{8} + \frac{5}{6} \times \frac{1}{8} = \frac{7}{48} + \frac{5}{48} = \frac{12}{48} = \frac{1}{4}$$

Example 147. A couple has 2 children. The probability that both children are boys if the older one is a boy is

(A) 1/4 (B) 1/3 (C) 1/2 (D) 1 [ME 2017]

Answer (C)

The probability of birth of a boy = probability of birth of a girl = 1/2

The couple will have both children boys if the second child is also a boy.

Probability = 1/2

Example 148. Two dice are thrown simultaneously. The probability that the product of the numbers appearing on the top faces of the dice is a perfect square is

(A) 1/9 (B) 2/9 (C) 1/3 (D) 4/9 [CE 2017]

Answer (B)

Total no. of cases $= n = 6^2 = 36$

Favourable cases: {(1, 1), (1, 4), (2, 2), (3, 3), (4, 4), (4, 1), (5, 5), (6, 6)}

No. of favourable cases $= m = 8$.   Probability $= m/n = 8/36 = 2/9$

Question of two marks:

Example 149. Shaquille O' Neal is a 60% career free throw shooter, meaning that he successfully makes 60 free throws out of 100 attempts on average. What is the probability that he will successfully make exactly 6 free throws in 10 attempts?

(A) 0.2508 (B) 0.2816 (C) 0.2934 (D) 0.6000   [EE 2016] Answer (A)

Let X represents the number of successful throws. $p = 0.6, q = 0.4, n = 10, x = 6$

$$P(X = 6) = {}^{10}C_6(0.6)^6.(0.4)^4 = 210 \times 0.00467 \times 0.0256 = 0.2508$$

$$\left[Binomial \ Distn : P(X = x) = {}^nC_x p^x q^{n-x} \right]$$

Example 150. The probabilities that a student passes Mathematics, Physics and Chemistry are m, p and c respectively. Of these subjects, the student has 75% chance of passing in at least one, a 50% chance   of passing in at least two and a 40% chance of passing in exactly two. Following relations are drawn in m, p, c :

(I) $p + m + c = 27/20$ (II) $p + m + c = 13/20$ (III) $(p) \times (m) \times (c) = 1/10$
(A) Only relation I is true   (B) Only relation II is true
(C) Relations II and III are true (D) Relations I and III are true[CS 2015]
Answer (D)

Let M, P and C denote the events of passing Mathematics, Physics and Chemistry respectively. $P(M) = m, P(P) = p, P(C) = c$

Chance of passing in at least two = chance of passing in exactly two + chance of passing in three + chance of passing in the three

Or, $50/100 = 40/100 + P(MPC) \Rightarrow P(M)P(P)P(C) = 10/100$ [As the events M, P and C are independent]

$\Rightarrow (p) \times (m) \times (c) = 1/10$ [Statement III]

$$\frac{75}{100} = p + m + c - \left(\frac{50}{100} + \frac{10}{100} \right) \Rightarrow p + m + c = \frac{135}{100} = \frac{27}{20}$$

Example 151. Out of all 2-digit integers between 1 and 100, a 2-digit number has to be selected at random. What is the probability that the selected number is not divisible by 7?

(A) 13/90 (B) 12/90 (C) 78/90 (D) 77/90 [ME 2013]
Answer (D)

The number of 2-digit integers between 1 and 100 is 90   (10 – 99)
Out of them 13 are divisible by 7, not divisible by 7 is 90 – 13 = 77
Probability = 77/90

Example 152. A cab was involved in a hit and run accident at night. You are given the following data about the cabs in the city and accident.
(i) 85% of the cabs in the city are green and the remaining cabs are blue

387

(ii) A witness identified the cab involved in the accident as blue.

(iii) It is known that a witness can correctly identify the cab color only 80% of the time.

Which one of the following options is closest to the probability that the accident was caused by a blue cab?

(A)   12%   (B)   15%   (C)   41%   (D)   80%   [EC 2018]

Answer (C)

Two mutually exclusive cases may arise:

(a) The accident was caused by a blue cab and the witness correctly

identified the cab color: Prob. $= \dfrac{15}{100} \times \dfrac{80}{100} = \dfrac{12}{100}$

(b) The accident was caused by a green cab and the witness cannot

correctly identify the cab color: Prob. $= \dfrac{85}{100} \times \dfrac{20}{100} = \dfrac{17}{100}$

Total Prob. $= \dfrac{29}{100} = 29\%$

Example 153. Two urns contain respectively 2 white, 1 black balls and 1white 5 black balls. One ball is transferred from the 1st to the 2nd urn and then a ball is drawn from the 2nd urn. The probability that the ball drawn is white:

(A) $\dfrac{16}{21}$ (B) $\dfrac{5}{21}$ (C) $\dfrac{3}{7}$ (D) $\dfrac{4}{7}$

Answer (B)

Probability = P(1 white ball from urn I and 1white ball from urn II) + P(1 black ball from urn I and 1white ball from urn II) $= \dfrac{2}{3} \times \dfrac{2}{7} + \dfrac{1}{3} \times \dfrac{1}{7} = \dfrac{5}{21}$

Example 154. A six sided unbiased die with four green faces and two red faces is rolled seven times. Which of the following combinations is the most likely outcome of the experiment?

(A) three green faces and four red faces

(B) four green faces and three red faces

(C) five green faces and two red faces

(D) six green faces and one red face   [CS/IT 2018]

Answer (C)

If getting a green face is 'success', then getting red face is 'failure'.

$p = 4/6 = 2/3, q = 1/3$

Probability of getting x green faces and $(7 - x)$ red faces in 7 throw is

$$^7c_x\left(\frac{2}{3}\right)^x\left(\frac{1}{3}\right)^{7-x} = {}^7c_x\frac{2^x}{3^7} = \frac{P}{3^7}, \quad P = {}^7c_x.2^x$$

Option (A): $x = 3, \ P = {}^7c_3.2^3 = 35 \times 8 = 280$

Option (B): $x = 4, \ P = {}^7c_4.2^4 = 35 \times 16 = 560$

Option (C): $x = 5, \ P = {}^7c_5.2^5 = 21 \times 32 = 6720$

Option (D): $x = 6, \ P = {}^7c_6.2^6 = 7 \times 64 = 448 \Rightarrow P$ is max. in Option (C)

Example 155. A coin is tossed thrice. Let X be the event that head occurs in each of the first two tosses. Let Y be the event that a tail occurs on the third toss. Let Z be the event that two tails occur in three tosses. Based on the above information, which one of the following statements is TRUE?

(A) X and Y are not independent   (B) Y and Z are dependent

(C) Y and Z are independent   (D) X and Z are independent

Answer: (B)

For X, the favourable cases {HHT, HHH} $\Rightarrow$ P(X) = 2/8 = 1/4

For Y, the favourable cases {HHT, HTT, THT, TTT} $\Rightarrow$ P(Y) = 4/8 = 1/2

For XY, the favourable case {HHT} $\Rightarrow$ P(XY) = 1/8

For Z, the favourable cases {HTT, THT, TTH} $\Rightarrow$ P(Z) = 3/8

For YZ, the favourable cases{ HTT, THT} $\Rightarrow$ P(YZ) = 2/8 = 1/4

For XZ, the favourable cases {ϕ} $\Rightarrow$ P(XZ) = 0/8 = 0

P(XY) = P(X).P(Y) $\Rightarrow$ X and Y are independent $\Rightarrow$ Not (A)

P(YZ) $\neq$ P(Y).P(Z) $\Rightarrow$ Y and Z are dependent $\Rightarrow$ (B) is answer.

V. Numerical Reasoning

Question of one mark:

Example 156. There are five buildings called V, W, X, Y and Z in a row (not necessarily in that order). V is to the west of W. Z is to the east of X and the west of V. W is to the west of Y. Which is the building in the middle?

(A)   V   (B)   W   (C)   X   (D)   Y   [CS/IT 2017]

Answer (A)

The possible arrangement is

west   *east*

389

V is to the west of W: V   W
Z is to the east of X and the west of V: X Z   V   W
W is to the west of Y: X Z   V   W   Y
So, V is in the middle.

Example 157. Michael lives 10 km away from where I live. Ahmed lives 5 km away and Susan lives 7 km away from where I live. Arun is farther away than Ahmed but closer than Susan from where I live. From the information provided here what is one possible distance (in km) at which I live from Arun's place?

(A)   3.00   (B)   4.99   (C) 6.02   (D)   7.01   [EC 2016]
Answer (C)

Michael →M, Ahmed →H, Susan →S, Arun →A,
IM = 10, IH = 5, IS = 7 Arun lives between H and S but closer than S
IA = 6.02 km

Example 158. Find the missing sequence in the letter series below:
A, CD, GHI, ?,   UVWXY
(A) LMN (B) MNO (C) MNOP (D) NOPQ [EC 2015]
Answer (C)

A   CD   GHI   <u>MNOP</u>   UVWXY

    B (1)   EF (2)   JKL (3)   QRST (4)

Example 159. Find the missing sequence in the letter series.

            B, FH, LNP, _ _ _ _.

(A) S U W Y (B) T U V W (C) T V X Z (D) T W X Z [ME 2016]
Answer (C)
If the position of letters in English alphabet system is considered

B	F	H		L	N	P
2	6	8		12	14	16
2		2			2	2

T   V   X   Z   So, (C) follows the same pattern as the given ones.

20   22   24   26

    2   2   2

Example 160. Which number does not belong in the series below?
2, 5, 10, 17, 26, 37, 50, 64

(A)   17   (B)   37   (C)   64   (D)   26   [CS/IT 2014]
Answer: (C)

	2	5	10	17	26	37	50	64
Difference:		3	5	7	9	11	13	14

The last difference should be 15. So, the last number should be 65, but 64 is given.

Example 161. What is the next number in the series?
8,   14,   26,   50,   98,   194,   __.
(A)   378   (B)   384   (C)   386   (D)   388
Answer: (C)
Next term = (previous term) × 2 - 2   Missing number = 2 × 194 – 2 = 386

Example 162. What is the missing number in the following sequence?
2, 12, 60, 240, 720, 1440, _____
(A)   2880   (B)   1440   (C)   720   (D)   0   [CS/IT 2018]
Answer: (B)

2   12   60   240   720   1440

  2 × 6   12 × 5   60 × 4   240 × 3   720 × 2

    *(1440)

    1440 × 1

Example 162. What is the next number in the series?
12   35   81   173   357   ____   [EC 2014]
Answer: 725

12   35   81   173   357   --
  1 × 23   2 × 23   4 × 23   8 × 23   16 × 23
Next number = 357 + 16 × 23 = 725

Example 163. What is the missing number in the following sequence?
    212   60   240   720   1440   ---   0
(A) 2880 (B)   1440   (C)   720   (D) 0   [CS/IT 2018]
Answer (B)
2   12   60   240   720   1440   ---   0
  2 × 6   12 × 5   60 × 4   240 × 3   720 × 2   1440 × 1
1440 × 0
Missing number = 1440 × 1 = 1440

Example 164. Fill in the missing number in the series.

| 2 | 3 | 6 | 15 | ___ | 157.5 | 630 |

Answer: 45

| 2 | 3 | 6 | 15 | 45 | 157.5 | 630 |

$\times 1.5$ $\times 2$ $\times 2.5$ $\times 3$ $\times 3.5$ $\times 4$

Example 165. Pick the odd one out in the following

13, 23, 33, 43, 53

(A)   23   (B)   33   (C)   43   (D)   53 [EE 2016]

Answer (B)

i) all numbers end with 5 ii) The sum of the digits = 5, 6, 7, 8

But 23, 43,53 are prime numbers (no factor other than 1 and the no. itself) while 33 = 11 × 3 – is a composite number.

Example 166. Find the odd one in the following group

Q,W,Z,B B,H,K,M W,C,G,J M,S,V,X

(A) Q,W,Z,B (B) B,H,K,M (C) W,C,G,J (D) M,S,V,X

Answer (C)

The difference between the consecutive alphabets are 521, 521,532,521.

Example 167. Find the odd one from the following group:

W,E,K,O   I,Q,W,A   F,N,T,X   N,V,B,D

(A) W,E,K,O (B) I,Q,W,A   (C) F,N,T,X (D) N,V,B,D   [EC 2014]

Answer (D)

The difference between the consecutive alphabets are 753, 753, 753, 751. So, D is odd.

Example 168. Based on the given statements, select the appropriate option with respect to grammar and usage.

Statements (i) The height of Mr. X is 6 feet.

(ii) The height of Mr. Y is 5 feet.

(A) Mr. X is longer than Mr. Y.

(B) Mr. X is more elongated than Mr. Y.

(C) Mr. X is taller than Mr. Y.

(D) Mr. X is lengthier than Mr. Y.   [EC 2016]

Answer (C)

Example 169. P, Q, and R talk about S's car collection. P states that S has at least 3 cars. Q believes that S has less than 3 cars. R indicates that to his knowledge, S has at least one car. Only one of P, Q and R is right. The number of cars owned by S is

(A)   0   (B)   1   (C)   3   (D)   Cannot be determined   [ME 2017]

Answer (A)

The statements of P, Q and R are contradictory.

Example 170. P looks at Q, while Q looks at R. P is married, R is not. The number of pairs of people in which a married person is looking at an unmarried person is

(A) 0   (B)   1   (C) 2   (D) Cannot be determined [ME 2017]

Answer (B)

Q may be i) married ii) unmarried For case i), $Q \rightarrow R$ is the answer. For case ii), $P \rightarrow Q$ is the answer. Required no. of pairs = 1 + 1 = 2

Example 171. If ROAD is written as URDG, then SWAN should be written as:

(A) VXDQ   (B) VZDQ   (C) VZDP   (D) UXDQ

Answer: (B)

R	O	A	D		S	W	A	N
+ 3	+ 3	+ 3	+ 3		+ 3	+ 3	+ 3	+ 3
U	R	D	G		V	Z	D	Q

Question of two marks:

Example 172. Fill in the missing value

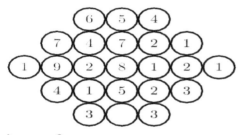

[EC 2015]

Answer: 3

The numbers appearing in the centre of each row is the average is the average of the sum of the numbers in the left and right.

Missing value = (3 + 3)/2 = 3

Example 173. If a and b are integers and $a - b$ is even, which of the following must be even?

(A) ab (B) $a^2 + b^2 + 1$ (C) $a^2 + b + 1$ (D) $ab - b$ [ME 2017]

Answer (D)

$a - b$ is even implies two cases:

i) Both a and b are even ii) Both a and b are odd.

Option (D)

In case i), ab is even and as b is even, $ab - b$ is even. (even–even=even)

393

In case ii), ab is odd and as b is odd, $ab - b$ is even.(odd-odd=even)
Option (A) is right for case i) only.
Option (B) is right neither for case i) nor for case ii). [It is always odd]
Option (C) is right neither for case i) nor for case ii). [It is always odd]

Example 174. Six people are seated around a circular table. There are at least two men and two women. There are at least three right handed persons. Every woman has a left handed person to her immediate right. None of the women are right-handed. The number of women at the table is
(A)   2 (B)   3 (C)   4   (D)   Cannot be determined   [EE 2017]
Answer (A)
The possible arrangements are
i) 2 M, 4 W   ii) 3 M, 3 W   iii) 4 M, 2W
Number of right handed persons a)   3   b)   4 and all are men.
As no woman is right handed and there are at least three right handed persons so, i) is rejected.
Option ii) or iii) is correct
As every woman has a left handed person to her immediate right ii) is rejected. Right option is iii).

Example 175. In a certain code, AMCF is written as EQGJ and NKUF is written as ROYJ. How will DHLP be written in that code?
(A)   RSTN   (B)   TLPH (C) HLPT (D) XSVR [EE 2018]
Answer (C)
From each letter 5 letters ahead including the base letter.
DHLP → HLPT

VI. Data Interpretation
Tables, Charts and Diagrams
Question of one mark:
Example 176. In a survey, 300 respondents were asked whether they own a vehicle or not. If yes, they were further asked to mention whether they own a car or scooter or both. Their responses are tabulated below. What percent of respondents do not own a scooter?

		Men	Women
Own vehicle	Car	40	34
	Scooter	30	20
	Both	60	46
Do not own vehicle		20	50

394

Answer: 48%

Percentage $= \left(\dfrac{300 - (30 + 60 + 20 + 46)}{300} \times 100 \right)\% = \left(\dfrac{144}{3} \right)\% = 48\%$

Example 177. The table below has question-wise data on the performance of students in an examination. The marks for each question are also listed. There is no negative or partial marking in   the examination.

Q. No.	Marks	Answered correctly	Answered wrongly	Not attempted
1	2	21	17	6
2	3	15	27	2
3	2	23	18	3

What is the average of the marks obtained by the class in the examination?
(A)   1.34   (B)   1.74   (C)   3.02   (D)   3.91
Answer (C)
Total marks obtained = 21 × 2 + 15 × 3 + 23 × 2 = 133
Total no. of students = 21 + 17 + 6 = 15 + 27 + 2 = 23 + 18 + 3 = 44
Average = 133/44 = 3.02

Example 178. The table below has question-wise data on the performance of students in an examination. The marks for each question are also listed. There is no negative or partial marking in the examination.

Q. No.	Marks	Answered Correctly	Answered Wrongly	Not Attempted
1	2	21	17	6
2	3	15	27	2
3	2	23	18	3

What is the average of the marks obtained by the class in the examination?
(A)   1.34   (B)   1.74   (C)   3.02   (D)   3.91 [CS/IT 2014]

Answer (C)

Total no. of questions in each category = 44

395

Total marks obtained = 2 × 21 + 3 × 15 + 2 × 23 = 133
Average marks = 133/44 = 3.02

Example 179. The following graph represents the installed capacity for cement production (in tones) and the actual production (in tones) of nine cement plants of a cement company Capacity utilization of a plant is defined as ratio of actual production of cement to installed capacity. A plant with installed capacity of at least 200 tonnes is called a large plant and a plant with lesser capacity is called a small plant. The difference between total production of large plants and small plant in tones is
_____.   [EE 2016]

Answer 120

Large Plants:

Plant No.	1	4	8	9	Total
Inst. Cap.	220	200	250	200	
Act. Prod.	160	190	230	190	770

Small Plants:

Plant No.	2	3	5	6	7	Total
Inst. Cap.	180	190	160	150	140	
Act.Prod.	150	160	120	100	120	650

Difference between total production of large plants and small plants (in tones) = 770 – 650 = 120

Example 180. An electric bus has on board instruments that report the total electricity consumed since the start of the trip as well as the total distance covered. During a single day of operation, the bus travels on stretches M, N, O and P, in that order. The cumulative distances travelled and the corresponding electricity consumption are shown in the Table below:

396

Stretch	Cumulative distance (Km)	Electricity used (KWh)
M	20	12
N	45	25
O	75	45
P	100	57

The stretch where the electricity consumption per km is minimum is
(A) M (B) N (C) O (D) P   [EC 2015]

Answer: (D)
Electricity consumption per km for
M $\rightarrow$ 12/20 =0.6   N $\rightarrow$ 13/25 = 0.52
O $\rightarrow$ 20/30 = 0.66   P $\rightarrow$ 12/25 = 0.48 $\Rightarrow$ Minimum is P.

Question of two marks:

Example 181. In the figure below, $\angle DEC + \angle BFC =$

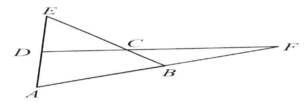

(A) $\angle BCD - \angle BAD$ (B) $\angle BAD + \angle BCF$
(C) $\angle BAD + \angle BCD$ (D) $\angle CBA + \angle ADC$ [CS/IT 2018]
Answer: (A)
$\angle DEC + \angle BFC = \angle ECF - \angle EDC + \angle BFC$
$= \angle BCD - (\angle EDC - \angle BFC) = \angle BCD - \angle BAD$

Example 182. The data given in the following table summarizes the monthly budget of an average household.

Category	Amount (Rs.)
Food	4000
Clothing	1200
Rent	2000
Savings	1500
Other expenses	1800

The approximate percentages of the monthly budget NOT spent on savings is

(A)   10%   (B)   14%   (C)   81%   (D)   86% [EC 2012]

Answer: (D)

Total monthly budget = 4000 + 1200 + 2000 +1500 + 1800 = 10 500

Amount spent on saving = 1500

The amount not spent on saving = 10,500 − 1500 = 9000

The percentage of the monthly budget NOT spent on savings

$$= \left(\frac{9000}{10500} \times 100 \right)\% = 86\%$$

Example 183. A shaving set company sells 4 different types of razors, Elegance, Smooth, Soft and Executive. Elegance sells at Rs. 48, Smooth at Rs. 63, Soft at Rs. 78 and Executive at Rs. 173 per piece.

The table below shows the numbers of each razor sold in each quarter of a year.

Quarter/Product	Elegance	Smooth	Soft	Executive
Q1	27300	20009	17602	9999
Q2	25222	19392	18445	8942
Q3	28976	22429	19544	10234
Q4	21012	18229	16595	10109

Which product contributes the greatest fraction to the revenue of the company in that year?

(A) Elegance (B) Executive (C) Smooth (D) Soft   [CS/IT 2016]

Answer (B)

Quarter/Product	Elegance	Smooth	Soft	Executive
Q1	27300	20009	17602	9999
Q2	25222	19392	18445	8942
Q3	28976	22429	19544	10234
Q4	21012	18229	16595	10109
TOTAL	102510	80059	72186	39284
Contribution (Rs.)	102510×48 = 4920480	80058×63 = 5043717	72186×8 =5630508	39284×173 = 6796132

Contribution of the 'Executive' brand is a maximum.

BAR Chart

Example 184. *Directions: (Q.I to Q.III) : Study the following bar chart carefully and answer the questions given below it.*

Students enrolled for their graduation in different disciplines in ABC college:

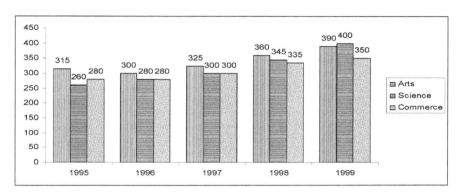

I. What is the ratio of the no. of Arts students and that of commerce students over the period ?

(A) 103 : 112   (B) 337 : 300   (C) 340 : 209   (D) 338 : 309

Answer (D)
Total no. of Arts students = 315 + 300 + 325 + 360 + 390 = 1690

Total no. of commerce students = 1545
Reqd. ratio = 1690 : 1545 = 338 : 309
II. What is the difference between the percentage of Arts students of 1996 and that of 1998 with respect to total students of the corresponding years ?
(A) 0.25%   (B) 0.37 %   (C) 0.27%   (D) 0.35%
Answer (C)
Difference = (300/860 - 360/1040) × 100 = 0.27%

III. What fractional part of the no. of Arts graduates in 1997 is equal to that of commerce graduates in 1997?

(A) 1.02   (B) 0.23   (C) 1.08   (D) 0.92

Answer (D)

Fractional part = (300/325) = 0.92

Example 185. *Directions (Q.I to Q.IV): Study the following graph carefully and answer the questions given below it.*

Income and expenditure of a company over the years (in lakh rupees):

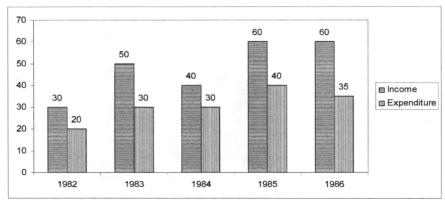

I. What was the difference in profit between 1983 and 1984 ?

(A) Rs. 5 lakhs   (B) Rs. 10 lakhs   (C) Rs. 15 lakhs   (D) No profit

Answer:(B)

Required difference in profit = (50 – 30) – (40 – 30) = 10 lakhs

II. In case of how many years, was the income more than average income of the given years?

(A) One   (B) Two   (C) Three   (D) Four

Answer (C)

Average income = Rs. (30 + 50 + 40 + 60 + 60)/5 = Rs. 48 Income is more than average income in the years 1983, 1985, 1986.

III. What was the percentage increase in expenditure from 1984 to 1985?

(A) 66.66   (B) 33.33   (C) 10   (D) 120

Answer (B)

IV. In which of the following years was the profit the maximum?

(A) 1982   (B) 1983   (C) 1985   (D) 1986

Answer (D)

Max.profit = Rs. (60 – 35) = 25 lakhs

Example 186. *Directions (Q.I to Q.III): Study the following graph carefully and answer the following questions.*

The graph shows the production of wheat by three states A, B and C (lakh tones) over the years 1994 through 1998.

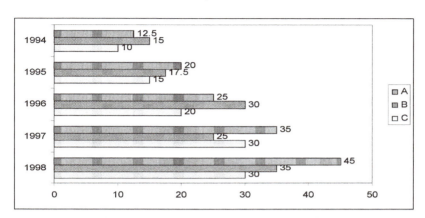

I. What is the difference between the production of state C in 1997 and production of state A in 1994 (in tones)?

(A) 1,80,000   (B) 1,75,000   (C) 17,50,000   (D) None of these

Answer (C)

II. The production of wheat by state A in 1998 is what percent more than the production of wheat by state B in 1996 ?

(A) 33.33%   (B) 40%   (C) 45 %   (D) 50%

Answer (D)

III. What is the difference between the average production of states B and A (in tones) over the period?   (A)

6,00,000   (B) 5,00,000   (C) 4,00,000   (D) None of these

Answer (D)

The difference is 3,00,000 tones.

PIE Chart

Example 187. 40% of deaths in the city road may be attributed to drunken driving. The number of degrees needed to represent it as a slice of a pie chart is

(A)   120   (B)   144   (C)   160   (D)   212 [EC 2017]

Answer (B)

In pie chart, $100\% \to 360^0 \Rightarrow 40\% \to \left(\dfrac{360}{100} \times 40\right)^0 = 144^0$

***Example* 188.** *Directions (Q.I to Q.II): The pie chart shows the distribution of detergent soaps stocked by a shop on a particular day.*
The total number of detergent soap is 225.

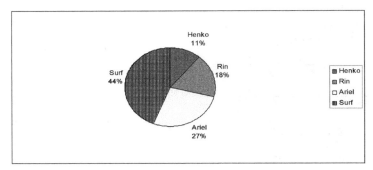

Q.I. If the average price of a detergent soap is Rs. 40, what is the value of the stock of Rin soaps in the shop ?
(A) Rs. 1800   (B) Rs. 1600 (C) Insufficient data (D) None of these
Answer (C)
As we don't have the price of a pack of Rin, we can't determine.
Q.II. If the average % profit made by sale of a detergent soap is 10%, what % of the profit earned by way of sale of soaps in this shop contributed by Ariel?

(A) 27 % (B) More than 27 % (C) Less than 27 % (D) Insufficient data
Answer (D)
The profit contributed by Ariel is not known. Therefore, we can't find out the %.

***Example* 189.** *Directions: (Q.I to Q.II): Use the information provided in the pie chart given below.*
The pie chart gives information about the region wise quarterly sales figures for the number of cars sold by XYZ company in the first quarter ending June 2006. The company sold 5,00,000 vehicles during this time frame.

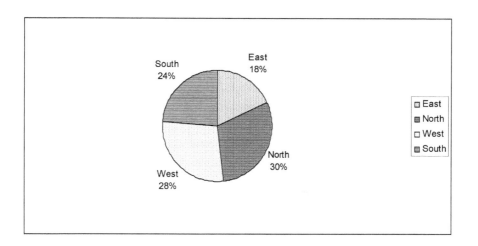

I. If the company is expecting to witness 0% growth in sale during next three quarters, how many vehicles is XYZ company likely to sell in the Eastern region during the next two quarters ?

(A) 180,000   (B) 270,000   (C) 360,000   (D) Can't be determined

Answer (D)

The overall growth in sales of the company remains flat (0% growth) during the next three quarters.

However, we do not know if the region wise distribution in sales is going to remain the same as in first quarter.

II. If the average profit on sale of a vehicle in the Eastern region is Rs. 2100 and that in the Western region is Rs. 1800, how much more profit (in crores of Rs.) did the company make on an average every month in the quarter ending June 2006, in Western region than Eastern region ?

(A)   6.3   (B)   0.63   (C)   2.1   (D)   8

Answer (C)

Total sales in Western region in quarter 1 = 28% of 5,00,000 = 1,40,000.

Profit per vehicle = 1800

Therefore, total profit in Western region = 25,20,00,000 = 25.2 crores

Total sales in Eastern region in quarter 1 = 18% of 5,00,000 = 90,000.

Profit per vehicle = 2100

Total profit in Eastern region in first quarter = 18,90,00,000 = 18.9 crores

Difference in profit = 6.2 crores for the quarter = 2.1 crores on an average per month.

Percentage of exports done by U.S.A = (233.6/777.6) × 100 = 30%

403

Example **190.** *Directions (Q.I to Q.IV) : The land cultivated under different crops in a district is given in the following chart. Study the chart, and on the basis of this chart answer the questions given below.*

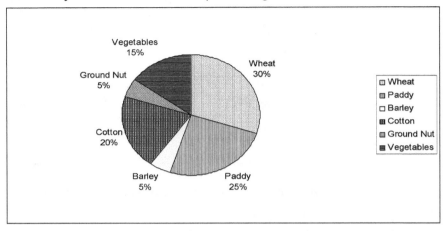

I. If the total area cultivated is 20,000 sq. meters, what is the area cultivated for vegetables (in sq. meters)?

(A) 15,000   (B) 30,000   (C) 20,000   (D) 35,000

Answer (B)

$(15/100) \times 20000 = 30,000$ sq. meters

II. How much more area in sq. meter is cultivated of the largest cultivated crop than to the second largest, if the total area cultivated be 2,00,000 sq. meter ?

(A) 10,000   (B) 20,000   (C) 5,000   (D) 25,000

Answer (A)

$[(30 - 25)/100] \times 200000 = 10,000$ sq. meter

III. If the area cultivated for paddy be 5,000 sq. meter, what is the area cultivated for cotton (sq. meter)?

(A) 2,000   (B) 1,500   (C) 3.400   (D) 4,000

Answer (D)

$(20/25) \times 5000 = 4,000$ sq. meter

IV. What is the ratio of area cultivated for barley and wheat together to the area cultivated for paddy, ground nut and vegetables together?

(A) 5:9   (B) 7:9   (C) 3:10   (D) 4:9

Answer (B)

Ratio = $(5 + 30):(25 + 5 +15) = 7:9$

Example 191. The pie chart below has the break-up of the number of students from different departments in an engineering college for the year 2012. The proportion of male to female students in each department is 5:4. There are 40 males in Electrical Engineering. What is the difference between the number of female students in the Civil department and the number of female students in the Mechanical department?   [CS/IT 2015]

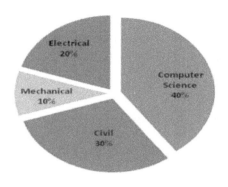

Answer: 32

The no. of males in EE is $40 \Rightarrow 20\% = 40$

The no. of males in ME is $10\% = 20$ females $\dfrac{4}{5} \times 20 = 16$

The no. of males in CE is $30\% = 60$ females $\dfrac{4}{5} \times 60 = 48$

Difference of no. of females in ME and CE = 48-16 = 32

TYPE -3: Two dimensional line graphs on X-Y plane

Example 192. *Directions (Q.I to Q.III): Study the following graph carefully and answer the questions given below.*

Major Exporters (in billion $ for the year 1998)

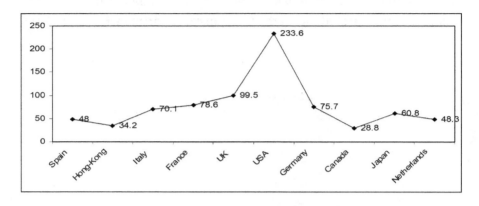

I. The ratio of exports between Spain and Japan is
(A) 3:4　　　　(B) 19:15　　　(C) 15:19　　　　(D) 4:3
Answer (C)
Required ratio = 48:60.8 = 15:19

II. If the exchange rate per dollar is Rs. 43, then find the difference between the values of exports by Hong-Kong and UK in terms of Rs.
(A) 28.179×10^9 (B) 28.179×10^{10} (C) 28.079×10^{10} (D) None of these
Answer (D)
Reqd. difference = $(99.5 – 34.2) \times 43 \times 10^9 = 28.079 \times 10^{11}$ (1 billion = 10^9)

III.Approximately what percent of total exports is done by USA alone ?

(A) 25　　　　(B) 30　　　　(C) 35　　　　(D) 40
Answer (B)

***Example* 193**. In the graph below the concentration of a particular pollutant in a lake is plotted over (alternate) days of a month in winter (average temperature $10\,^\circ C$) and a month in summer (average temperature $30\,^\circ C$)

406

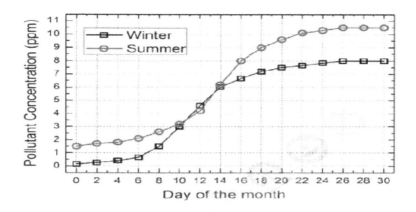

Consider the following statements based on the following data:

i. Over the given months, the difference between the maximum and minimum pollutant concentrations is the same in both the winter and the summer.

ii. There are at least four days in the summer month such that the pollutant concentrations in those days are within 1 ppm of the pollution concentrations on the corresponding days of the winter month.

Which one of the following options is correct?

(A) Only I (B) Only ii (C) Both I and ii (D) Neither i nor ii [ME 2017]

Answer (B)

In the summer, the maximum and minimum pollutant concentrations are 10.5 and 1.5 ppm, difference = 9 ppm

In the winter, the maximum and minimum pollutant concentrations are 8 and 0 ppm, difference = 8 ppm

so, I is not correct.

On 10, 12, 14, 16 the pollution concentrations are within a difference limit of 1 ppm. So ii. is correct.

EXERCISE

Question of one mark:

1. Hema's age is 5 years more than twice Hari's age. Suresh's age is 13 years less than 10 times Hari's age. If Suresh is 3 times as old as Hema, how old is Hema?

(A)   14   (B)   17   (C)   18   (D)   19

2. For $0 \leq x \leq 2\pi$, sin x and cos x are both decreasing functions in the interval _____.

(A) $(0,\pi/2)$ (B) $(\pi/2, \pi)$ (C) $(\pi, 3\pi/2)$ (D) $(3\pi/2, 2\pi)$

3. If $x > y > 1$, which one of the following must be TRUE?

i. $\ln x > \ln y$ ii. $e^x > e^y$ iii. $y^x > x^y$ iv. $\cos x > \cos y$
(A) i & ii (B) i & iii (C) iii & iv (D) ii & iv

4. A function $f(x)$ is linear and has a value 29 at $x = -2$ and 39 at $x = 3$. Find its value at $x = 5$.
(A) 59 (B) 45 (C) 43 (D) 35

5. Arrange the following three-dimensional objects in the descending order of their volumes:

(i) A cuboid with dimensions 10 cm, 8 cm and 6 cm

(ii) A cube of side 8 cm

(iii) A cylinder with base radius 7 cm and height 7 cm

(iv) A sphere of radius 7 cm

(A) (i), (ii), (iii), (iv)   (B) (ii), (i), (iv), (iii)

(C) (iii), (ii), (i), (iv)   (D) (iv), (iii), (ii), (i)

Question of two marks:

6. A fruit seller sold a basket of fruits at 12.5% loss. Had he sold for Rs. 108 more, he would have made a 10% gain. What is the loss in Rupees incurred by the fruit seller?
(A) 48 (B) 52 (C) 60 (D) 108 (C)

7. An automobile travels from city A to city B and returns to city A by the same route. The speed of the vehicle during the onward and return journeys were constant at 60 km/h and 90 km/h, respectively. What is the

average speed in km/h for the entire journey?

(A)   72   (B)   73   (C)   74   (D)   75

8. To pass a test, a candidate needs to answer at least 2 out of 3 questions correctly. A total of 6,30,000 candidates appeared for the test. Question A was correctly answered by 3,30,000 candidates. Question B was answered correctly by 2,50,000 candidates. Question C was answered correctly by 2,60,000 candidates. Both questions A and B were answered correctly by 1,00,000 candidates. Both questions B and C were answered correctly by 90,000 candidates. Both questions A and C were answered correctly by 80,000 candidates. If the number of students answering all questions correctly is the same as the number answering none, how many candidates failed to clear the test?

(A)   30,000   (B)   2,70,000   (C)   3,90,000   (D)   4,20,000

9. The price of a wire made of super alloy material is proportional to the square of its length. The price of length 10 m of the wire is Rs. 1600. What would be the total price (in Rs.) of two wires of lengths 4 m and 6m?

(A)   768   (B)   832   (C)   1440   (D)   1600

10. An unbiased coin is tossed six times in a row and four different such trials are conducted.
One trial implies six tosses of the coin. If H stands for head and T stands for tail, the following are the observations from the four trials:
(1) HTHTHT (2) TTHHHT (3) HTTHHT (4) HHHT__ __.
Which statement describing the last two coin tosses of the fourth trial has the highest probability of being correct?

(A) Two T will occur.   (B) One H and one T will occur.

(C) Two H will occur.   (D) One H will be followed by one T.

Answers:

1. (D) 2. (B) 3. (A) 4. (C) 5. (D) 6. (C) 7. (A) 8. (D) 9. (B) 10. (B)